AF426640

Appreciating Geometry Through Extraordinary Relationships

Problem Solving in Mathematics and Beyond

Print ISSN: 2591-7234
Online ISSN: 2591-7242

Series Editor: Dr. Alfred S. Posamentier
Distinguished Lecturer
New York City College of Technology - City University of New York

There are countless applications that would be considered problem solving in mathematics and beyond. One could even argue that most of mathematics in one way or another involves solving problems. However, this series is intended to be of interest to the general audience with the sole purpose of demonstrating the power and beauty of mathematics through clever problem-solving experiences.

Each of the books will be aimed at the general audience, which implies that the writing level will be such that it will not be engulfed in technical language — rather the language will be simple everyday language so that the focus can remain on the content and not be distracted by unnecessarily sophiscated language. Again, the primary purpose of this series is to approach the topic of mathematics problem-solving in a most appealing and attractive way in order to win more of the general public to appreciate this most important subject rather than to fear it. At the same time we expect that professionals in the scientific community will also find these books attractive, as they will provide many entertaining surprises for the unsuspecting reader.

Published

Vol. 40 *Appreciating Geometry Through Extraordinary Relationships*
by Alfred S Posamentier

Vol. 39 *Fun Math: Problem Solving Beyond the Classroom*
by Alfred S Posamentier

Vol. 38 *Mathematics: The Quest for Truth and Beauty*
by James D Stein

Vol. 37 *Mathematics in Architecture, Art, Nature, and Beyond*
by Alfred S Posamentier and Günter J Maresch

Vol. 36 *Geometric Gems: An Appreciation for Geometric Curiosities*
Volume III: The Wonders of Circles
by Alfred S Posamentier and Robert Geretschläger

For the complete list of volumes in this series, please visit www.worldscientific.com/series/psmb

Appreciating Geometry Through Extraordinary Relationships

Alfred S. Posamentier

The City University of New York, USA

NEW JERSEY · LONDON · SINGAPORE · BEIJING · SHANGHAI · HONG KONG · TAIPEI · CHENNAI · TOKYO

Published by

World Scientific Publishing Co. Pte. Ltd.

5 Toh Tuck Link, Singapore 596224

USA office: 27 Warren Street, Suite 401-402, Hackensack, NJ 07601

UK office: 57 Shelton Street, Covent Garden, London WC2H 9HE

British Library Cataloguing-in-Publication Data
A catalogue record for this book is available from the British Library.

Problem Solving in Mathematics and Beyond — Vol. 40
APPRECIATING GEOMETRY THROUGH EXTRAORDINARY RELATIONSHIPS

Copyright © 2026 by World Scientific Publishing Co. Pte. Ltd.

All rights reserved. This book, or parts thereof, may not be reproduced in any form or by any means, electronic or mechanical, including photocopying, recording or any information storage and retrieval system now known or to be invented, without written permission from the publisher.

For photocopying of material in this volume, please pay a copying fee through the Copyright Clearance Center, Inc., 222 Rosewood Drive, Danvers, MA 01923, USA. In this case permission to photocopy is not required from the publisher.

ISBN 978-981-98-1514-2 (hardcover)
ISBN 978-981-98-1578-4 (paperback)
ISBN 978-981-98-1515-9 (ebook for institutions)
ISBN 978-981-98-1516-6 (ebook for individuals)

For any available supplementary material, please visit
https://www.worldscientific.com/worldscibooks/10.1142/14366#t=suppl

Desk Editors: Kannan Krishnan/Gabriel Rawlinson

Typeset by Stallion Press
Email: enquiries@stallionpress.com

To my wonderful family: Lisa, David, Daniel, Max, Samuel, Jack, and Charles . . . . . and to Barbara for her continued love and support.

About the Author

Alfred S. Posamentier is currently a Distinguished Lecturer at New York City College of Technology of the City University of New York. Prior to that he was Executive Director for Internationalization and Funded Programs at Long Island University, New York. This was preceded by 5 years as Dean of the School of Education and Professor of Mathematics Education at Mercy University, New York. Before that he was for 40 years at The City College of the City University of New York, at which he is now Professor Emeritus of Mathematics Education and Dean Emeritus of the School of Education. He is the author and co-author of more than 90 mathematics books for teachers, secondary and elementary school students, as well as the general readership. Dr. Posamentier is also a frequent commentator in newspapers and journals on topics related to education and mathematics.

After completing his B.A. degree in mathematics at Hunter College of the City University of New York, he took a position as a teacher of mathematics at Theodore Roosevelt High School (Bronx, New York), where he focused his attention on improving the students' problem-solving skills and at the same time enriching their instruction far beyond what the traditional textbooks offered. During his six-year tenure there, he also developed the school's first mathematics teams (both at the junior and senior level). He is still involved in working with mathematics teachers and supervisors, nationally and internationally, to help them maximize their effectiveness.

Immediately upon joining the faculty of the City College of New York in 1970 (after having received his master's degree there in 1966), he began to develop in-service courses for secondary school mathematics teachers, including such special areas as recreational mathematics and problem solving in mathematics. As Dean of the City College School of Education for 10 years, his scope of interest in educational issues covered the full gamut of educational issues. During his tenure as Dean, he took the School of Education from the bottom of the New York State rankings to the top with a perfect NCATE accreditation assessment in 2009. He also raised more than $12,000,000 from the private sector for educational innovative programs. Posamentier repeated this successful transition at Mercy College, where he enabled it to become the only college to have received both NCATE and TEAC accreditation simultaneously.

In 1973, Dr. Posamentier received his Ph.D. from Fordham University (New York) in mathematics education, and he has since extended his reputation in mathematics education to Europe. He has been visiting professor at several European universities in Austria, England, Germany, the Czech Republic, Turkey and Poland. In 1990, he was Fulbright Professor at the University of Vienna.

In 1989 he was awarded an Honorary Fellow position at the South Bank University (London, England). In recognition of his outstanding teaching, the City College Alumni Association named him Educator of the Year in 1994, and in 2009. New York City had the day, May 1, 1994, named in his honor by the President of the New York City Council. In 1994, he was also awarded the *Grosse Ehrenzeichen für Verdienste um die Republik Österreich* (Grand Medal of Honor from the Republic of Austria), and in 1999, upon approval of Parliament, the President of the Republic of Austria awarded him the title of University Professor of Austria. In 2003, he was awarded the title of *Ehrenbürgerschaft* (Honorary Fellow) of the Vienna University of Technology, and in 2004 he was awarded the *Österreichisches Ehrenkreuz für Wissenschaft & Kunst 1. Klasse* (Austrian Cross of Honor for Arts and Science,

First Class) from the President of the Republic of Austria. In 2005, he was inducted into the Hunter College Alumni Hall of Fame, and in 2006 he was awarded the prestigious Townsend Harris Medal by the City College Alumni Association. He was inducted into the New York State Mathematics Educator's Hall of Fame in 2009, and in 2010 he was awarded the coveted Christian-Peter-Beuth Prize from the Technische Fachhochschule – Berlin. In 2017, Posamentier was awarded *summa cum laude nemine discrepante* by the Fundacion Sebastian, A.C., Mexico City, Mexico.

He has taken on numerous important leadership positions in mathematics education locally. He was a member of the New York State Education Commissioner's Blue Ribbon Panel on the Math-A Regents Exams, and the Commissioner's Mathematics Standards Committee, which redefined the Mathematics Standards for New York State, and he also served on the New York City Schools' Chancellor's Math Advisory Panel.

Dr. Posamentier is still a leading commentator on educational issues and continues his long-time passion of seeking ways to make mathematics interesting to both teachers, students and the general public – as can be seen from some of his more recent books.

For more information and a list of his publications see: https://en.wikipedia.org/wiki/Alfred_S._Posamentier.

Contents

Introduction

Presented properly, the most appealing field of mathematics is geometry. Geometry is visual, contains many unusual relationships, and intrigues teachers, students, and amateurs alike. There are many unexpected geometric principles that are simple to demonstrate and easy to appreciate. A number of special lines in triangles, namely, the medians, the angle bisectors, the altitudes, and the perpendicular bisectors of the sides, all have unique concurrencies. Even otherwise unremarkable quadrilaterals generate unexpected relationships: when the midpoints of the sides are joined consecutively, no matter the shape of the quadrilateral, the result is always a parallelogram.

Geometry has attracted many famous names not known for their mathematical pursuits. Napoleon Bonaparte is said to have discovered a beautiful relationship where the centers of equilateral triangles placed on the sides of any random triangle will generate another equilateral triangle. The 20th president of the United States, James A. Garfield, published an original proof of the Pythagorean theorem. Recently, two American high school students proved the Pythagorean theorem using trigonometry, which to this time was believed to be impossible.

What makes geometry so special is the boundless new relationships that evolve over time. Of course, there are many complicated geometric configurations that provide curious challenges to solve. Although this book will present a number of unusual geometric relationships, the reader requires very little information beyond the high school geometry course to understand and appreciate the solutions. In some of these examples, the elegance of the proof, especially when compared to traditional methods, should be impressive. However, many proofs rely on basic geometric theorems that, unfortunately, are not presented in the typical high school geometry course. Yet they ought to be, as they are quite simple and extremely helpful tools for better understanding geometric relationships. These will be presented in the first chapter so that the reader can benefit from these powerful procedures right from the start. Some of these "tools" include Ptolemy's theorem, Ceva's theorem, Menelaus' theorem, the angle trisection relationships, Fibonacci applications, Desargues' theorem, Stewart's theorem, Apollonius' theorem, Simson's theorem, and Heron's formula, among others.

The plethora of unusual geometric relationships presented in this book will awe motivated students of geometry and lay readers. Furthermore, these relationships, or geometric problems, could inspire readers to pursue other fascinating relationships. Join us now as we take this extraordinary tour through some of the most unusual geometric problems that can be appreciated without "higher mathematics."

Chapter 1

The Basic Elements of Geometry

In the interest of providing the readership with all the necessary "equipment" for the proofs in this book, and cognizant of the fact that some readers may have forgotten a few of the basics from the secondary school geometry curriculum, we present this geometry primer. First, we will review some of the facts and principles you should recognize from your high school geometry class. These are presented without proofs, as they are intended to refresh the reader's memory. Naturally, any high school geometry book will furnish further understanding of these basic concepts. Next, we present a number of more sophisticated theorems from basic trigonometry. Readers may refer back to this chapter when encountering unfamiliar concepts in the remaining chapters of the book.

Congruence of Triangles

Triangles can be proved congruent by showing that their corresponding parts are equal:

- Two sides and their included angle (SAS)
- Two angles and their included side (ASA)
- Two angles and a side not included (AAS)
- Three sides (SSS)
- For right triangles: the hypotenuse and a leg (HL)

Similarity of Triangles

Triangles can be proved similar by showing any of the following equivalent properties:

- The corresponding angles are equal (AAA).
- The corresponding sides are proportional.
- Two pairs of corresponding sides are proportional, and the included angles are equal.

Right Triangle Properties

For the following properties, we refer to a right triangle ABC with its right angle at C, shown in Figure 1-1. The lengths of the sides opposite vertices A, B, and C are named a, b, and c, respectively. The altitude CD to the hypotenuse AB has length h and divides the hypotenuse into segments AD and DB with lengths p and q, respectively.

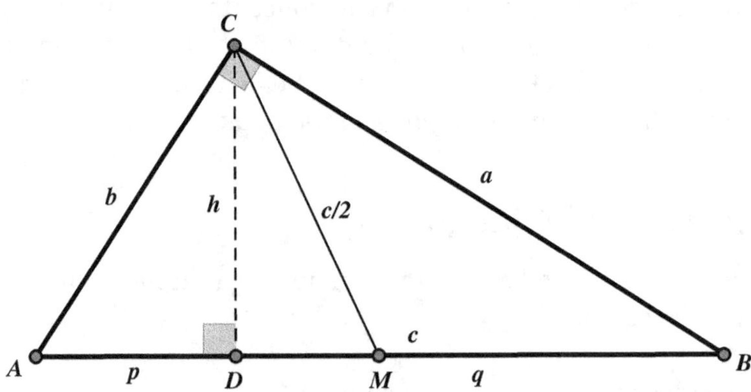

Figure 1-1

- The Pythagorean theorem: the sum of the squares of the legs equals the square of the hypotenuse: $a^2 + b^2 = c^2$.
- The square of the altitude h on the hypotenuse is equal to the product of the lengths of the segments p and q on the hypotenuse: $h^2 = p \cdot q$.
- The square of the length a of a leg is equal to the product of the length c of the hypotenuse and the adjoining altitude segment q: $a^2 = c \cdot q$, also $b^2 = c \cdot p$.
- The median CM to the hypotenuse of a right triangle is one-half the length c of the hypotenuse.

Angles Related to a Circle

These relationships are given in two common notations, as different notations are suited for different applications. Readers may also be more familiar with one notation than the other. Referring to Figure 1-2, we have the following circle-related relationships:

- An inscribed angle: $\angle A = \frac{1}{2}\overset{\frown}{BAC}$ (or $\angle CAB = \frac{1}{2} \cdot \angle COB$)
- An angle formed by 2 secants: $\angle F = \frac{1}{2}(\overset{\frown}{BC} - \overset{\frown}{DE})$ (or $\angle CFB = \frac{1}{2} \cdot (\angle COB - \angle EOD)$)
- An angle formed by 2 tangents: $\angle J = \frac{1}{2}(\overset{\frown}{HAK} - \overset{\frown}{HCK}) = 180° - \overset{\frown}{HCK}$ (or $\angle HJK = \frac{1}{2} \cdot (180° - \angle KOH)$)
- An angle formed by a secant and a tangent: $\angle L = \frac{1}{2}(\overset{\frown}{CK} - \overset{\frown}{NK})$ (or $\angle KLC = \frac{1}{2} \cdot (\angle KOC - \angle NOK)$)
- An angle formed by 2 chords intersecting inside the circle: $\angle AME = \frac{1}{2}(\overset{\frown}{AE} + \overset{\frown}{BC})$ (or $\angle AME = \frac{1}{2} \cdot (\angle AOE + \angle COB)$)
- An angle formed by a chord and a tangent: $\angle JKC = \frac{1}{2}\overset{\frown}{KC} = \angle KAC$ (or $\angle JKC = \frac{1}{2}\angle KOC = \angle KAC$)
- Opposite angles in a circle (i.e., opposite angles of an inscribed quadrilateral): $\angle BCK = 180° - \angle KAB$

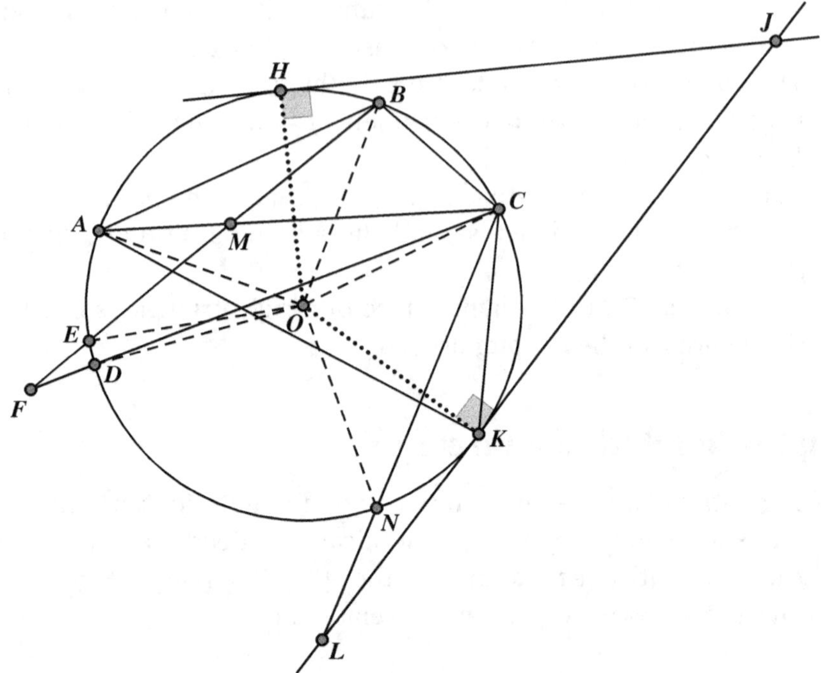

Figure 1-2

Tangents, Secants, and Chords of a Circle

The following axioms refer to Figure 1-3:

- From the same external point, the tangent is a mean proportional between the entire secant and its external segment:

$$\frac{e+f}{h} = \frac{h}{f}.$$

- For two secants from the same external point, the product of the length of a secant and its external segment is equal to the product of the other secant and its external segment:

$$(e+f) \cdot f = (g+k) \cdot k.$$

- For two chords intersecting inside a circle, the product the segments of one chord is equal to the product or segments of the other chord:

$$a \cdot b = c \cdot d.$$

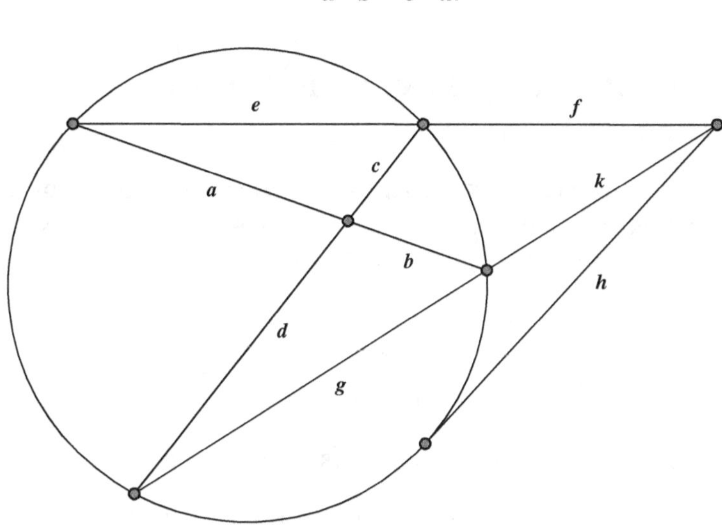

Figure 1-3

The Law of Sines and the Law of Cosines

Although the laws of sines and cosines are not used very much in this book, we offer them here for the few times that they do occur. In triangle *ABC*, in Figure 1-4, we let *a*, *b*, and *c* denote the lengths of

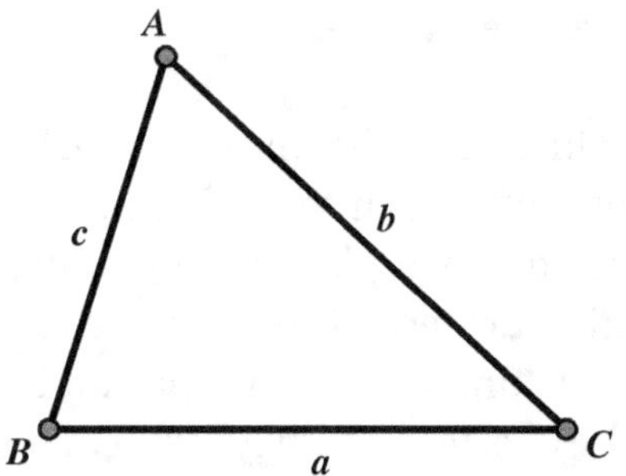

Figure 1-4

the sides opposite the vertices A, B, and C, respectively. Then $\frac{a}{\sin \angle A} = \frac{b}{\sin \angle B} = \frac{c}{\sin \angle C}$. This is known as the *law of sines*.

Also, $c^2 = a^2 + b^2 - 2ab \cos \angle ACB$, and analogously $a^2 = b^2 + c^2 - 2bc \cos \angle BAC$ and $b^2 = c^2 + a^2 - 2ca \cos \angle CBA$. This is known as the *law of cosines.*

Trigonometry: Angle Sum and Difference Identities

It is often useful to express the sine or cosine of the sum or difference of two angles in terms of the individual sines and cosines. To this purpose, we can use the following identities:

$$\sin (\alpha + \beta) = \sin \alpha \cos \beta + \cos \alpha \sin \beta$$

$$\sin (\alpha - \beta) = \sin \alpha \cos \beta - \cos \alpha \sin \beta$$

$$\cos (\alpha + \beta) = \cos \alpha \cos \beta - \sin \alpha \sin \beta$$

$$\cos (\alpha - \beta) = \cos \alpha \cos \beta + \sin \alpha \sin \beta$$

Interior Angle Bisector in a Triangle

This relationship, involving a triangle's angle bisectors, is not usually taught in high school geometry. However, it can be very useful, and so we offer this simple proof. The interior angle bisector of a triangle partitions the side to which it is drawn proportional to the two adjacent sides of the bisected angle. In Figure 1-5, line AD is a bisector of angle $\angle BAC$, and therefore, $\frac{DC}{DB} = \frac{AC}{AB}$.

Proof: In Figure 1-5, line AD is the bisector of $\angle BAC$ of $\triangle ABC$. The line parallel to AD through point B intersects the extension of side CA at point E. To make matters simpler, we have marked four angles in the figure, all of which are equal. We have $\angle ABE = \angle BAD$, as these are alternate-interior angles of the parallel lines AD and BE. Furthermore, $\angle BEA = \angle DAC$, as they are corresponding angles of these parallel lines, and since AD is the bisector of $\angle BAC$, we also have $\angle BAD = \angle DAC$. By substitution, we get $\angle BEA = \angle ABE$. This yields isosceles triangle ABE, where $AE = AB$. Since triangles EBC and ADC

are similar, we have $\frac{DC}{DB}=\frac{AC}{AE}$. Now we substitute AB for AE to get the desired result, namely, $\frac{DC}{DB}=\frac{AC}{AB}$.

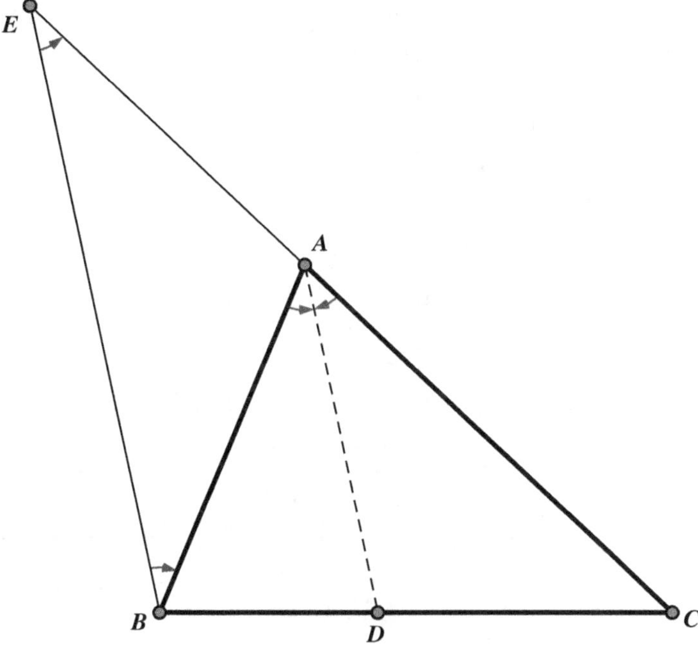

Figure 1-5

Exterior Angle Bisector in a Triangle

The exterior angle bisector of a triangle partitions the line containing the opposite side to which it is drawn into segments proportional to the two adjacent sides of the bisected angle. In Figure 1-6, line AD is a bisector of angle $\angle EAB$, and therefore, $\frac{DC}{DB}=\frac{AC}{AB}$. Once again, a proof is provided to justify this powerful theorem.

Proof: In Figure 1-6, line AD is the bisector of exterior angle $\angle FAB$ of $\triangle ABC$. The line parallel to AD through point B intersects side CA at point E. As in the previous proof, we have marked four angles in the figure, all of which are equal.

We have $\angle ABE = \angle DAB$, as these are alternate-interior angles of the parallel lines AD and BE. Furthermore, $\angle FAD = \angle AEB$ because they are corresponding angles of these parallel lines, and since AD is the bisector of $\angle FAB$, we also have $\angle FAD = \angle DAB$. Therefore, we can substitute to get $\angle AEB = \angle EBA$. This yields isosceles triangle ABE, where $AE = AB$. Since triangles EBC and ADC are similar, we have $\frac{DC}{DB} = \frac{AC}{AE}$. We substitute AB for AE and once again get the desired result, namely, $\frac{DC}{DB} = \frac{AC}{AB}$.

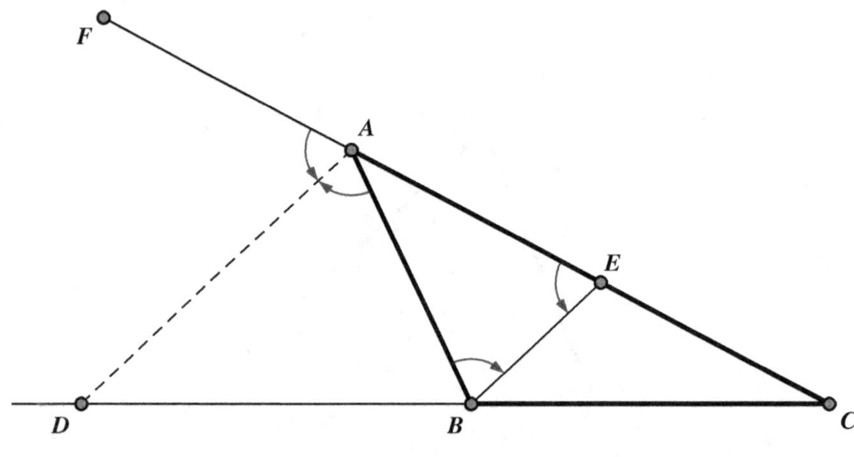

Figure 1-6

The Minimum Distance Point of an Equilateral Triangle

The equilateral triangle contains a host of unusual and unexpected relationships. A number of them will be presented throughout the book, but one such relationship will be useful as we begin our tour through geometry. It is known as Viviani's theorem, and was originally developed by the Italian mathematician Vincenzo Viviani (1622–1703).

Theorem: The sum of the distances from any point in the interior of an equilateral triangle to the sides of the triangle is constant, which is the length of the altitude of the triangle.

Two proofs of this interesting property are provided here. The first compares the length of each perpendicular segment to a portion of the altitude, and the second involves area comparisons.

Proof I: In Figure 1-7, we have equilateral $\triangle ABC$, with $PR \perp AC$, $PQ \perp BC$, $PS \perp AB$, and $AD \perp BC$. Draw a line through P parallel to BC, meeting AD, AB, and AC at G, E, and F, respectively.

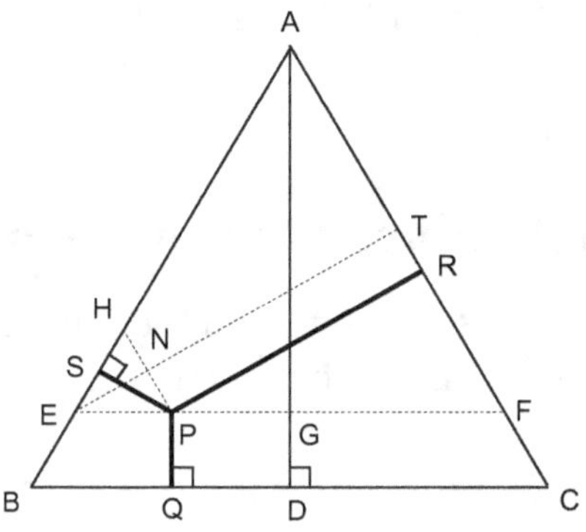

Figure 1-7

Since $PGDQ$ is a rectangle, $PQ = GD$. Draw $ET \perp AC$. Since $\triangle AEF$ is equilateral, $AG = ET$, as all the altitudes of an equilateral triangle are equal. Draw $PH \parallel AC$ to intersect ET at N so that $NT = PR$. Since $\triangle EHP$ is equilateral, altitudes PS and EN are equal. Therefore, we have shown that $PS + PR = ET = AG$. Since $PQ = GD$, we have $PS + PR + PQ = AG + GD = AD$, which is a constant for the given triangle.

Proof II: In Figure 1-8, the equilateral $\triangle ABC$ has $PR \perp AC$, $PQ \perp BC$, $PS \perp AB$, and $AD \perp BC$. Then we draw PA, PB, and PC.

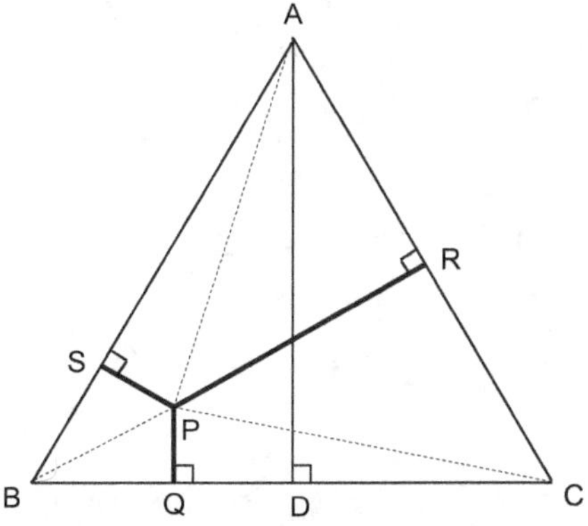

Figure 1-8

The area of $\triangle ABC$ = area of $\triangle APB$ + area of $\triangle BPC$ + area of $\triangle CPA = \frac{1}{2}(AB)(PS) + \frac{1}{2}(BC)(PQ) + \frac{1}{2}(AC)(PR)$. Since $AB = BC = AC$, the area of $\triangle ABC = \frac{1}{2}(BC)[PS + PQ + PR]$. However, the area of $\triangle ABC = \frac{1}{2}(BC)(AD)$. Therefore, $PS + PQ + PR = AD$, a constant for the given triangle.

The Minimum Distance Point of a Quadrilateral

There are times when it is useful to find the point in the quadrilateral where the sum of the distances to the vertices is a minimum.

Theorem: The minimum-distance point of a quadrilateral is the point of intersection of the diagonals.

To prove that, among the interior points of a quadrilateral, the point where the diagonals intersect has the smallest sum of distances to the vertices, we choose any other interior point and compare its sum of distances to the vertices to that of the diagonal-intersection point.

Consider quadrilateral $ABCD$ with diagonals AC and BD intersecting at Q, shown in Figure 1-9.

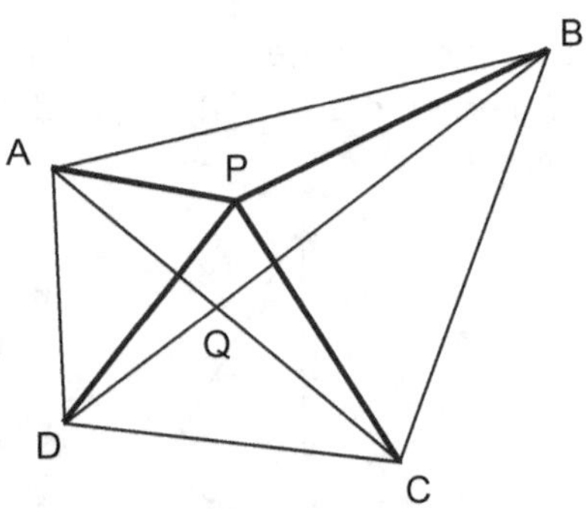

Figure 1-9

Select any point P (not at Q) in the interior of quadrilateral $ABCD$. Then $PA + PC > QA + QC$, since the sum of the lengths of two sides of a triangle is greater than the length of the third. Similarly, $PB + PD > QB + QD$. By addition, $PA + PB + PC + PD > QA + QB + QC + QD$, which shows that the sum of the distances from the point of intersection of the diagonals of a quadrilateral to the vertices is less than the sum of the distances from *any other* interior point of the quadrilateral to the vertices. Therefore, the intersection of the diagonals of a quadrilateral is the minimum distance point in the quadrilateral.

The Minimum Distance Point of a Triangle

Next, we will consider where the minimum distance point of a general triangle would be. We shall use $\triangle ABC$ with no angle measuring greater than 120°, as shown in Figure 1-10.

Theorem: The minimum distance point of a triangle (with no angle of measure greater than 120°) is the equiangular point, which is the point at which the sides of the triangle subtend equal angles.

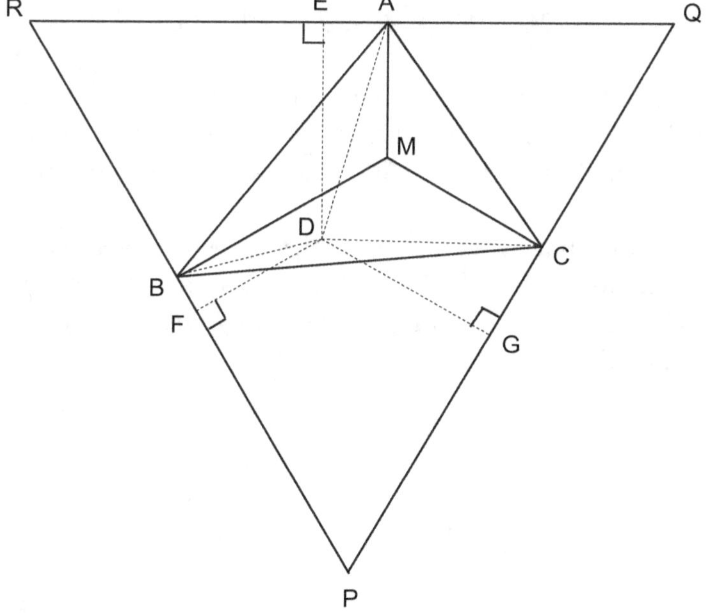

Figure 1-10

Proof: Let M be the point in the interior of $\triangle ABC$ where $\angle AMB = \angle BMC = \angle AMC = 120°$. Draw lines through A, B, and C perpendicular to AM, BM, and CM, respectively. These lines meet to form equilateral $\triangle PQR$. To prove that $\triangle PQR$ is equilateral, notice that each angle has measure $60°$. This can be shown by considering, for example, quadrilateral $AMBR$. Since $\angle RAM = \angle RBM = 90°$, and $\angle AMB = 120°$, it follows that $\angle ARB = 60°$. Let D be *any other* point in the interior of $\triangle ABC$. We must show that the sum of the distances from M to the vertices is less than the sum of the distances from D to the vertices.

Since we proved above that the sum of the distances from any point in the interior of an equilateral triangle to the sides of the triangle is a constant, we know that $MA + MB + MC = DE + DF + DG$, where DE, DF, and DG are the perpendiculars to REQ, RBP, and QGP, respectively.

But $DE + DF + DG < DA + DB + DC$. Recall that the shortest distance from an external point to a line is the length of the perpendicular segment from the point to the line. By substitution:

$$MA + MB + MC < DA + DB + DC.$$

We chose to restrict our discussion to triangles with angles of measure less than $120°$. If you try to construct the point M in a triangle with one angle measuring $150°$, the reason for our restriction will become obvious.

Now that we have covered the basics of geometry and then some, we are ready to appreciate the geometry that is perhaps beyond the typical high school curriculum, but certainly within easy reach of a motivated reader.

Chapter 2

Spectacular Named
Relationships in Geometry

Many theorems in geometry are known by the name of their discoverer. These theorems are usually quite spectacular and very useful in solving geometric problems. They provide a much more glamorous view of Euclidean geometry than is presented in a high school geometry course. We present them with an introduction, as well as a proof to justify their veracity. Enjoy!

The Ubiquitous Pythagorean Theorem

Perhaps the best-known theorem in mathematics is the Pythagorean theorem. Therefore, it is only appropriate that we begin the discussion of named theorems with this relationship. Whenever a right triangle appears in a geometric problem, chances are that the Pythagorean theorem can be called on to help solve it.

The Pythagorean theorem is among the most popular geometric theorems to prove. For example, *The Pythagorean Proposition*, a book by Elisha Scott Loomis (1852–1940) first published in 1927, presents 367 proofs of the Pythagorean theorem.[1] Since that time, many additional proofs of this famous theorem have been developed.

---

[1] Elisha Scott Loomis. *The Pythagorean Proposition*, 2nd ed., National Council of Teachers of Mathematics, 1940.

Some proofs of the Pythagorean theorem were published by people not known for their mathematical prowess. Prior to his political career, 20th President of the United States James A. Garfield taught Greek and Latin, and dabbled in geometry. In 1876, while serving in the House of Representatives, Garfield produced the following proof of the Pythagorean theorem.[2] Let's take a look at the proof he discovered.

President Garfield's Proof of the Pythagorean Theorem

In Figure 2-1 we have triangles *ABC*, *EAD*, and *ABE*. All the triangles in the diagram are right triangles, and $\triangle ABC \cong \triangle EAD$. Recall that the area of the trapezoid *DCBE* is half the product of the altitude *CD* $(a + b)$ and the sum of the bases *CB* (a) and *DE* (b), which we can write as $\frac{1}{2}(a+b)^2$.

We can also get find area of the trapezoid *DCBE* by adding the areas of each of the three right triangles: $\triangle ABC, \triangle ABE, \triangle ADE$.

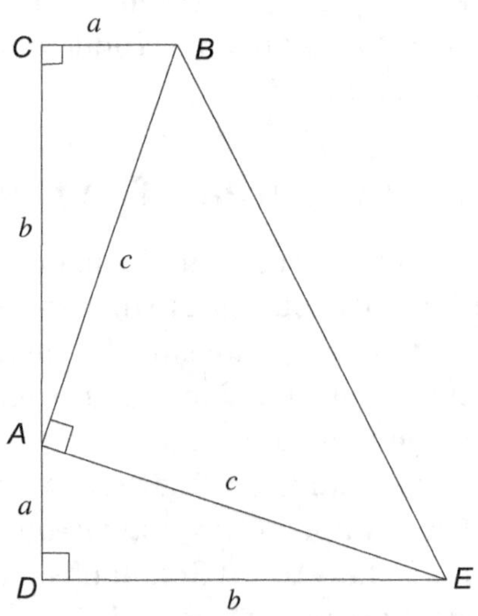

Figure 2-1

---

[2] James A. Garfield. "Pons Asinorum." *New England Journal of Education* 3, no. 14 (1876): 161.

$$\frac{1}{2}ab+\frac{1}{2}ab+\frac{1}{2}c^2$$

$$2\left(\frac{1}{2}ab\right)+\frac{1}{2}c^2.$$

We can then equate the two expressions representing the area of the entire trapezoid:

$$2\left(\frac{1}{2}ab\right)+\frac{1}{2}c^2=\frac{1}{2}(a+b)^2.$$

This can be simplified to:

$$2ab+c^2=(a+b)^2$$

$$2ab+c^2=a^2+2ab+b^2$$

$$c^2=a^2+b^2.$$

This is the Pythagorean theorem as applied to the right triangle *ABC*.

An astute reader may notice that Garfield's proof is somewhat similar to the one believed to be used by Pythagoras. If we "complete" a square from the given trapezoid in Figure 2-1, we get a configuration similar to that in Figure 2-2, which we can call a "completed square." Therefore, the area of this newly formed square with sides of

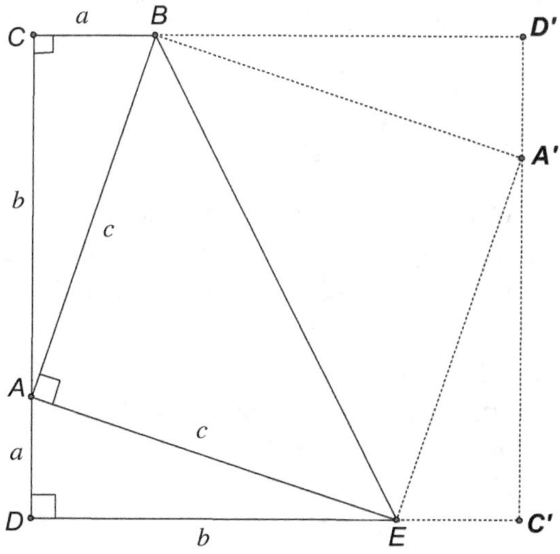

Figure 2-2

length $a + b$ is $(a+b)^2 = a^2 + b^2 + 2ab$. The area could also be obtained by adding the four small rectangles to the two larger right triangles: $4\frac{ab}{2} + 2\frac{c^2}{2} = 2ab + c^2$. When we equate these two area representations, we get $a^2 + b^2 + 2ab = 2ab + c^2$, which is the Pythagorean theorem $a^2 + b^2 = c^2$.

Jamie deLemos, a student at Framingham High School in Framingham, MA, further manipulated this configuration and came up with yet another proof of the Pythagorean theorem.[3] She flipped[4] the trapezoid in Garfield's proof (Figure 2-1) to get the shape shown in Figure 2-3.

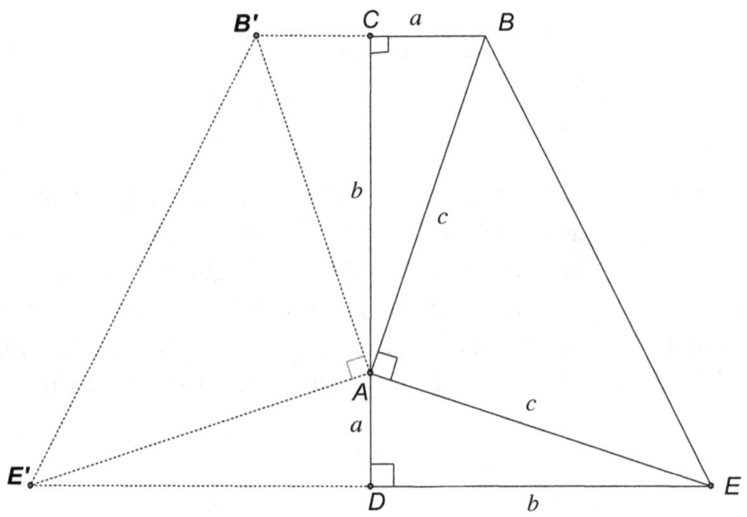

Figure 2-3

She simply found the area of the trapezoid as the sum of the areas of the various right triangles. First, the area of the trapezoid $BB'E'E$ is equal to one-half the product of its altitude and the sum of its bases, namely, $\frac{1}{2}(a+b)(2a+2b)$. The area of this trapezoid is also the sum of the six right triangles that comprise it. That is, four times the area of triangle ABC and twice the area of triangle ABE, or $4(\frac{1}{2}ab) + 2(\frac{1}{2}c^2)$.

Equating these two expressions for the same area (trapezoid $BB'E'E$), we get:

---

[3] Published in *The Mathematics Teacher* 88, no. 1 (January 1995): 79.

[4] This is also called a reflection in line *CAD*.

$$\frac{1}{2}(a+b)(2a+2b)=4\left(\frac{1}{2}ab\right)+2\left(\frac{1}{2}c^2\right).$$

Then simplifying, we get:

$$(a+b)^2 = 2ab+c^2$$
$$a^2+2ab+b^2 = 2ab+c^2$$
$$a^2+b^2 = c^2.$$

As deLemos herself says, "Who would have guessed that an average geometry student could come up with an original derivation of this famous theorem? Logic and creativity are the keys, plus a little push and some thinking." This might motivate the reader to seek other proofs of the Pythagorean theorem.

Apollonius's Theorem Relating the Median of a Triangle to the Side Lengths

The theorem of Apollonius states that the square of a triangle's median equals half the sum of the squares of the two adjacent sides minus the square of the third side. In triangle ABC, where AD is the median to side BC, this can be stated as $AD^2 = \frac{AB^2+AC^2}{2} - BD^2$.

Proof: In Figure 2-4, we begin by drawing the altitude from vertex A to point E on base BC. We will let $BD = CD = x$, $DE = y$, and $AE = h$, so that $BE = (x-y)$ and $CE = (x+y)$.

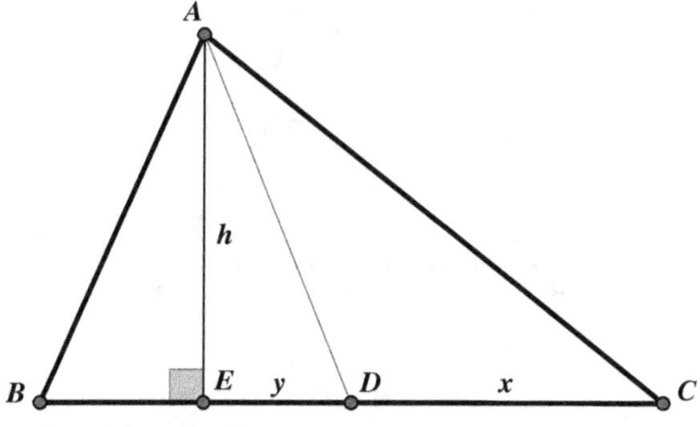

Figure 2-4

We apply the Pythagorean theorem to right triangles ABE and ACE to get $AB^2 = AE^2 + BE^2$ and $AC^2 = AE^2 + CE^2$. Adding these equations, we get

$$AB^2 + AC^2 = AE^2 + BE^2 + AE^2 + CE^2 = h^2 + (x - y)^2 + h^2 + (x + y)^2,$$

or

$$AB^2 + AC^2 = h^2 + x^2 - 2xy + y^2 + h^2 + x^2 + 2xy + y^2 = 2(h^2 + x^2 + x^2).$$

Since $AD^2 = h^2 + y^2$, and $BD^2 = x^2$, this yields $AB^2 + AC^2 = 2(AD^2 + BD^2)$, which is

$$AD^2 = \frac{AB^2 + AC^2}{2} - BD^2.$$

Heron's Formula for the Area of a Triangle

The lengths of the three sides of a triangle determine a unique triangle. Therefore, in such a case we should be able to establish the area of the triangle. Heron of Alexandria (ca. 62 C.E.) developed a nifty formula that enables us to find the area of a triangle when the only information that we are given is the lengths of its sides. This formula for the area of triangle ABC shown Figure 2-5 is $area \triangle ABC = \sqrt{s(s-a)(s-b)(s-c)}$, where $s = \frac{a+b+c}{2}$, which is the semi-perimeter of the triangle.

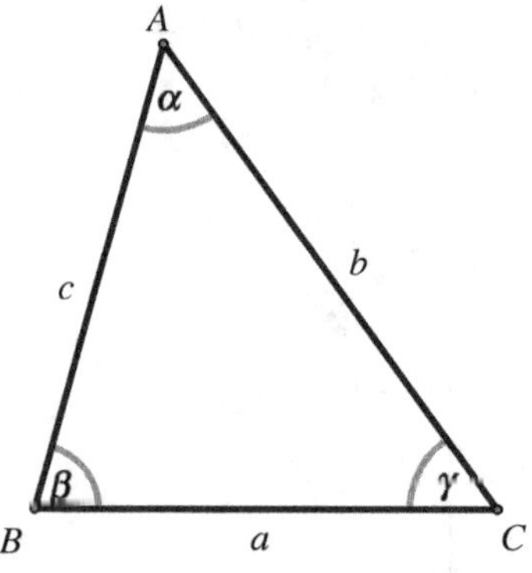

Figure 2-5

This formula is explained further on in this chapter. Applying this formula is quite simple. For practice, we shall apply it to a triangle whose sides have lengths 13, 14, and 15 units, which then has a semi-perimeter of 21. Thus, with Heron's formula, we get the area of the triangle as

$$\sqrt{21\cdot(21-13)\cdot(21-14)\cdot(21-15)} = \sqrt{21\cdot 8\cdot 7\cdot 6} = \mathbf{84}.$$

There are lots of other formulas for finding the area of a triangle, each requiring the measure of various parts of the triangle. We offer just some of these triangle-area formulas here, referring to Figure 2-6:

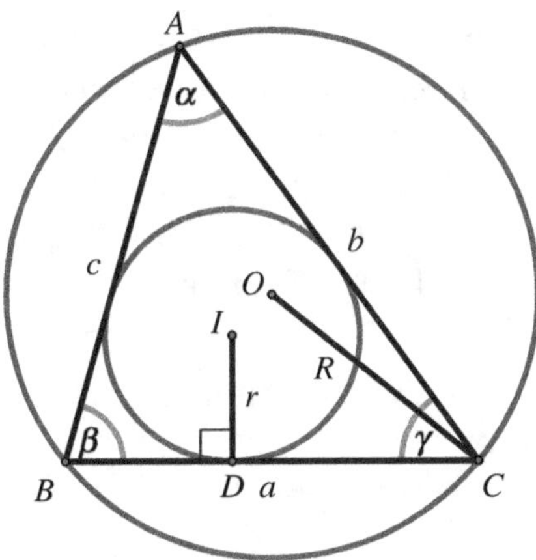

Figure 2-6

$$area\triangle ABC = \frac{abc}{4R} \quad (R = \text{radius of the circumscribed circle})$$

$$area\triangle ABC = r\cdot s \quad (r = \text{radius of the inscribed circle, and}$$
$$s = \text{semi-perimeter})$$

$$area\triangle ABC = \frac{\tan\alpha}{4}\,(b^2 + c^2 - a^2) \quad (\text{when } \alpha \neq 90°)$$

$$area\triangle ABC = \frac{a^2}{2}\cdot\frac{\sin\beta\cdot\sin\gamma}{\sin\alpha}$$

$$area\triangle ABC = \frac{1}{4}\cdot\frac{a^2 + b^2 + c^2}{\cot\alpha + \cot\beta + \cot\gamma}$$

$$area\triangle ABC = \frac{h_a h_b}{2\sin\gamma}$$

$$area\triangle ABC = \frac{R\cdot h_a h_b}{c}$$

$$area\triangle ABC = \frac{3}{4}\sqrt{m(m-m_a)(m-m_b)(m-m_c)}, \text{ where } m_a, m_b \text{ and } m_c$$
$$\text{are the medians of } \triangle ABC, \text{ and } m = \frac{m_a + m_b + m_c}{2}.$$

Derivation of Heron's Formula for the Area of a Circle

$$area\triangle ABC = \sqrt{s(s-a)(s-b)(s-c)}, \text{ where } s = \frac{a+b+c}{2}.$$

There is a nice geometric proof—attributed to Heron—provided in Thomas Heath's *A Manual of Greek Mathematics* (Dover, 1963). To conserve space, we will provide a much shorter trigonometric derivation building on two high school-level relationships: the Law of Cosines ($c^2 = a^2 + b^2 - 2ab \cos C$) and the Pythagorean identity ($\sin^2 \theta + \cos^2 \theta = 1$). The formula for the area of a triangle is $area\triangle ABC = \frac{1}{2}ab\sin C$.

With $\cos^2 C = \frac{(a^2+b^2-c^2)^2}{4a^2b^2}$, we substitute $\sin C$ in this equation for the area of a triangle to get:

$$area\triangle ABC = \frac{1}{2}ab\sin C = \frac{1}{2}ab\sqrt{1-\cos^2 C} = \frac{1}{2}ab\sqrt{1-\frac{(a^2+b^2-c^2)^2}{4a^2b^2}}$$
$$= \frac{1}{2}ab\sqrt{\frac{4a^2b^2-(a^2+b^2-c^2)^2}{4a^2b^2}} = \frac{1}{4}\sqrt{4a^2b^2-(a^2+b^2-c^2)^2}.$$

We now must factor the term under the radical sign:

$$4a^2b^2 - (a^2+b^2-c^2)^2 = -(a+b+c)\cdot(a+b-c)\cdot(a-b-c)\cdot(a-b+c)$$
$$= -(a+b+c)\cdot(a+b-c)\cdot[-(-a+b+c)]\cdot(a-b+c)$$
$$= (a+b+c)\cdot(a+b-c)\cdot(-a+b+c)\cdot(a-b+c).$$

With $a + b + c = 2s$, $a + b - c = 2(s-c)$, $-a + b + c = 2(s-a)$, and $a - b + c = 2(s-b)$, we get $area\triangle ABC = \frac{1}{4}\sqrt{2s\cdot 2(s-c)\cdot 2(s-a)\cdot 2(s-b)} = \sqrt{s\cdot(s-a)\cdot(s-b)\cdot(s-c)}.$

Brahmagupta's Theorem for the Area of a Quadrilateral

Now that you are familiar with Heron's formula for finding the area of any triangle given only the lengths of its three sides, it is natural to try to extend this formula to quadrilaterals. One common way is to consider the triangle as a quadrilateral with a zero-length side. Such an extension is credited to Brahmagupta,[5] an Indian mathematician who lived in the early part of the seventh century. Brahmagupta used the following formula to find the area of a *cyclic quadrilateral*, that is, a quadrilateral that may be inscribed in a circle, with sides of lengths a, b, c, and d, and where s is the semiperimeter: Area $= \sqrt{(s-a)(s-b)(s-c)(s-d)}$. You will notice that Brahmagupta considered Heron's formula to treat the triangle as a quadrilateral with $d = 0$.

Proof of Brahmagupta's Formula: First consider the case where quadrilateral $ABCD$ is a rectangle with $a = c$ and $b = d$. Using Brahmagupta's formula, we have area of rectangle $ABCD$:

$$= \sqrt{(s-a)(s-b)(s-c)(s-d)}$$
$$= \sqrt{(a+b-a)(a+b-b)(a+b-a)(a+b-b)}$$
$$= \sqrt{a^2 b^2} = ab,$$

which is the area of the rectangle as found by the traditional methods.

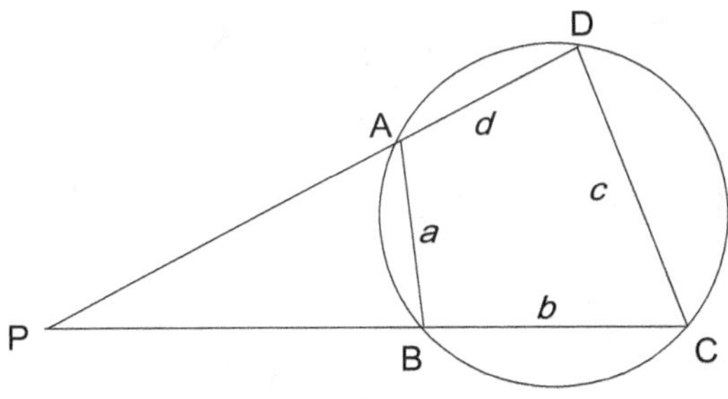

Figure 2-7

---

[5] In 628, Brahmagupta (born 598) wrote his *Brahma-sphuta-siddhānta* ("The Revised System of Brahma"), the twelfth and thirteenth chapters of which are devoted to mathematics.

Consider any non-rectangular cyclic quadrilateral *ABCD*, as shown in Figure 2-7. Extend *DA* and *CB* to meet at *P*, forming $\triangle DCP$. Let $PC = x$ and $PD = y$. By Heron's formula,

$$\text{area of } \triangle DCP = \frac{1}{4}\sqrt{(x+y+c)(y-x+c)(x+y-c)(x-y+c)}. \quad \text{(I)}$$

Since $\angle CDA$ is supplementary to $\angle CBA$, and $\angle ABP$ is also supplementary to $\angle CBA$,

$$\angle CDA = \angle ABP. \text{ Then, } \triangle BAP \sim \triangle DCP. \quad \text{(II)}$$

From (II) we get

$$\frac{area\triangle BAP}{area\triangle BAP} = \frac{a^2}{c^2},$$

$$\frac{area\triangle DCP}{area\triangle DCP} - \frac{area\triangle BAP}{area\triangle DCP} = \frac{c^2}{c^2} - \frac{a^2}{c^2},$$

$$\frac{area\triangle DCP - area\triangle BAP}{area\triangle DCP} = \frac{areaABCD}{area\triangle DCP} = \frac{c^2 - a^2}{c^2}. \quad \text{(III)}$$

From (II) we also get

$$\frac{x}{c} = \frac{y-d}{a}, \quad \text{(IV)}$$

and

$$\frac{y}{c} = \frac{x-b}{a}. \quad \text{(V)}$$

We can add (IV) and (V) to get $\frac{x+y}{c} = \frac{x+y-b-d}{a}$, and $x+y = \frac{c}{c-a}(b+d)$, so that

$$x+y+c = \frac{c}{c-a}(b+c+d-a). \quad \text{(VI)}$$

The following relationships are found by using similar methods:

$$y-x+c = \frac{c}{c+a}(a+c+d-b). \quad \text{(VII)}$$

$$x + y - c = \frac{c}{c-a}(a+b+d-c). \qquad\qquad \text{(VIII)}$$

$$x - y + c = \frac{c}{c+a}(a+b+c-d). \qquad\qquad \text{(IX)}$$

Substitute (VI), (VII), (VIII), and (IX) into (I) to get the area of

$$\Delta DCP = \frac{c^2}{4(c^2-a^2)}\sqrt{(b+c+d-a)(a+c+d-b)(a+b+d-c)(a+b+c-d)}.$$

Since (III) may be read as the area of $\Delta DCP = \frac{c^2}{c^2-a^2}$ (area $ABCD$), the area of cyclic quadrilateral $ABCD = \sqrt{(s-a)(s-b)(s-c)(s-d)}$.

An interesting extension of Brahmagupta's formula to the general quadrilateral is given without proof. The area of any (convex) quadrilateral $= \sqrt{(s-a)(s-b)(s-c)(s-d) - abcd \cdot \cos^2\left(\frac{\alpha+\gamma}{2}\right)}$, where a, b, c, and d are the lengths of the sides, $s = \frac{a+b+c+d}{2}$, and α and γ are the measures of a pair of opposite angles of the quadrilateral. This formula shows that of all the quadrilaterals that can be formed from four sides with given lengths, the one with the maximum area is the cyclic quadrilateral. The maximum area is achieved when $abcd \cdot \cos^2\left(\frac{\alpha+\gamma}{2}\right) = 0$, which occurs when $\alpha + \gamma = 180°$, a fact true only in cyclic quadrilaterals.

Brahmagupta also found that for a cyclic quadrilateral of consecutive sides lengths a, b, c, and d, where m and n are the lengths of the quadrilateral's internal diagonals, the following relationship holds true:

$$m^2 = \frac{(ab+cd)(ac+bd)}{ad+bc}$$

$$n^2 = \frac{(ac+bd)(ad+bc)}{ab+cd}.$$

Ptolemy's Theorem for Cyclic Quadrilaterals

Perhaps the most famous theorem involving cyclic quadrilaterals is that attributed to Claudius Ptolemaeus of Alexandria (c. 100–170 CE), commonly known as Ptolemy. In his major astronomical work the

Almagest[6] (ca. 150 CE), Ptolemy states this theorem on cyclic quadrilaterals.

Ptolemy's Theorem: The product of the lengths of the diagonals of a cyclic quadrilateral equals the sum of the products of the lengths of the pairs of opposite sides.

Two methods for proving Ptolemy's theorem are provided. The second method incorporates the proof of the converse as well.

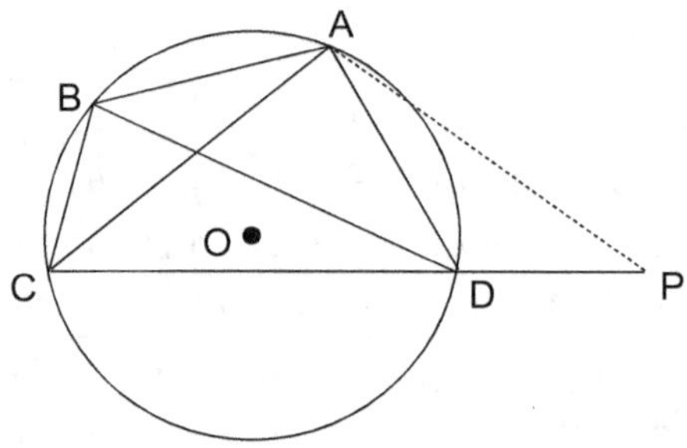

Figure 2-8

Ptolemy's Theorem Proof 1: In Figure 2-8, quadrilateral *ABCD* is inscribed in circle *O*. A line is drawn through *A* to meet *CD* at *P* so that

$$\angle BAC = \angle DAP. \tag{I}$$

Since quadrilateral *ABCD* is cyclic, $\angle ABC$ is supplementary to $\angle ADC$. However, $\angle ADP$ is also supplementary to $\angle ADC$.
Therefore

$$\angle ABC = \angle ADP. \tag{II}$$

---

[6] The Greek title, *Syntaxis Mathematica*, means mathematical (or astronomical) compilation. The Arabic title, *Almagest*, is a renaming meaning "great collection (or compilation)." The book is a manual of all of the mathematical astronomy that the ancients knew at that time. Book I of the thirteen books which comprise this monumental work contains the theorem (6.11) that now bears Ptolemy's name.

Thus,

$$\triangle BAC \sim \triangle DAP \text{ (AA)}, \tag{III}$$

and

$$\frac{AB}{AD} = \frac{BC}{DP}, \quad \text{or} \quad DP = \frac{(AD)(BC)}{AB}. \tag{IV}$$

From (I), $\angle BAD = \angle CAP$, and from (III), $\frac{AB}{AD} = \frac{AC}{AP}$.
Therefore, $\triangle ABD \sim \triangle ACP$ (SAS), and

$$\frac{BD}{CP} = \frac{AB}{AC}, \quad \text{or} \quad CP = \frac{(AC)(BD)}{AB}. \tag{V}$$

So that

$$CP = CD + DP. \tag{VI}$$

Substituting (IV) and (V) into (VI),

$$\frac{(AC)(BD)}{AB} = CD + \frac{(AD)(BC)}{AB}.$$

Thus, $(AC)(BD) = (AB)(CD) + (AD)(BC)$.

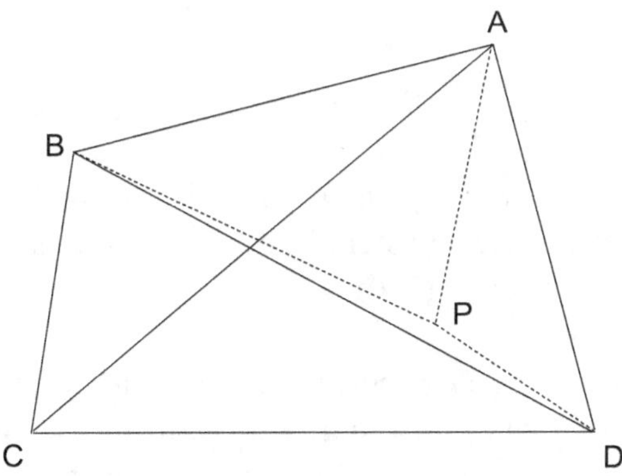

Figure 2-9

Ptolemy's Theorem Proof 2: In quadrilateral $ABCD$, as shown in Figure 2-9, we draw $\triangle DAP$ on side AD similar to $\triangle CAB$. Thus,

$$\frac{AB}{AP} = \frac{AC}{AD} = \frac{BC}{PD} \qquad \text{(I)}$$

and

$$(AC)(PD) = (AD)(BC). \qquad \text{(II)}$$

Since $\angle BAC = \angle PAD$, then $\angle BAP = \angle CAD$. Therefore, from (I), $\triangle BAP \sim \triangle CAD$ (SAS), and $\frac{AB}{AC} = \frac{BP}{CD}$, or

$$(AC)(BP) = (AB)(CD). \qquad \text{(III)}$$

Adding (II) and (III), we have

$$(AC)(BP + PD) = (AD)(BC) + (AB)(CD). \qquad \text{(IV)}$$

Now $BP + PD > BD$ (triangle inequality), unless P is on DB. However, P will be on DB if and only if $\angle ADP = \angle ADB$. But we already know that $\angle ADP = \angle ACB$ (similar triangles). And if $ABCD$ were cyclic, then $\angle ADB$ would equal $\angle ACB$, and $\angle ADB$ would equal $\angle ADP$. Therefore, we can state that if (and only if) $ABCD$ is cyclic, P lies on BD.

This tells us that

$$BP + PD = BD. \qquad \text{(V)}$$

Substituting (V) into (IV), we get $(AC)(BD) = (AD)(BC) + (AB)(CD)$.

Notice we have proved Ptolemy's theorem and its converse, which we now state as our next theorem.

Converse of Ptolemy's Theorem: If the product of the lengths of the diagonals of a quadrilateral equals the sum of the products of the lengths of the pairs of opposite sides, then the quadrilateral is cyclic.

Assume quadrilateral $ABCD$ is not cyclic (see Figure 2-7). If there is a line CDP, then $\angle ADP \neq \angle ABC$. If C, D, and P are not collinear, then it is possible to have $\angle ADP = \angle ABC$. However, then $CP < CD + DP$, and from steps (IV) and (V) in Proof I, above, $(AC)(BD) < (AB)(CD) + (AD)(BC)$. But this contradicts the given information that $(AC)(BD) = (AB)(CD) + (AD)(BC)$. Therefore, quadrilateral $ABCD$ is cyclic.

Ceva's Theorem on Concurrency in a Triangle

We now consider one of the strongest theorems determining concurrency in a triangle. The theorem is often attributed to Giovanni Ceva (1647–1734), who published it in his 1678 work *De lineis rectis se invicem secantibus statica constructio*. However, it is believed to have been proven much earlier by Yusuf Al-Mu'taman ibn Hűd, an eleventh-century king of Zaragoza.

Ceva's Theorem: The three lines containing the vertices A, B, and C of $\triangle ABC$ and intersecting the opposite sides at points L, M, and N, respectively, are concurrent if and only if $\frac{AN}{NB} \cdot \frac{BL}{LC} \cdot \frac{CM}{MA} = 1$.

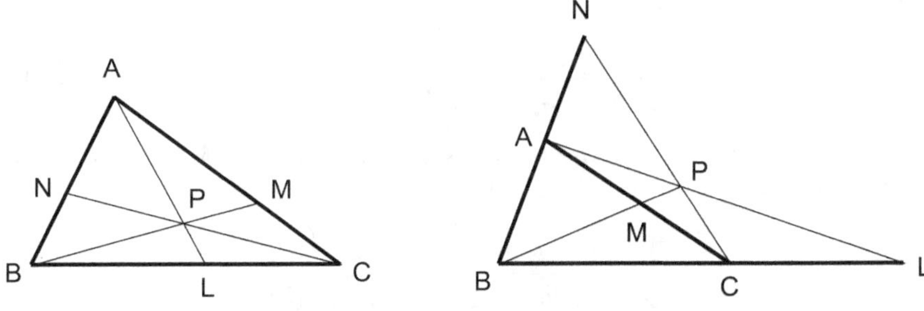

Figure 2-10

To prove Ceva's theorem, we must note that there are two possible situations in which the three lines drawn from the vertices may intersect the sides and still be concurrent, as shown in Figure 2-10. It is perhaps easier to follow the proof with the left-side diagram and verify the validity of the statements in the right-side diagram. In any case, the statements made in the proof hold for *both* diagrams.

Ceva's theorem is an equivalence (or biconditional) and therefore requires two proofs (one the converse of the other). We offer three proofs. We shall first prove that if the three lines containing the vertices of $\triangle ABC$ and intersecting the opposite sides in points L, M, and N, respectively, are concurrent, then $\frac{AN}{NB} \cdot \frac{BL}{LC} \cdot \frac{CM}{MA} = 1$. This first proof, although not the simplest, requires no auxiliary lines.

Ceva's Theorem Proof 1: In Figure 2-10, AL, BM, and CN meet in point P. Since $\triangle ABL$ and $\triangle ACL$ share the same altitude (from point A),

$$\frac{area\,\triangle ABL}{area\,\triangle ACL} = \frac{BL}{LC}. \tag{I}$$

Similarly,

$$\frac{area\,\triangle PBL}{area\,\triangle PCL} = \frac{BL}{LC}. \tag{II}$$

From (I) and (II): $\dfrac{area\,\triangle ABL}{area\,\triangle ACL} = \dfrac{area\,\triangle PBL}{area\,\triangle PCL}.$

A basic property of proportions $\left(\frac{w}{x} = \frac{y}{z} = \frac{w-y}{x-z}\right)$ provides that:

$$\frac{BL}{LC} = \frac{area\,\triangle ABL - area\,\triangle PBL}{area\,\triangle ACL - area\,\triangle PCL} = \frac{area\,\triangle ABP}{area\,\triangle ACP}. \tag{III}$$

We now repeat the same process as above using BM instead of AL. Here:

$$\frac{CM}{MA} = \frac{area\,\triangle BMC}{area\,\triangle BMA} = \frac{area\,\triangle PMC}{area\,\triangle PMA}.$$

It follows that:

$$\frac{CM}{MA} = \frac{area\,\triangle BMC - area\,\triangle PMC}{area\,\triangle BMA - area\,\triangle PMA} = \frac{area\,\triangle BCP}{area\,\triangle BAP}. \tag{IV}$$

Once again, we repeat the process, now using CN where AL was used earlier:

$$\frac{AN}{NB} = \frac{area\,\triangle ACN - area\,\triangle APN}{area\,\triangle BCN - area\,\triangle BPN}.$$

This permits us to get:

$$\frac{AN}{NB} = \frac{area\,\triangle ACN - area\,\triangle APN}{area\,\triangle BCN - area\,\triangle BPN} = \frac{area\,\triangle ACP}{area\,\triangle BCP}. \qquad (V)$$

We now simply multiply (III), (IV), and (V) to get the desired result:

$$\frac{BL}{LC} \cdot \frac{CM}{MA} \cdot \frac{AN}{NB} = \frac{area\,\triangle ABP}{area\,\triangle ACP} \cdot \frac{area\,\triangle BCP}{area\,\triangle BAP} \cdot \frac{area\,\triangle ACP}{area\,\triangle BCP} = 1.$$

By introducing an auxiliary line, we can produce a simpler proof.

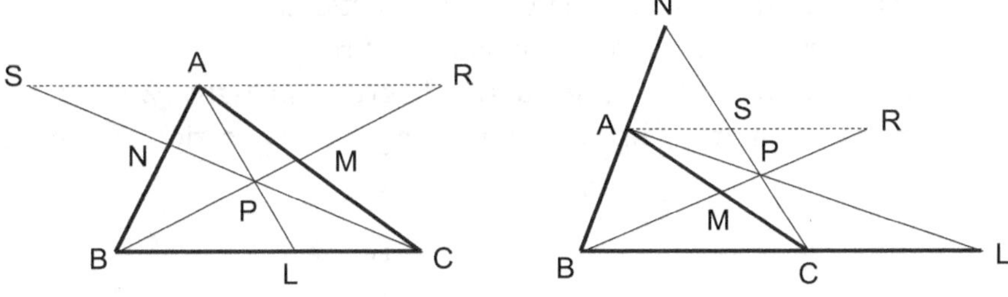

Figure 2-11

Ceva's Theorem Proof 2: Consider Figure 2-10 but include a line parallel to BC that contains vertex A and intersects CP at S and BP at R, as shown in Figure 2-11. The parallel lines enable us to establish the following pairs of similar triangles:

$$\triangle AMR \sim \triangle CMB, \text{ therefore } \frac{AM}{MC} = \frac{AR}{CB} \qquad (I)$$

$$\triangle BNC \sim \triangle ANS, \text{ therefore } \frac{BN}{NA} = \frac{CB}{SA} \qquad (II)$$

$$\triangle CLP \sim \triangle SAP, \text{ therefore } \frac{CL}{SA} = \frac{LP}{AP} \qquad (III)$$

$$\triangle BLP \sim \triangle RAP, \text{ therefore } \frac{BL}{RA} = \frac{LP}{AP}. \qquad (IV)$$

From (III) and (IV) we get

$$\frac{CL}{SA} = \frac{BL}{RA}.$$

This can be rewritten as

$$\frac{CL}{BL} = \frac{SA}{RA}. \tag{V}$$

Now by multiplying (I), (II), and (V), we obtain our desired result:

$$\frac{AM}{MC} \cdot \frac{BN}{NA} \cdot \frac{CL}{BL} = \frac{AR}{CB} \cdot \frac{CB}{SA} \cdot \frac{SA}{RA} = 1.$$

We rearrange the terms and invert the fractions to get $\frac{AN}{NB} \cdot \frac{BL}{LC} \cdot \frac{CM}{MA} = 1$ (which reads the same as the conclusion of Proof 1).

By adding two auxiliary lines to the diagrams in Figure 2-9, we are able to produce another proof, again using the properties of similar triangles.

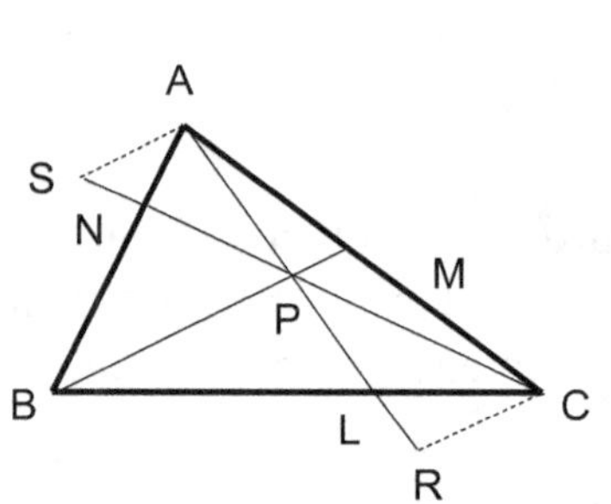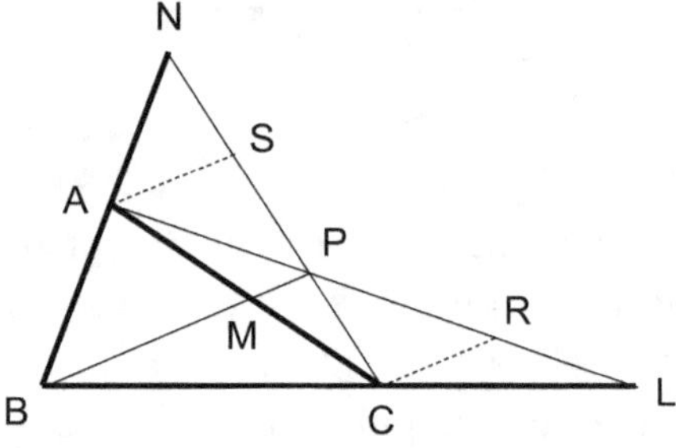

Figure 2-12

Ceva's Theorem Proof 3: We begin with the diagrams shown in Figure 2-9 but add two lines to each. We draw a line through A and a line through C, each parallel to BP and intersecting CP and AP at points S and R, respectively, as seen in Figure 2-12.

$$\triangle ASN \sim \triangle BPN, \text{ which yields } \frac{AN}{NB} = \frac{AS}{BP}. \tag{I}$$

$$\triangle BPL \sim \triangle CRL, \text{ which yields } \frac{BL}{LC} = \frac{BP}{CR}. \tag{II}$$

$\triangle PAM \sim \triangle RAC$, which yields $\frac{CA}{MA} = \frac{RC}{PM}$. This can be written as

$$CA = \frac{(RC)(MA)}{PM}. \tag{III}$$

$\triangle PCM \sim \triangle SCA$, which yields $\frac{CM}{CA} = \frac{PM}{AS}$. This can be written as

$$CA = \frac{(AS)(CM)}{PM}. \tag{IV}$$

From (III) and (IV):

$$\frac{(RC)(MA)}{PM} = \frac{(AS)(CM)}{PM},$$

which then can be written as

$$\frac{CM}{MA} = \frac{RC}{AS}. \tag{V}$$

This gives us

$$\frac{AN}{NB} \cdot \frac{BL}{LC} \cdot \frac{CM}{MA} = \frac{AS}{BP} \cdot \frac{BP}{CR} \cdot \frac{RC}{AS} = 1.$$

To complete the proof of Ceva's theorem, we must now prove the converse of the implication proved above. That is, we shall now prove that, if the lines containing the vertices of $\triangle ABC$ intersect the opposite sides in points L, M, and N, respectively, so that $\frac{AN}{NB} \cdot \frac{BL}{LC} \cdot \frac{CM}{MA} = 1$, then the lines AL, BM, and CN are concurrent.

Suppose BM and AL intersect at P. Draw PC and call its intersection with AB point N'. Now that AL, BM, and CN' are concurrent, we can use the part of Ceva's theorem proved earlier to state the following: $\frac{AN'}{N'B} \cdot \frac{BL}{LC} \cdot \frac{CM}{MA} = 1$. But our hypothesis stated that $\frac{AN}{NB} \cdot \frac{BL}{LC} \cdot \frac{CM}{MA} = 1$. Therefore, $\frac{AN'}{N'B} = \frac{AN}{NB}$, so that N and N' must coincide and, thereby proving concurrency.

The Concept of Duality

Many statements in geometry involve relationships between points and lines. When, in a statement concerning points and lines in a plane, the word *point* is replaced by the word *line* and the word *line* is replaced by the word *point* each time these words are used, the newly-formed statement is the *dual* of the original statement. (Occasionally, some other modifications may be necessary in order to preserve proper sentence structure.) The principle of duality relates concurrency of lines and the collinearity of points. Let us first consider the principle of duality with the following examples of dual statements:

STATEMENT	DUAL STATEMENT
1. Two distinct points determine a unique line.	1. Two distinct lines determine a unique point.
2. Any point contains an infinite number of lines.	2. Any line contains an infinite number of points.
3. Only one triangle is determined by three non-collinear points.	3. Only one trilateral is determined by three non-concurrent lines.

This third example of duality implies that other related words need also be changed when forming the dual of a statement. Specifically, notice that *collinear* and *concurrent* are dual words, as are *triangle* and *trilateral*.

Recall Ceva's theorem, where the three lines containing the vertices A, B, and C of $\triangle ABC$ and intersecting the opposite sides at points L, M, and N, respectively, are concurrent if and only if $\frac{AN}{NB} \cdot \frac{BL}{LC} \cdot \frac{CM}{MA} = 1$. (See Figure 2-13.) For the most part, the dual of a postulate is also a postulate; the dual of a definition is itself a definition. Thus, if a statement is a theorem, its dual is also likely to be a theorem,[7] although a valid proof would be needed to establish the statement as a theorem.

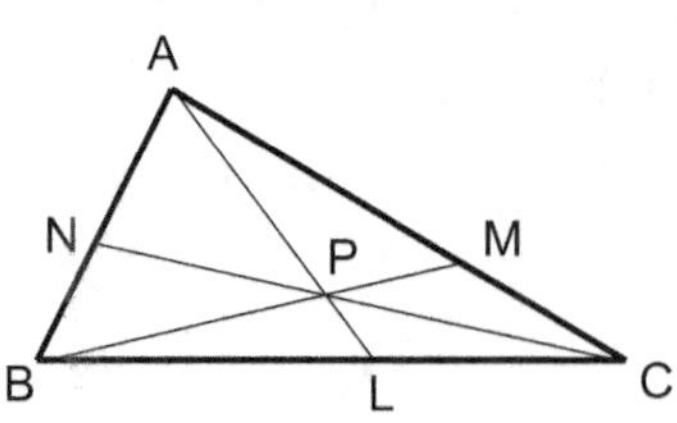
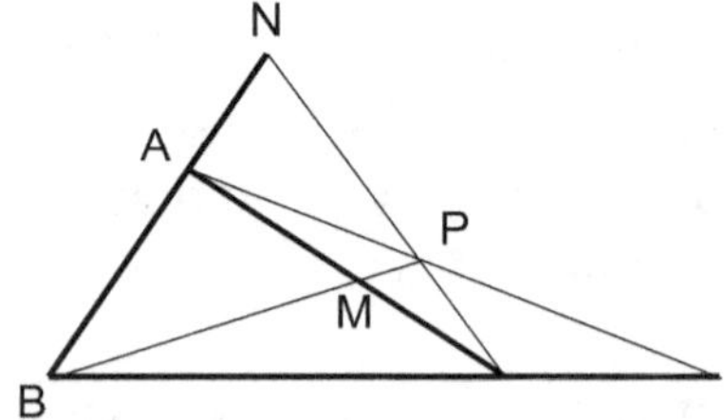

Figure 2-13

---

[7] In a geometric system, exclusively based on postulates and definitions whose respective duals are all true, the dual of every theorem is also a theorem. This claim is easily justified by realizing that the proof of the dual of a theorem can be produced by simply replacing each statement of the proof of the original theorem by its dual.

This is precisely what we will now investigate. With our knowledge of duality, we shall form the dual statement of Ceva's theorem. Actually, it was the rediscovery of Menelaus of Alexandria's famous but forgotten[8] theorem, which will be presented next, that led Giovanni Ceva in 1678 to publish is first book, *De lineis rectis se invicem secantibus statica constructio* to produce his theorem by the principle of duality. Notice the duality relationship between these two theorems.

Menelaus' Theorem on Collinearity

Greek mathematician Menelaus of Alexandria (70–130 CE) published a work entitled *Sphaerica* circa 100 CE, in which he produces the well-known theorem presented here. Menelaus used the plane version to develop a spherical analogue,[9] which was the true purpose of his treatise. This theorem, which today bears Menelaus' name, did not become popular until it was rediscovered by Giovanni Ceva as a part of his work in 1678.

Menelaus' Theorem: In Figure 2-14, the three points P, Q, and R on the sides AC, AB, and BC, respectively, of $\triangle ABC$, are collinear if and only if $\frac{AQ}{QB} \cdot \frac{BR}{RC} \cdot \frac{CP}{PA} = -1$.[10]

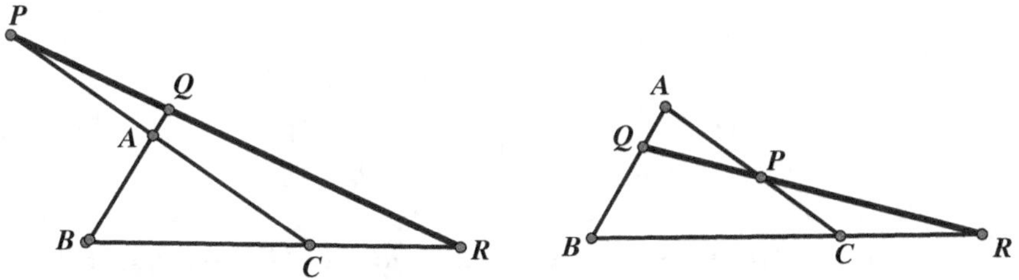

Figure 2-14

---

[8] During the Dark Ages, much of classical Greek mathematics was lost and forgotten.

[9] The spherical analogue to Menelaus' theorem for spherical triangle ABC is

$$\frac{\sin \widehat{AQ}}{\sin \widehat{QB}} \cdot \frac{\sin \widehat{BR}}{\sin \widehat{RC}} \cdot \frac{\sin \widehat{CP}}{\sin \widehat{PA}} = -1.$$

[10] The reason for the negative sign is explained in the proof of this theorem.

Like Ceva's theorem, Menelaus' theorem is an equivalence and, therefore, requires proofs for each of the two statements (converses of each other) that comprise the entire theorem. We shall first prove that if the three points P, Q, and R on the sides AC, AB, and BC, respectively, are collinear, then $\frac{AQ}{BQ} \cdot \frac{BR}{RC} \cdot \frac{CP}{PA} = -1$.

Two proofs of this part of Menelaus' Theorem follow.

Proof of Menelaus' Theorem 1: Draw a line containing C, parallel to AB, and intersecting PQR or QPR at point D, as can be seen in Figure 2-15.

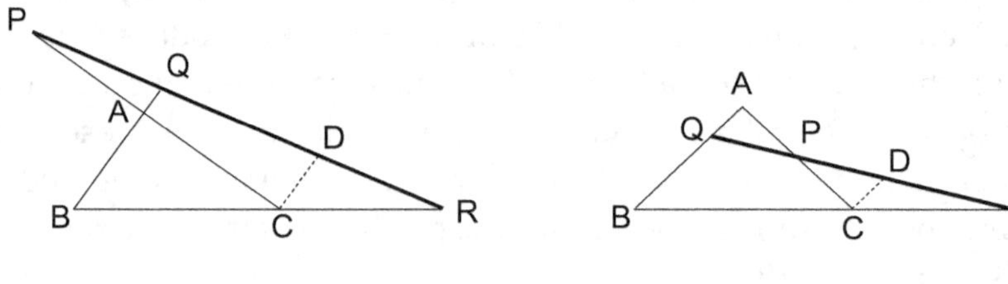

Figure 2-15

We then have $\triangle DCR \sim \triangle QBR$, therefore

$$\frac{DC}{QB} = \frac{RC}{BR}, \quad \text{or} \quad DC = \frac{(QB)(RC)}{BR}. \tag{I}$$

Also, $\triangle PDC \sim \triangle PQA$, therefore

$$\frac{DC}{AQ} = \frac{CP}{PA}, \quad \text{or} \quad DC = \frac{(AQ)(CP)}{PA}. \tag{II}$$

From (I) and (II), $\frac{(QB)(RC)}{BR} = \frac{(AQ)(CP)}{PA}$, and $(QB)(RC)(PA) = (AQ)(CP)(BR)$, or $\frac{AQ}{QB} \cdot \frac{BR}{RC} \cdot \frac{CP}{PA} = 1$.

By taking direction into account in the left-hand diagram of Figure 2-15, we see that $\frac{AQ}{QB}$, $\frac{BR}{RC}$, and $\frac{CP}{PA}$ are each negative ratios, and in the right-hand diagram of Figure 2-15, $\frac{BR}{RC}$ is a negative ratio, while $\frac{AQ}{QB}$ and $\frac{CP}{PA}$ are positive ratios. Therefore, $\frac{AQ}{QB} \cdot \frac{BR}{RC} \cdot \frac{CP}{PA} = -1$, since in each case there is an odd number of negative ratios.

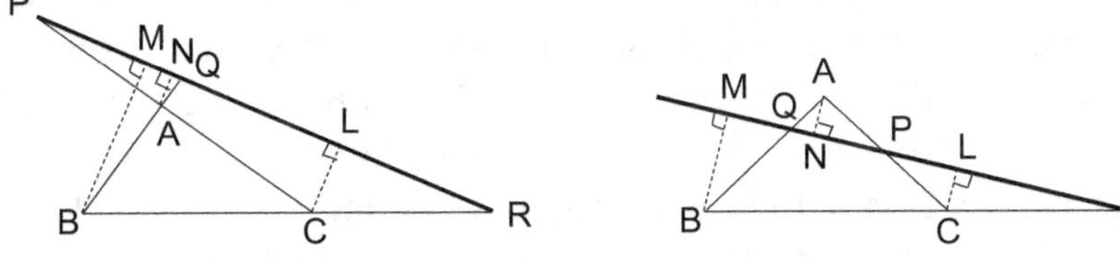

Figure 2-16

Proof of Menelaus' Theorem 2: Once again, we begin by assuming collinearity of P, Q, and R in Figure 2-16.

Draw $BM \perp PR$, $AN \perp PR$, and $CL \perp PR$. Since $\triangle BMQ \sim \triangle ANQ$,

$$\frac{AQ}{QB} = \frac{AN}{BM}. \tag{I}$$

Also $\triangle LCP \sim \triangle NAP$, and $\dfrac{CP}{PA} = \dfrac{LC}{AN}$; $\tag{II}$

and $\triangle MRB \sim \triangle LRC$, and $\dfrac{BR}{RC} = \dfrac{BM}{LC}$. $\tag{III}$

By multiplying (I), (II), and (III), we get, numerically:

$$\frac{AQ}{QB} \cdot \frac{CP}{PA} \cdot \frac{BR}{RC} = \frac{AN}{BM} \cdot \frac{LC}{AN} \cdot \frac{BM}{LC} = 1.$$

In the left-hand diagram of Figure 2-16, $\frac{AQ}{QB}$ is negative, $\frac{CP}{PA}$ is negative, and $\frac{BR}{RC}$ is negative.

Therefore,

$$\frac{AQ}{QB} \cdot \frac{CP}{PA} \cdot \frac{BR}{RC} = -1.$$

In the right-hand diagram of Figure 2-16, $\frac{AQ}{QB}$ is positive, $\frac{CP}{PA}$ is positive, and $\frac{BR}{RC}$ is negative.

Therefore,

$$\frac{AQ}{QB} \cdot \frac{CP}{PA} \cdot \frac{BR}{RC} = -1.$$

To complete the proof of Menelaus' theorem, we must now prove the *converse* of the above theorem. We shall prove that if the three points P, Q, and R are on the sides AC, AB, and BC, respectively, and if $\frac{AQ}{QB} \cdot \frac{BR}{RC} \cdot \frac{CP}{PA} = -1$, then points P, Q, and R are collinear.

In Figure 2-14, let the line containing R and Q intersect AC at P'. Using the portion of the theorem just proved, we know that $\frac{AQ}{QB} \cdot \frac{BR}{RC} \cdot \frac{CP'}{P'A} = -1$. However, our hypothesis tells us that $\frac{AQ}{QB} \cdot \frac{BR}{RC} \cdot \frac{CP}{PA} = -1$. Therefore, $\frac{CP'}{P'A} = \frac{CP}{PA}$, which indicates that P and P' must coincide. This proves the collinearity. Menelaus' theorem provides us with a useful method for proving points collinear.

Napoleon's Theorem

We will introduce this theorem by discussing a relevant point within the triangle, namely, the point inside the triangle at which congruent angles are formed by drawing rays from this point to the vertices. This is the foundation of the Napoleon theorem.

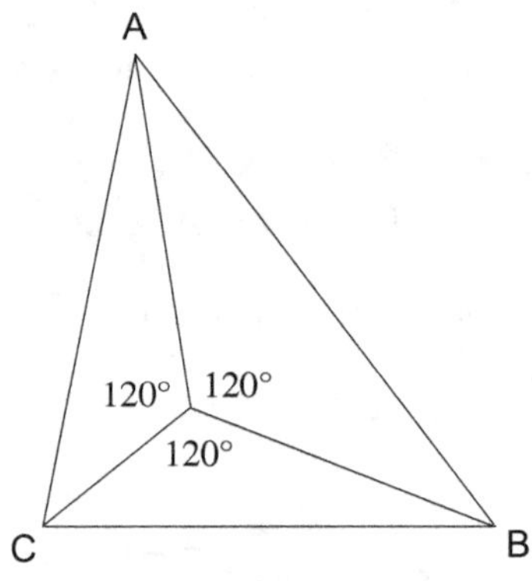

Figure 2-17

We aim to locate the point in Figure 2-17, which indicates the equal angles created in a triangle. To get this point we must first find another point with an interesting property. Begin by constructing an equilateral triangle externally on each of the three sides of the given triangle, as seen in Figure 2-18. Draw segments joining each vertex of the given triangle with the remote vertex of the equilateral triangle on the opposite side. The following theorem presents an astonishing property of these three line segments. After we prove this property, we can return to our original problem.

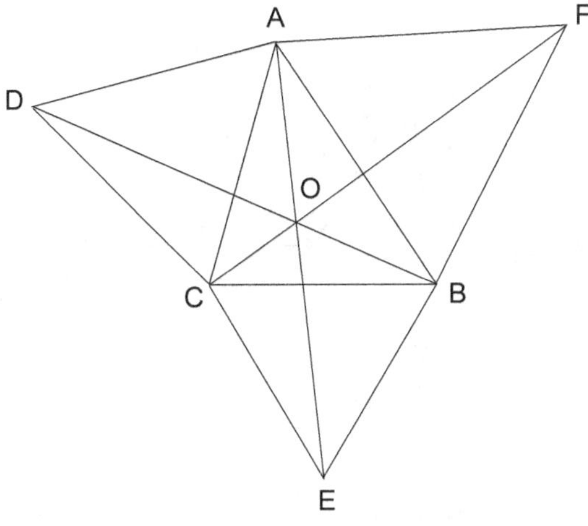

Figure 2-18

Pre-Napoleon's Theorem 1: The segments joining each vertex of a given triangle to the remote vertex of the equilateral triangle drawn externally on the opposite side of the given triangle are congruent.

Proof of Pre-Napoleon's Theorem 1: The plan of the proof is to prove $DB = AE$ and $AE = CF$ by first proving $\triangle DCB \cong \triangle ACE$ and then $\triangle EBA \cong \triangle CBF$. We notice that since $\angle DCA = \angle ECB = 60°$, $\angle DCB = \angle ACE$ (by addition). Also, since we have equilateral triangles, $DC = AC$ and $CB = CE$. Therefore, $\triangle DCB \cong \triangle ACE$ (SAS) and $DB = AE$. In a similar manner, we may prove that $\triangle EBA \cong \triangle CBF$. This enables us to conclude that $AE = CF$. Thus, $DB = AE = CF$. From the diagram in Figure 2-18, it appears that DB, AE and CF are concurrent. This observation gives us our next theorem.

Pre-Napoleon's Theorem 2 (Fermat's Theorem): The segments joining each vertex of a given triangle with the remote vertex of the equilateral triangle drawn externally on the opposite side of the given triangle are concurrent. This point of concurrency is called the *Fermat point* of the triangle, named after the French mathematician Pierre de Fermat (1601–1665).

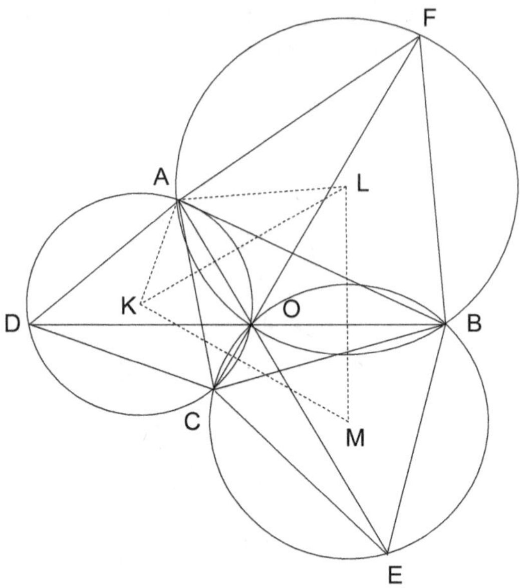

Figure 2-19

Proof of Pre-Napoleon's Theorem 2: Begin by constructing the circumcircle of each of the three equilateral triangles. We will show that the three circles are concurrent at O. Essentially, we will prove that the six segments from O to A, B, C, D, E, and F will determine the three concurrent lines. Consider the circumcircles of the three equilateral triangles $\triangle ACD$, $\triangle ABF$, and $\triangle BCE$.

Let K, L, and M be the centers of these circles shown in Figure 2-19. Circles K and L meet at points O and A. Since $\overset{\frown}{ADC} = 240°$, and $\angle AOC = \frac{1}{2}(\overset{\frown}{ADC})$, therefore, $\angle AOC = 120°$. Similarly, $\angle AOB = \frac{1}{2}(\overset{\frown}{AFB}) = 120°$. Then $\angle COB = 120°$, since a complete revolution $= 360°$. Since, $\overset{\frown}{CEB} = 240°$, $\angle COB$ is an inscribed angle and point O must lie on circle M. Thus, we see that the three circles are concurrent, intersecting at point O. We connect point O with points A, B, C, D, E, and F to find $\angle DOA = \angle AOF = \angle FOB = 60°$, and therefore, $\angle DOB = 180°$, as do angles COF and AOE. Thus it has been proved that DB, AE, and CF are concurrent and intersect at point O, which is also the point of intersection of circles K, L, and M.

The point O is called the *equiangular point* of $\triangle ABC$ since $\angle AOB = \angle AOC = \angle BOC = 120°$.

Before continuing with the equiangular point, let us take advantage of another interesting property. Sources indicate that the following theorem was developed by Napoleon Bonaparte, who took pride in his mathematical talents. Thus, the resulting equilateral triangle is often called the *Napoleon triangle*.

Napoleon's Theorem: The circumcenters of the three equilateral triangles drawn externally on the sides of a given triangle determine an equilateral triangle.

Proof of Napoleon's Theorem: We will prove that the sides of $\triangle KLM$ are proportional to AE, BD, and CF. Begin by considering $\triangle DAC$, shown in Figure 2-19. Since K is the centroid, which is the point of intersection of the medians of $\triangle DAC$, then AK is two-thirds of the length of the altitude (or median). Using the relationships in a 30-60-90 triangle, we find that $AC : AK = \sqrt{3} : 1$. Similarly, in equilateral $\triangle AFB$, $AF : AL = \sqrt{3} : 1$. Therefore, $AC : AK = AF : AL$. Also, $\angle KAC = \angle LAF = 30°$, $\angle CAL = \angle CAL$ (reflexive), and $\angle KAL = \angle CAF$ (addition). Therefore, $\triangle KAL \sim \triangle CAF$. Thus, $CF : KL = CA : AK = \sqrt{3} : 1$.

Similarly, we can prove $DB : KM = \sqrt{3} : 1$, and $AE : ML = \sqrt{3} : 1$. Therefore, $DB : KM = AE : ML = CF : KL$. But since $DB = AE = CF$, as proven earlier, we obtain

$KM = ML = KL$. Therefore, $\triangle KML$ is equilateral.

The Simson Line (or, the Pedal Line)

One of the great injustices in the history of mathematics involves a theorem originally published by William Wallace in Thomas Leybourn's *Mathematical Repository* (1799–1800). Through careless misquotations, this theorem has been erroneously attributed to Robert Simson (1687–1768), a famous Scottish mathematician who produced the first translation of Euclid's *Elements* into English. Despite this slight against Wallace, we shall use the popular name *Simson's theorem*.

Simson's Theorem: The feet of the perpendiculars drawn from any point on the circumcircle of a triangle to the sides of the triangle are collinear.

In Figure 2-20, point P is on the circumcircle of $\triangle ABC$. $PY \perp AC$ at Y, $PZ \perp AB$ at Z, and $PX \perp BC$ at X. According to Simson's (*i.e.,* Wallace's) theorem, points X, Y, and Z are collinear. This line connecting X, Y, and Z is usually referred to as the *Simson Line*.

Although not necessarily the simplest proof, we shall, for the sake of consistency, prove Simson's theorem using Menelaus' theorem, as it is useful for proofs involving collinearity. A second proof will be provided to demonstrate the independence of Simson's theorem.

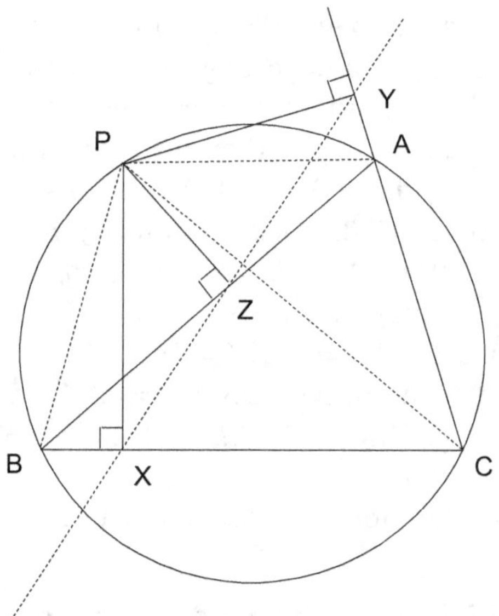

Figure 2-20

Proof of Simson's Theorem 1: In Figure 2-20, draw PA, PB, and PC. We then have $\angle PBA = \frac{1}{2}\overset{\frown}{AP}$ and $\angle PCA = \frac{1}{2}\overset{\frown}{AP}$. Therefore, $\angle PBA = \angle PCA = a$. Thus, $\frac{BZ}{PZ} = Cot\,a = \frac{CY}{PY}$ (in $\triangle PZB$ and $\triangle PYC$), or $\frac{BZ}{PZ} = \frac{CY}{PY}$, which implies

$$\frac{BZ}{CY} = \frac{PZ}{PY}. \tag{I}$$

Similarly, $\angle PAB = \angle PCB$, since both $= \frac{1}{2}\overset{\frown}{PB}$. Therefore, $\frac{AZ}{PZ} = Cot\ \angle PAZ = \frac{CX}{PX}$ (in $\triangle PAZ$ and $\triangle PCX$) or $\frac{AZ}{PZ} = \frac{CX}{PX}$, which implies

$$\frac{CX}{AZ} = \frac{PX}{PZ}. \tag{II}$$

Since $\angle PBC$ and $\angle PAC$ are opposite angles of an inscribed (cyclic) quadrilateral, they are supplementary. However, $\angle PAY$ is also supplementary to $\angle PAC$. Therefore, $\angle PBC = \angle PAY = c$. Thus, $\frac{BX}{PX} = Cot\ \angle PBX = \frac{AY}{PY}$ (in $\triangle PBX$ and $\triangle PAY$) or $\frac{BX}{PX} = \frac{AY}{PY}$, which implies

$$\frac{AY}{BX} = \frac{PY}{PX}. \tag{III}$$

By multiplying (I), (II), and (III) we obtain $\frac{BZ}{CY} \cdot \frac{CX}{AZ} \cdot \frac{AY}{BX} = \frac{PZ}{PY} \cdot \frac{PX}{PZ} \cdot \frac{PY}{PX} = 1$ (or -1, had we considered direction). Thus, by Menelaus' theorem, X, Y, and Z are collinear. These three points determine the Simson line of $\triangle ABC$ with respect to point P.

Proof of Simson's Theorem 2: Consider Figure 2-20 and draw PA, PB, and PC. Since $\angle PYA$ is supplementary to $\angle PZA$, quadrilateral $PZAY$ is cyclic. Therefore,

$$\angle PYC = \angle PZA. \tag{I}$$

Similarly, since $\angle PYC$ is supplementary to $\angle PXC$, quadrilateral $PXCY$ is cyclic, and

$$\angle PYX = \angle PCB. \tag{II}$$

However, quadrilateral $PACB$ is also cyclic, since it is inscribed in the given circumcircle, and therefore

$$\angle PAZ = \angle PCB. \tag{III}$$

From (I), (II), and (III), $\angle PYZ = \angle PYX$, and thus, points X, Y, and Z are collinear.[11]

---

[11] For other proofs of Simson's theorem, see *Challenging Problems in Geometry*, by A.S. Posamentier and C.T. Salkind (Dover, 1996), pp. 43–45.

Stewart's Theorem

A nice challenge in geometry is to find the length of "any" Cevian. A cevian is a line segment that has one endpoint on a vertex of a given triangle and the other endpoint on the opposite side. That is, for $\triangle ABC$, shown in Figure 2-21, if we know the lengths of, AC, BC, AD, and BD our challenge could be to find the length of CD.

This problem was first solved by the Robert Simson, who taught the theorem in his lectures. Simson allowed his prized student Matthew Stewart (1717–1785) to publish in 1746 the theorem in Stewart's *General Theorems of Considerable Use in the Higher Parts of Mathematics*. Simson's generosity helped Stewart to obtain the chair of mathematics at the University of Edinburgh.

It is ironic that Simson was credited with "Simson's theorem," which he did not originate, yet not credited with "Stewart's theorem," which he did. We shall still refer to the theorem as Stewart's, the author of the book in which it was first published.

Simson deserves particular note for his definitive book *The Elements of Euclid*, which was published in 1756 and was in print for over 150 years. This book is the basis for all subsequent study of Euclid's *Elements*, and had a profound influence on the typical high school geometry course currently taught in the United States. We shall first state Stewart's theorem, then prove it. Some applications are provided in the next chapter.

Stewart's Theorem: Using the letter designation in Figure 2-21, Stewart's theorem states the following relationship: $a^2 n + b^2 m = c(d^2 + mn)$.

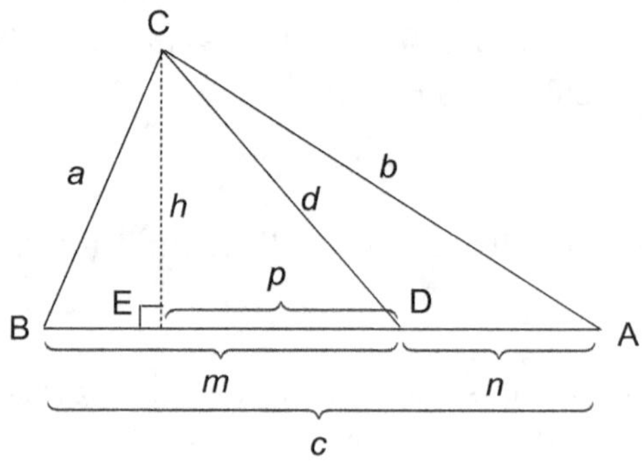

Figure 2-21

Proof of Stewart's Theorem: In $\triangle ABC$, let $BC = a$, $AC = b$, $AB = c$, and $CD = d$. Point D divides AB into two segments: $BD = m$ and $DA = n$. Draw altitude $CE = h$ and let $ED = p$. In order to proceed with the proof of Stewart's theorem, we must first employ two necessary formulas.

The first one is applicable to $\triangle CBD$. We apply the Pythagorean theorem to $\triangle CEB$ to obtain $(CB)^2 = (CE)^2 + (BE)^2$. Since

$$BE = m - p, \quad a^2 = h^2 + (m - p)^2. \tag{I}$$

However, by applying the Pythagorean theorem to $\triangle CED$, we have $(CD)^2 = (CE)^2 + (ED)^2$, or $h^2 = d^2 - p^2$. Replacing h^2 in equation (I), we obtain $a^2 = d^2 - p^2 + (m - p)^2$, which then becomes $a^2 = d^2 - p^2 + m^2 - 2mp + p^2$.

Thus,

$$a^2 = d^2 + m^2 - 2mp. \tag{II}$$

A similar argument is applicable to $\triangle CDA$. Applying the Pythagorean theorem to $\triangle CEA$, we find that $(CA)^2 = (CE)^2 + (EA)^2$. Since

$$EA = (n + p), \quad b^2 = h^2 + (n + p)^2. \tag{III}$$

However, $h^2 = d^2 - p^2$, so we substitute for h^2 in (III) as follows:

$b^2 = d^2 - p^2 + (n + p)^2$, which becomes $b^2 = d^2 - p^2 + n^2 + 2np + p^2$.

Thus,

$$b^2 = d^2 + n^2 + 2np. \tag{IV}$$

Equations (II) and (IV) give us the formulas we need. Now multiply equation (II) by n to get

$$a^2 n = d^2 n + m^2 n - 2mnp, \tag{V}$$

and multiply equation (IV) by m to get

$$b^2 m = d^2 m + n^2 m + 2mnp. \tag{VI}$$

Adding (V) and (VI), we have:

$$a^2 n + b^2 m = d^2 n + d^2 m + m^2 n + n^2 m + 2mnp - 2mnp.$$

Therefore, $a^2 n + b^2 m = d^2(n + m) + mn(m + n)$. Since $m + n = c$, we have $a^2 n + b^2 m = d^2 c + mnc$, or $a^2 n + b^2 m = c(d^2 + mn)$, which is the relationship we set out to prove.

Miquel's Theorem

Draw any convenient triangle and select a point on each side. Now construct three circles, each containing a vertex of the triangle and the two points on the vertex's adjacent sides. Although you draw this figure on paper with the aid of a pair of compasses, it is particularly easy to construct with a computer application, such as Geometer's Sketchpad or GeoGebra.

What relationship do you notice about these three circles? Your observation should lead you to a theorem published by in 1838 by the French mathematician Auguste Miquel (1816–1851). We shall state *Miquel's theorem* as follows:

Miquel's Theorem: If a point is selected on each side of a triangle, then the circles determined by each vertex and the points on the adjacent sides pass through a common point.

This theorem may be viewed in two ways. The expected form is shown in Figure 2-22. However, when two of the selected points are on the extensions of the sides, as shown in Figure 2-23, the theorem still holds.

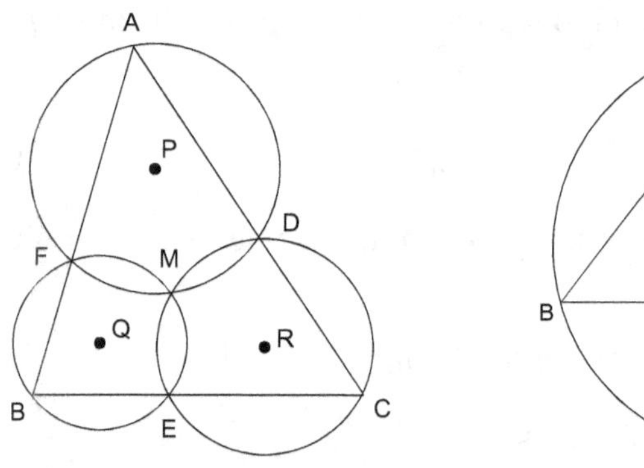

Figure 2-22

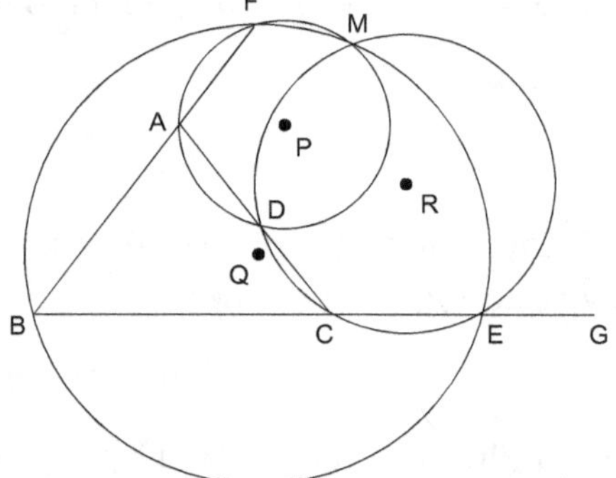

Figure 2-23

Proof of Miquel's Theorem—Case 1: Consider the problem when *M* is inside △*ABC*, as shown in Figure 2-24. Points *D*, *E*, and *F* are any points on sides *AC*, *BC*, and *AB*, respectively, of △*ABC*. Let circles *Q* and *R*, determined by points *F*, *B*, *E* and *D*, *C*, *E*, respectively, meet at *M*. Draw *FM*, *ME*, and *MD*.

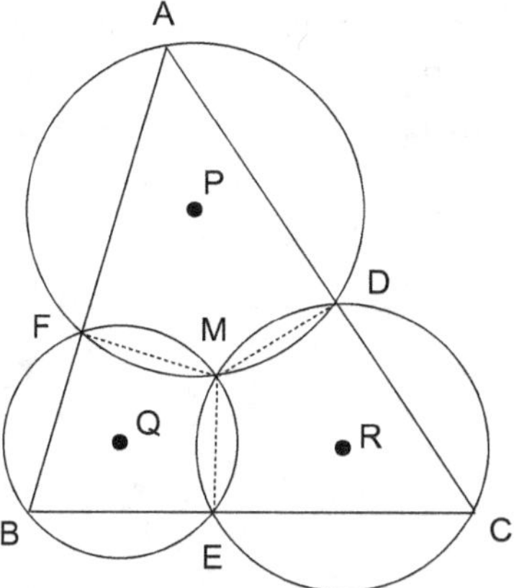

Figure 2-24

In cyclic quadrilateral *BFME*, we have $\angle FME = 180° - \angle B$. Similarly, in cyclic quadrilateral *CDME*, we have $\angle DME = 180° - \angle C$. By addition, $\angle FME + \angle DME = 360° - (\angle B + \angle C)$. Therefore, $\angle FMD = \angle B + \angle C$. However, in $\triangle ABC$, $\angle B + \angle C = 180° - \angle A$. Therefore, we get $\angle FMD = 180° - \angle A$, and quadrilateral *AFMD* is cyclic. Thus, point *M* lies on all three circles.

Proof of Miquel's Theorem—Case 2: Figure 2-25 illustrates the problem when *M* is outside $\triangle ABC$.

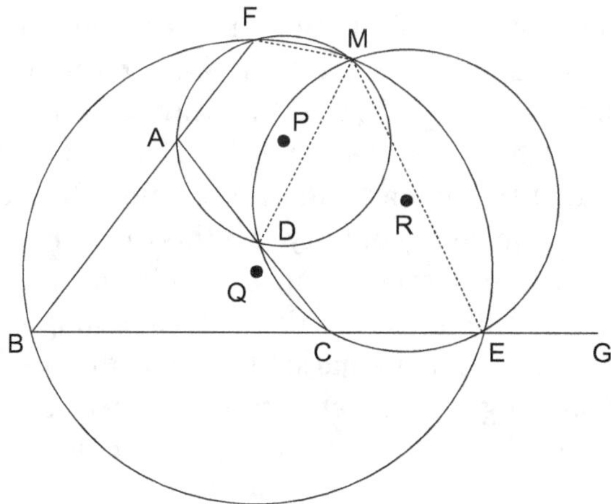

Figure 2-25

Again, let circles Q and R meet at M. Since quadrilateral $BFME$ is cyclic, $\angle FME = 180° - \angle B$.

Similarly, since quadrilateral $CDME$ is cyclic, $\angle DME = 180° - \angle DCE$. By subtraction,

$$\angle FMD = \angle FME - \angle DME = \angle DCE - \angle B. \tag{I}$$

However,

$$\angle DCE = \angle BAC + \angle B. \tag{II}$$

By substituting (II) into (I), $\angle FMD = BAC = 180° - \angle FAD$. Therefore, quadrilateral $ADMF$ is also cyclic, and point M lies on all three circles. Point M is called the *Miquel point* of $\triangle ABC$. The points F, D, and E determine the *Miquel triangle*, $\triangle FDE$.

Desargues' Theorem

During his lifetime, Girard Desargues (1591–1661) did not enjoy the fame that he attained after his death. This lack of popularity was due in part to the recent development of analytic geometry by René Descartes (1596–1650), as well as Desargues' introduction of many new and unfamiliar terms.

In 1648, Desargues' pupil Abraham Bosse (1604–1676), a master engraver, published a book entitled *Manière universelle de M. Desargues, pour pratiquer la perspective.* Although the book was not popularized until about two centuries later, it contained a theorem that became one of the fundamental propositions in the nineteenth-century study of projective geometry. It is this theorem that is of interest to us here. Desargues' theorem involves placing any two triangles in a position that will enable the three lines joining corresponding vertices to be concurrent. Remarkably, when this is achieved, the pairs of corresponding sides meet in three collinear points. As would be expected, we shall prove Desargues' theorem by using Menelaus' theorem.

Desargues' Theorem: If $\triangle A_1B_1C_1$ and $\triangle A_2B_2C_2$ are situated so that the lines joining the corresponding vertices, A_1A_2, B_1B_2, and C_1C_2, are concurrent, then the pairs of corresponding sides intersect in three collinear points.

In Figure 2-26, lines A_1A_2, B_1B_2, and C_1C_2 all meet at P, by the hypothesis. Lines B_2C_2, and B_1C_1 meet at A'; lines A_2C_2, and A_1C_1 meet at B'; and lines B_2A_2 and B_1A_1 meet at C'. Desargues' theorem states that points A', B', and C' are concurrent.

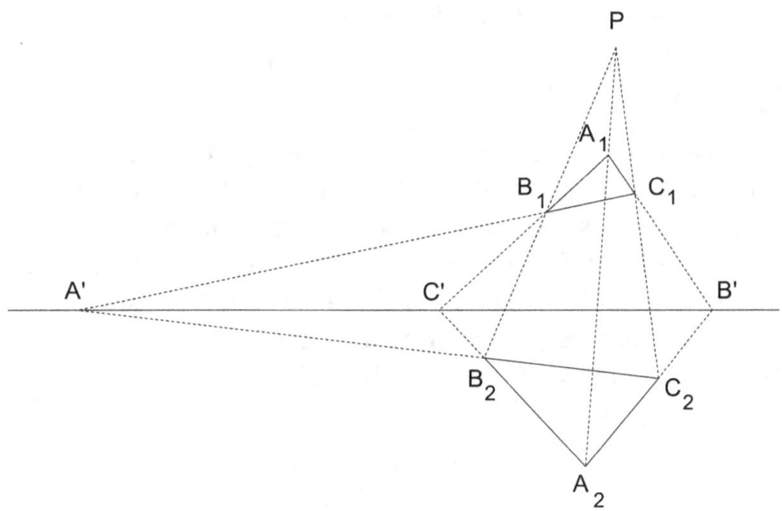

Figure 2-26

Proof of Desargues' Theorem: Consider $A'B_1C_1$ to be a transversal of $\triangle PB_2C_2$.

Therefore,

$$\frac{PB_1}{B_1B_2} \cdot \frac{B_2A'}{A'C_2} \cdot \frac{C_2C_1}{C_1P} = -1 \text{ (Menelaus' theorem).} \qquad \text{(I)}$$

Similarly, considering $C'B_1A_1$ as a transversal of $\triangle PB_2A_2$, we then have

$$\frac{PA_1}{A_1A_2} \cdot \frac{A_2C'}{C'B_2} \cdot \frac{B_2B_1}{B_1P} = -1 \text{ (Menelaus' theorem).} \qquad \text{(II)}$$

Now taking $B'A_1C_1$ as a transversal of $\triangle PA_2C_2$, we get

$$\frac{PC_1}{C_1C_2} \cdot \frac{C_2B'}{B'A_2} \cdot \frac{A_2A_1}{A_1P} = -1 \text{ (Menelaus' theorem).} \qquad \text{(III)}$$

By multiplying (I), (II), and (III), we get $\frac{B_2A'}{A'C_2} \cdot \frac{A_2C'}{C'B_2} \cdot \frac{C_2B'}{B'A_2} = -1$.

Thus, by applying Menelaus' theorem to $\triangle A_2B_2C_2$, we have points A', B', and C' collinear.

It should be noted that the converse of Desargues' theorem is also true. It is the dual of the original theorem.

Pappus' Theorems

Suppose we consider the vertices of a hexagon $AB'CA'BC'$, shown in Figure 2-27, being located alternately on two lines, shown in Figure 2-28. Suppose we now draw the lines that were the opposite sides of the hexagon to locate their point of intersection. We find that the three points of intersection of these pairs of "opposite sides" are collinear. This conclusion was first published by Pappus of Alexandria (290–350 CE) in his *Mathematical Collection* circa 300 CE.

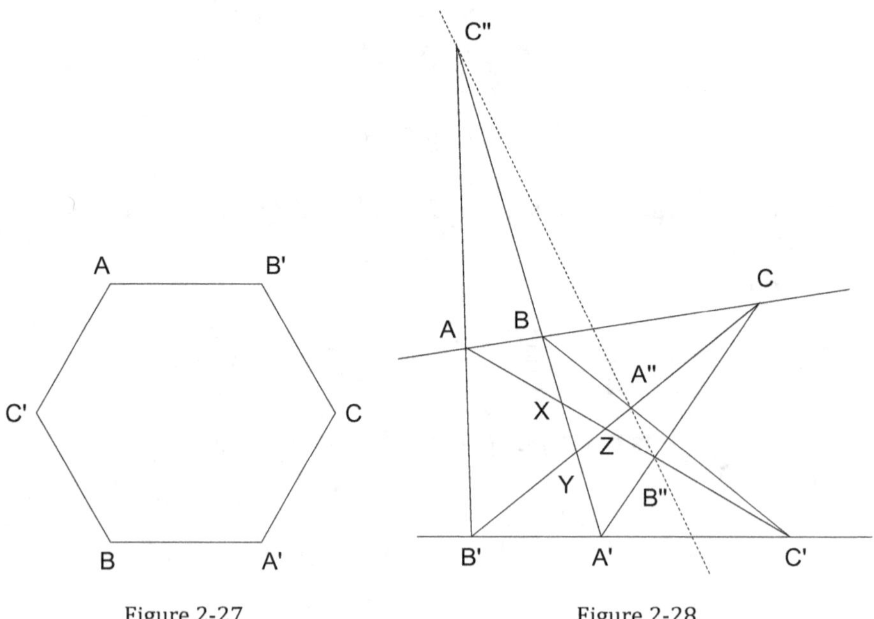

Figure 2-27   Figure 2-28

We shall first restate Pappus' theorem, then prove it, making liberal use of Menelaus' theorem.

Pappus' Theorem 1: Points A, B, and C are on one line and points A', B', and C' are on another line (in any order). If AB' and $A'B$ meet at C'', while AC' meet at B'', and BC' and $B'C$ meet at A'', then points A'', B'', and C'' are collinear.

Proof of Pappus' Theorem 1: In Figure 2-28, $B'C$ meets $A'B$ at Y, AC' meets $A'B$ at X, and $B'C$ meets AC' at Z. Consider $C''AB'$ as a transversal of ΔXYZ.

Then,

$$\frac{ZB'}{YB'} \cdot \frac{XA}{ZA} \cdot \frac{YC''}{XC''} = -1 \text{ (Menelaus' theorem).} \tag{I}$$

Now taking $A'B''C$ as a transversal of ΔXYZ, we have

$$\frac{YA'}{XA'} \cdot \frac{XB''}{ZB''} \cdot \frac{ZC}{YC} = -1 \text{ (Menelaus' theorem).} \tag{II}$$

$BA''C'$ is also a transversal of ΔXYZ, so that we get

$$\frac{YB}{XB} \cdot \frac{ZA''}{YA''} \cdot \frac{XC'}{ZC'} = -1 \text{ (Menelaus' theorem).} \tag{III}$$

Multiplying (I), (II), and (III) gives us equation (IV):

$$\frac{YC''}{XC''} \cdot \frac{XB''}{ZB''} \cdot \frac{ZA''}{YA''} \cdot \frac{ZB'}{YB'} \cdot \frac{YA'}{XA'} \cdot \frac{XC'}{ZC'} \cdot \frac{XA}{ZA} \cdot \frac{ZC}{YC} \cdot \frac{YB}{XB} = -1. \tag{IV}$$

Since points A, B, C are collinear, and A', B', C' are collinear, we obtain the following two relationships by Menelaus' theorem (when we consider each line a transversal of ΔXYZ):

$$\frac{ZB'}{YB'} \cdot \frac{YA'}{XA'} \cdot \frac{XC'}{ZC'} = -1 \tag{V}$$

$$\frac{XA}{ZA} \cdot \frac{ZC}{YC} \cdot \frac{YB}{XB} = -1. \tag{VI}$$

Substituting (V) and (VI) into (IV), we get $\frac{YC''}{XC''} \cdot \frac{XB''}{ZB''} \cdot \frac{ZA''}{YA''} = -1$. Thus, points A'', B'', and C'' are collinear, by Menelaus' theorem.

Pappus' Theorem 2: In Figure 2-29, parallelograms *ACDE* and *BCFG* are on sides *AC* and *BC* of triangle *ABC*. Sides *ED* and *GF* are extended to meet at point *H*. On side *AB*, line *AK* is produced parallel and equal to *HC* to draw parallelogram *ABLK*. The result is that area of parallelogram *ACDE* + area of parallelogram *BCFG* = area of parallelogram *ABLK*.

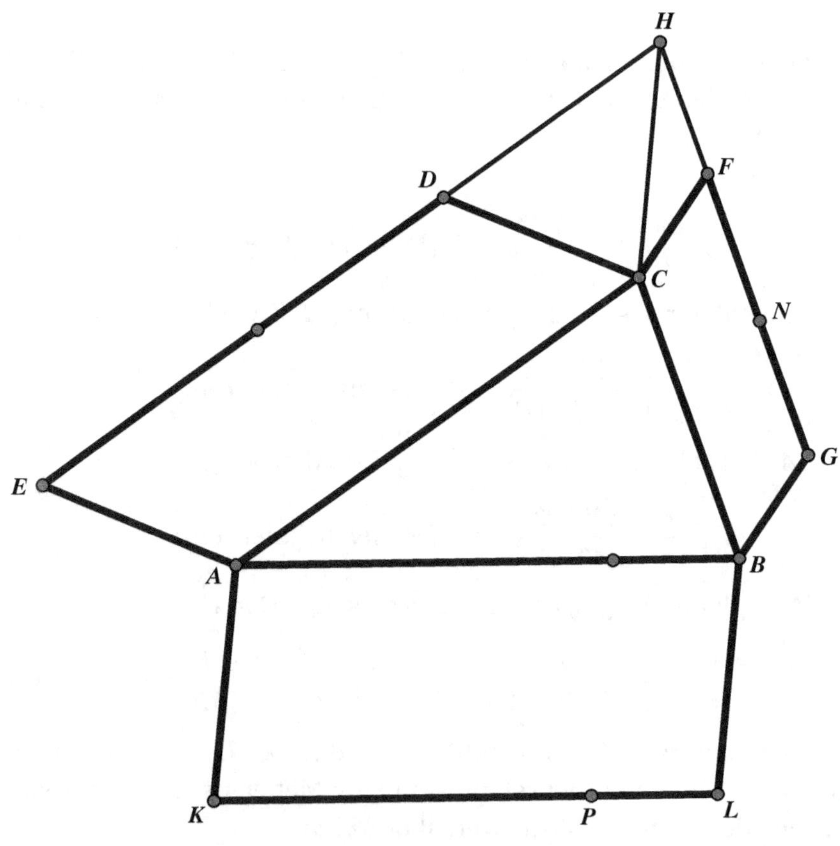

Figure 2-29

Proof of Pappus' Theorem 2: In Figure 2-30, we extend *KA* to intersect *ED* at point *J* and extend *LB* to intersect *GF* at point *N*. We then extend *HC* to intersect *KL* at point *P*. We find that $areaACDE = areaACHJ$, and $areaBCFG = areaBCHN$, since in both cases the two parallelograms have the same altitude and base. Once again, a similar reason justifies that $areaACHJ = areaAKPM$, and

$areaBCHN = areaBLPM.$ Since $areaAKPM + areaBLPM = areaABLK$, we have $areaACDE + areaBCFG = areaABLK$.

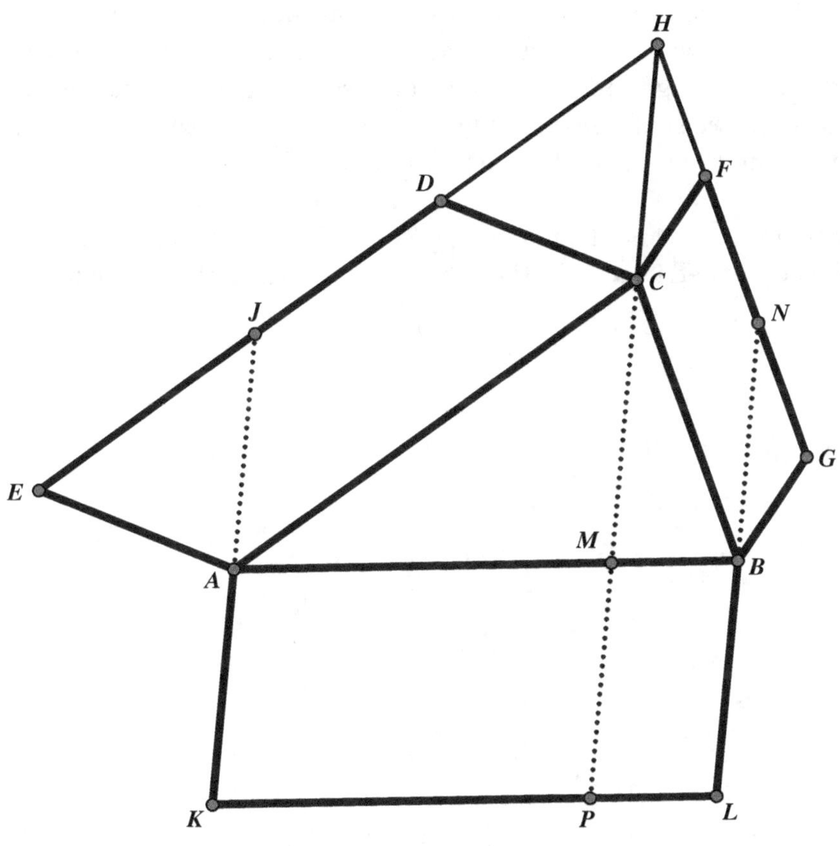

Figure 2-30

Pascal's Theorem on the Inscribed Hexagon

Blaise Pascal (1623–1662), a contemporary of Desargues, is regarded today as one of the true geniuses in the history of mathematics. Although eccentricities kept him from achieving his full potential, he is considered one of the originators of the formalized study of probability (a product of his correspondences with Fermat), and he made many important contributions to other branches of mathematics. Our concern here is one of his contributions to geometry.

In 1640, at the age of sixteen, Pascal published a one-page paper entitled *Essay pour les coniques*. It contained a theorem that Pascal referred to as *mysterium hexagrammicum*. The work highly impressed Descartes, who could not believe it was the work of a boy. The *mysterium hexagrammicum* states that the intersections of the opposite sides of a hexagon inscribed in a conic section are collinear. For our purposes, we shall consider only the case where the conic section is a circle, and the hexagon has no pair of opposite sides parallel.

Pascal's Theorem: If a hexagon, with no pair of opposite sides parallel, is inscribed in a circle, then the intersections of the opposite sides are collinear.

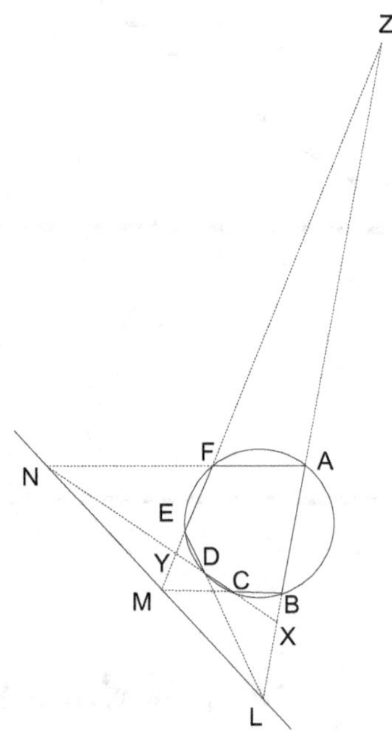

Figure 2-31

Proof of Pascal's Theorem: Hexagon *ABCDEF* is inscribed in a circle. In Figure 2-31, the pairs of opposite sides *AB* and *DE* meet at *L*, *CB* and *EF* meet at *M*, and *CD* and *AF* meet at *N*. Furthermore, *AB* meets *CN* at *X*, *EF* meets *CN* at *Y*, and *EF* meets *AB* at *Z*. Consider *BC* to be a transversal of $\triangle XYZ$.

Then:

$$\frac{ZB}{BX} \cdot \frac{XC}{CY} \cdot \frac{YM}{MZ} = -1 \text{ (Menelaus' theorem).} \tag{I}$$

Taking AF to be a transversal of $\triangle XYZ$:

$$\frac{ZA}{AX} \cdot \frac{YF}{FZ} \cdot \frac{XN}{NY} = -1, \text{ by Menelaus' theorem.} \tag{II}$$

Also, since DE is a transversal of $\triangle XYZ$:

$$\frac{XD}{DY} \cdot \frac{YE}{EZ} \cdot \frac{ZL}{LX} = -1, \text{ by Menelaus' theorem.} \tag{III}$$

By multiplying (I), (II), and (III), we get

$$\frac{YM}{MZ} \cdot \frac{XN}{NY} \cdot \frac{ZL}{LX} \cdot \frac{(ZB)(ZA)}{(EZ)(FZ)} \cdot \frac{(XD)(XC)}{(AX)(BX)} \cdot \frac{(YE)(YF)}{(DY)(CY)} = -1. \tag{IV}$$

When two secant segments are drawn to a circle from an external point, the product of the lengths of one secant and its external segment equals the product of the lengths of the other secant and its external segment. Therefore,

$$\frac{(ZB)(ZA)}{(EZ)(FZ)} = 1, \tag{V}$$

$$\frac{(XD)(XC)}{(AX)(BX)} = 1, \tag{VI}$$

and

$$\frac{(YE)(YF)}{(DY)(CY)} = 1. \tag{VII}$$

By substituting (V), (VI), and (VII) into (IV), we get

$$\frac{YM}{MZ} \cdot \frac{XN}{NY} \cdot \frac{ZL}{LX} = -1.$$

Thus, by Menelaus' theorem, points M, N, and L must be collinear.

Brianchon's Theorem

In 1806 France, a twenty-one year-old student at the École Polytechnique, published an article in the *Journal de L'École Polytechnique* that would become one of the fundamental contributions to the study of conic sections in projective geometry. The student's name was Charles Julien Brianchon (1785–1864). He restated of Pascal's overlooked theorem, then extended and developed it into a new theorem, which now bears his name. Brianchon's theorem states "In any hexagon circumscribed about a conic section, the three diagonals cross each other in the same point."[12] Brianchon's theorem bears a curious resemblance to Pascal's theorem. They are, in fact, duals of one another. This can be easily seen by comparing the following version of each.

Pascal's Theorem	**Brianchon's Theorem**
The points of intersection of the opposite sides of a hexagon inscribed in a conic section are collinear.	*The lines joining the opposite vertices of a hexagon circumscribed about a conic section are concurrent.*

Notice that the two statements above are alike except for the underlined words, which are geometric duals. As with Pascal's theorem, we shall consider only the conic section that is a circle.

Brianchon's Theorem: If a hexagon is circumscribed about a circle, the lines containing opposite vertices are concurrent, as shown in Figure 2-32.

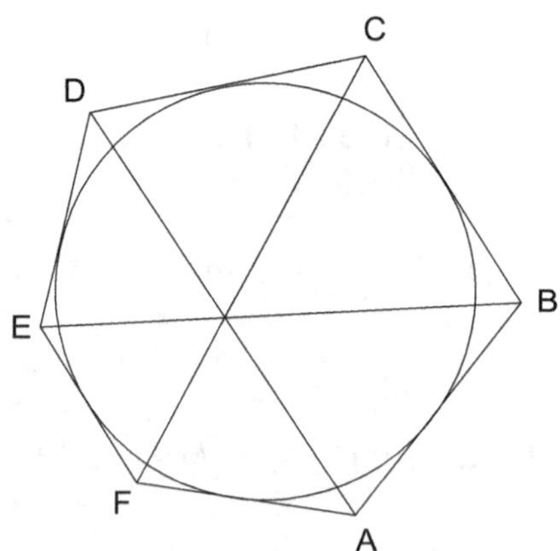

Figure 2-32

---

[12] *Source Book in Mathematics*, edited by D.E. Smith (Dover, 1929), p. 336.

The simplest proofs of this theorem require a knowledge of projective geometry. Although we are prepared at this point to prove this theorem by Euclidean methods, our proof would be more concise if we consider the concept of radical axes.

Radical Axes

Earlier we stated that Brianchon's theorem is the dual of Pascal's theorem. At that juncture we deferred the proof, since we needed knowledge about a radical axis to prove the theorem elegantly. We shall now establish some important properties of radical axes and use them to prove Brianchon's theorem.

In Figure 2-33, we have two circles, R and Q, that intersect at points A and B. Point P is any point on line AB not between points A and B. Lines PT and PS are tangent to circles R and Q at points T and S, respectively.

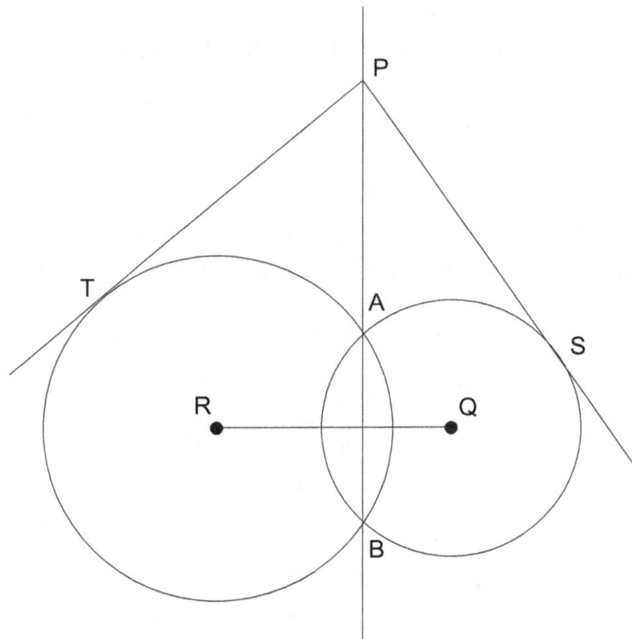

Figure 2-33

From elementary geometry, we know that PT is the mean proportional between PB and PA. Therefore, $(PT)^2 = (PB)(PA)$. Similarly for circle Q, we have $(PS)^2 = (PB)(PA)$. It then follows that $PT = PS$. Since P was selected as *any* external point on AB, we can conclude that, from any external point on AB, tangent segments to circles R and Q are congruent. Before we can state this as a locus theorem, we must prove that any point P, which generates congruent tangents to circles R and Q, must lie on AB.

Let us therefore suppose that P is any point where tangent segments PT and PS are congruent. Let PA intersect circle R at B and circle Q at B'. As before, $(PB)(PA) = (PT)^2$ and $(PB')(PA) = (PS)^2$. Because $PT = PS$, $PB = PB'$. Therefore, B and B' must coincide, and P lies on the common secant PA of the two circles. We call this line consisting of common endpoints of congruent tangent segments to two circles the *radical axis* of the two circles. We state this result as our next theorem.

Definition of Radical Axis: The radical axis of two intersecting circles is their common secant. It follows immediately that the radical axis of two tangent circles is their common tangent. Before we can investigate the radical axis of two non-intersecting circles, we need to consider the following theorem.

Radical Axis Theorem 1: The locus of a point, the difference of whose distances squared from two fixed points is a constant, is a line perpendicular to the segment determined by the two fixed points.

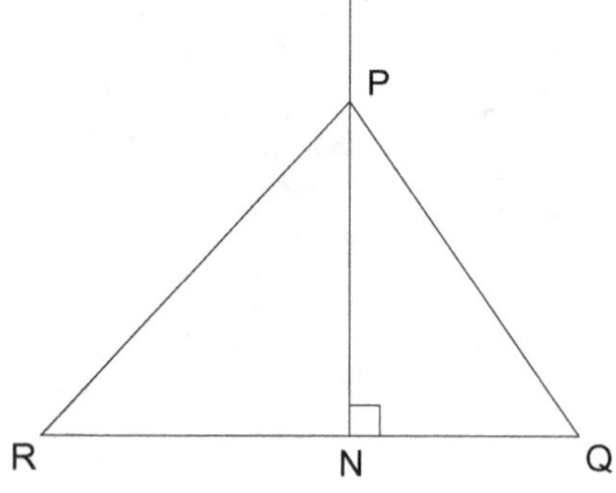

Figure 2-34

Proof of Radical Axis Theorem 1: In Figure 2-34, we let R and Q be the fixed points and P be a point on the locus. Draw PR and PQ. Construct $PN \perp RQ$. We use the Pythagorean theorem to get $(PR)^2 - (RN)^2 = (PN)^2$ and $(PQ)^2 - (QN)^2 = (PN)^2$.

Therefore, $(PR)^2 - (RN)^2 = (PQ)^2 - (QN)^2$, or $(PR)^2 - (PQ)^2 = (RN)^2 - (QN)^2 = k$. Let $RQ = d$. Then from the right side of the previous equation we have:

$$(RN + QN)(RN - QN) = k, \text{ then } d\,(RN - QN) = k, \text{ and } RN - QN = \frac{k}{d}. \quad \text{(I)}$$

Remember that

$$RN + QN = d. \qquad \text{(II)}$$

Solving equations (I) and (II) simultaneously, we get $RN = \frac{d^2+k}{2d}$ and $QN = \frac{d^2-k}{2d}$.

This fixes the position of N. Since d and k are constant for any given situation, P must lie on the line perpendicular to RQ at N, which divides RQ in the ratio $\frac{RN}{QN} = \frac{d^2+k}{d^2-k}$.

We may conclude this locus proof by showing that any point on PN satisfies the given conditions. This exercise is left to the reader.

We must now determine the radical axis of two non-intersecting circles. You can probably predict the next theorem.

Radical Axis Theorem 2: The radical axis of two non-intersecting circles is a line perpendicular to their line of centers.

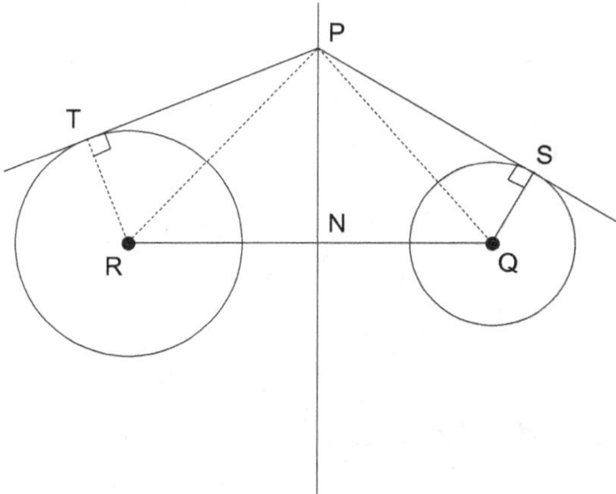

Figure 2-35

Proof of Radical Axis Theorem 2: Begin by letting r and q be the radii of circles R and Q, respectively. In Figure 2-35, let P be a point on the required locus so that tangent segments PT and PS are congruent. By applying the Pythagorean theorem to $\triangle PTR$ and $\triangle PSQ$ we get $(PR)^2 - r^2 = (PT)^2$ and $(PQ)^2 - q^2 = (PS)^2$, respectively. But $PT = PS$, and therefore $(PR)^2 - r^2 = (PQ)^2 - q^2$, or $(PR)^2 - (PQ)^2 = r^2 - q^2$.

Since the right side of this equality is a constant, we can conclude (by employing radical axis theorem 1) that the locus of P is the line containing P, which is perpendicular to the line of centers RQ. In a manner similar to that used in the previous proof, we can determine the location of N in terms of the radii and the distance between the centers. As a direct consequence of the previous theorem, we have the following.

Radical Axis Theorem 3: The radical axes of three given circles, whose centers are not collinear, are concurrent.

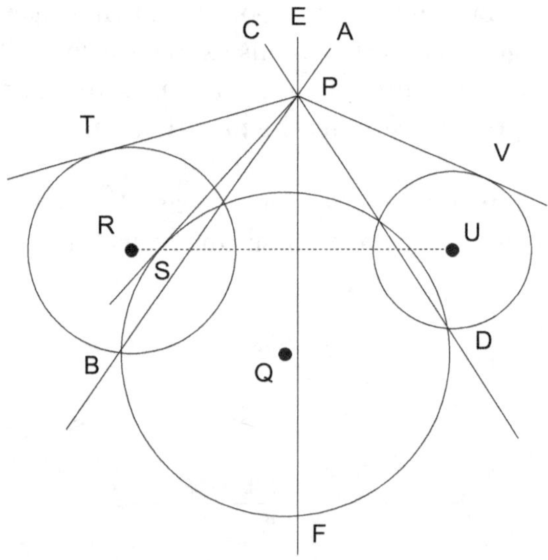

Figure 2-36

Proof of Radical Axis Theorem 3: Let us consider circles R, Q, and U whose radical axes are AB, CD, and EF, as shown in Figure 2-36. Let P be the intersection of AB, and CD. Using radical axis AB of circles R

and *Q*, we have *PT* = *PS*. Using radical axis *CD* of circles *Q* and *U*, we have *PV* = *PS*. (Note that *PT*, *PS*, and *PV* are tangents to the given circles.) Thus, *PT* = *PV*, which indicates that *P* must lie on the radical axis *EF* of circles *R* and *U*. This proves that the radical axes are concurrent at *P*.

Equipped with this knowledge of radical axes, we are now ready to prove Brianchon's theorem.

Brianchon's Theorem: If a hexagon is circumscribed about a circle, the lines containing the opposite vertices are concurrent.

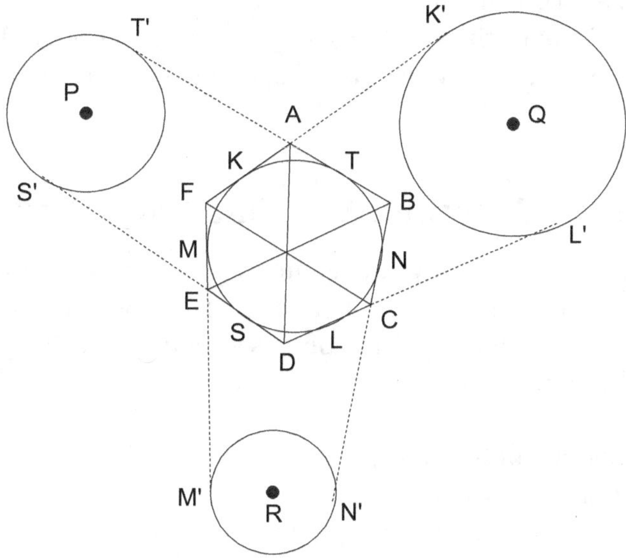

Figure 2-37

Proof of Brianchon's Theorem: As seen in Figure 2-37, the sides of hexagon *ABCDEF* are tangent to a circle at points *T*, *N*, *L*, *S*, *M*, and *K*. Points *K′*, *L′*, *N′*, *M′*, *S′*, and *T′* are chosen on *FA*, *DC*, *BC*, *FE*, *DE*, and *BA*, respectively, so that *KK′* = *LL′* = *NN′* = *MM′* = *SS′* = *TT′*. Now construct circle *P* tangent to *BA* and *DE* at points *T′* and *S′*, respectively (the existence of this circle is easily justified).

Likewise, construct circle *Q* tangent to *FA* and *DC* at points *K′* and *L′*, respectively. Then construct circle *R* tangent to *FE* and *BC* at points *M′* and *N′*, respectively. Since two tangent segments to a circle from an external point have the same length, *FM = FK*. We already know that *MM′ = KK′*. Therefore, by addition, *FM′ = FK′*. Similarly, *CL = CN* and *LL′ = NN′*. By subtraction, *CL′ = CN′*.

We now notice that points *F* and *C* are each endpoints of a pair of congruent tangent segments to circles *R* and *Q*. Thus, these points determine the radical axis, *CF*, of circles *R* and *Q*.

Using the same technique, we can easily show that *AD* is the radical axis of circles *P* and *Q*, and that *BE* is the radical axis of circles *P* and *R*. Earlier we proved that the radical axes of three circles with non-collinear centers (taken in pairs) are concurrent. Therefore *CF*, *AD*, and *BE* are concurrent.

We should note that the only way in which these circles would have had collinear centers is if the diagonals were to have coincided. This is impossible!

Feuerbach's Theorem: The Nine-Point Circle

Perhaps one of the true joys in geometry is to observe how one configuration can produce a seemingly endless array of properties and relationships. One such situation begins with nine specific points of a triangle. These points, for any given triangle, are:

- the midpoints of the sides,
- the feet of the altitudes,
- and the midpoints of the segments from the orthocenter to the vertices.

They have the surprising relationship of all being on the same circle. This circle is called the *nine-point circle* of the triangle. This fantastic relationship was popularized by the work of German mathematician Karl Wilhelm Feuerbach (1800–1834), who built on this relationship to develop the even more astonishing *Feuerbach theorem*, which will be presented after we establish the nine-point circle.

In 1765, Leonhard Euler showed that six points in a triangle (the midpoints of the sides and the feet of the altitudes) determine a unique circle. Yet it was not until 1820 that three other points (the midpoints of the segments from the triangle's orthocenter to its vertices) were found to be on this circle, when Charles Julien Brianchon and Jean-Victor Poncelet (1788–1867) published a paper[13] containing the first complete proof of the theorem the first usage of the name *"nine-point circle."*

The Nine-Point Circle Theorem: In any triangle, the midpoints of the sides, the feet of the altitudes, and the midpoints of the segments from the orthocenter to the vertices lie on a circle (Figure 2-38).

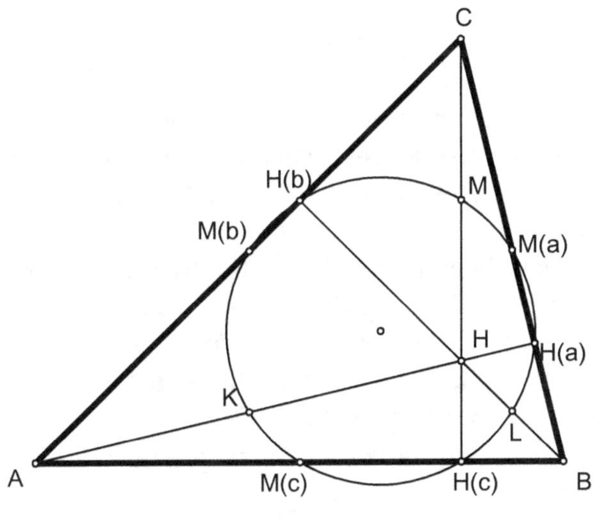

Figure 2-38

Proof of the Nine-Point Circle Theorem: In order to simplify the discussion of this proof, we shall consider each part with a separate

---

[13] Charles Julien Brianchon and Jean-Victor Poncelet. *Recherches sur la determination d'une hyperbole équilatère moyen de quartes conditions données* (Paris, 1820).

diagram. Figures 2-39 through 2-42 are all extracted from Figure 2-43, which is the complete diagram.

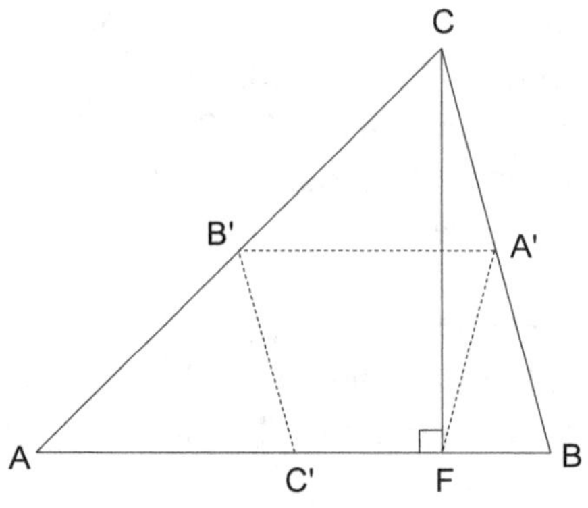

Figure 2-39

In Figure 2-39, points A', B', and C' are the midpoints of the three sides of $\triangle ABC$ opposite their respective vertices. is an altitude of $\triangle ABC$. Since $A'B'$ is a midline of $\triangle ABC$, $A'B' \parallel AB$. Therefore, quadrilateral $A'B'C'F$ is a trapezoid, where $B'C'$ is also a midline of $\triangle ABC$, so that $B'C' = \frac{1}{2}BC$. Since $A'F$ is the median to the hypotenuse of right $\triangle BCF$, $A'F = \frac{1}{2}BC$. Therefore, $B'C' = A'F$ and trapezoid $A'B'C'F$ is isosceles.

Recall that when the opposite angles of a quadrilateral are supplementary, as in the case of an isosceles trapezoid, the quadrilateral is cyclic. Therefore, quadrilateral $A'B'C'F$ is cyclic.

So far, we have four of the nine points on one circle.

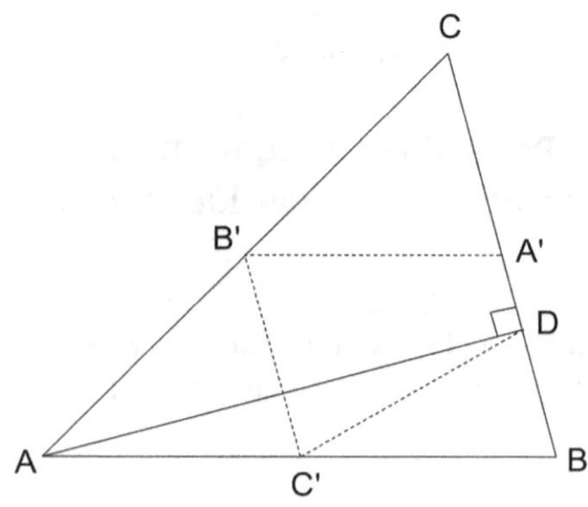

Figure 2-40

To avoid any confusion, we redraw $\triangle ABC$ as in Figure 2-40, and include altitude. Using the same argument as before, we find that quadrilateral $A'B'C'D$ is an isosceles trapezoid and a cyclic quadrilateral. So, we now have five of the nine points on one circle: midpoints A', B', and C', and altitude bases F and D.

By repeating the same argument for altitude BE, we can then state that points D, F, and E lie on the same circle as points A', B', and C'. These six points are as far as Euler got with this configuration.

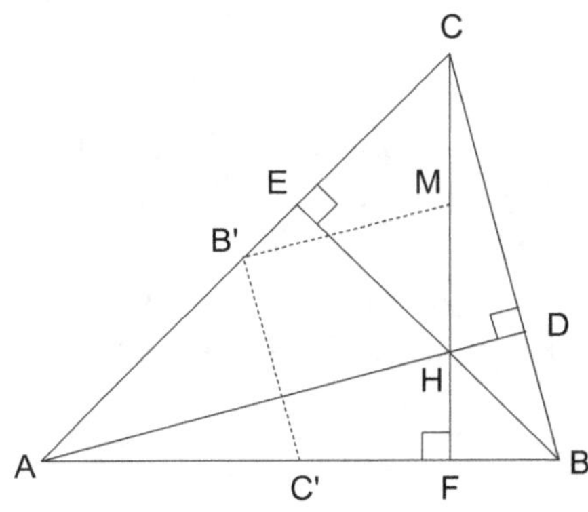

Figure 2-41

With H as the orthocenter (the point of intersection of the altitudes), M is the midpoint of CH, as can be seen in Figure 2-41. Therefore, $B'M$, a midline of $\triangle ACH$, is parallel to AH, or altitude AD. Since $B'C'$ is a midline of $\triangle ABC$, $B'C' \parallel BC$. Because $\angle ADC$ is a right triangle, $\angle MB'C'$ is also a right angle. Thus, quadrilateral $MB'C'F$ is cyclic because its opposite angles are supplementary. This places point M on the circle determined by points B', C', and F. We now have a seven-point circle.

We repeat this procedure with point L, the midpoint of BH, as shown in Figure 2-42. As before, $\angle B'A'L$ is a right angle, as is $\angle B'EL$. Therefore, points B', E, A', and L are concyclic (opposite angles supplementary). We now have L as an additional point on our circle, making it an eight-point circle.

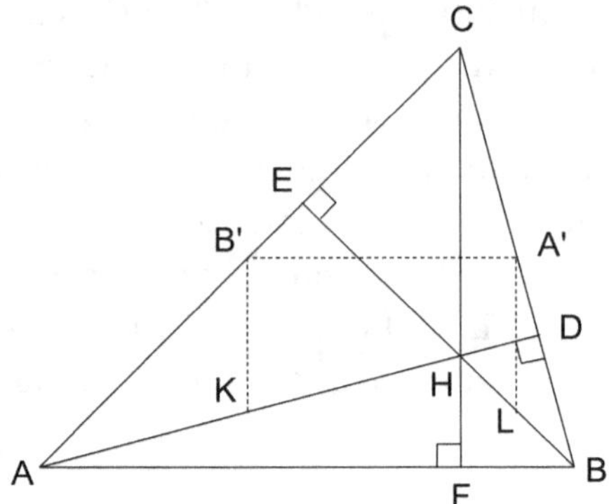

Figure 2-42

To locate our final point on the circle, consider point K, the midpoint of AH. As we did earlier, we find $\angle A'B'K$ to be a right angle, as is $\angle A'DK$. Therefore, quadrilateral $A'DKB'$ is cyclic and point K is on the same circle as points B', A', and D. We have thus proved that nine specific points lie on this circle, as can be seen in Figure 2-43.

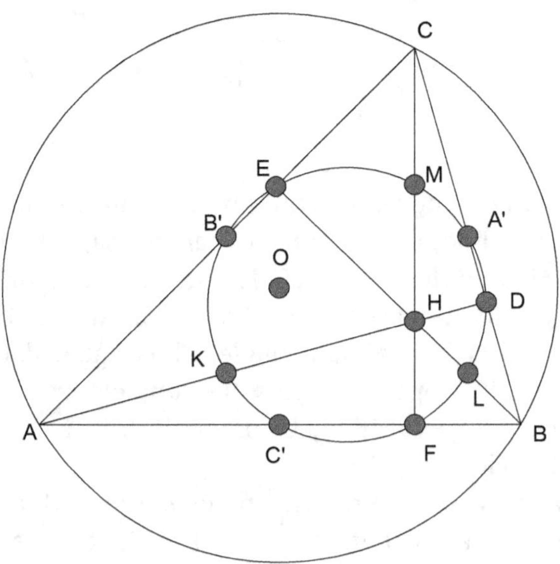

Figure 2-43

Fuerbach's Theorem Relating the Nine-Point Circle to Other Circles

As alluded to earlier, Karl Wilhelm Feuerbach owes much of his fame to an innovative 1822 paper,[14] wherein he states that "the circle which passes through the feet of the altitudes of a triangle touches all four of the circles which are tangent to the three sides of the triangle; it is internally tangent to the inscribed circle and externally tangent to each of the circles which touch the sides of the triangle externally." This is pictured in Figure 2-44. As a result of this work, the theorem is referred to as the Feuerbach theorem and the nine-point circle theorem is also frequently called the Feuerbach theorem.

Feuerbach's Theorem: The nine-point circle of a triangle is tangent to the incircle and the excircles of the triangle.

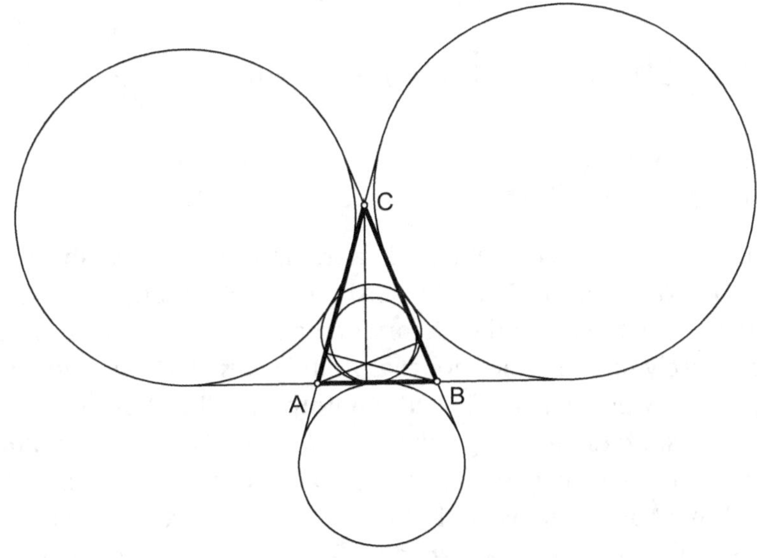

Figure 2-44

---

[14] Karl Wilhelm Feuerbach. *Eigenschaften einiger merkwürdigen Punkte des geradlinigen Dreiecks und mehrere durch sie bestimten Linien und Figuren. Eine analytische-trigonometrische Abhandlung* (Nürnberg, 1822).

The proof of this property is quite complex and time consuming. The curious reader will find four different proofs of Feuerbach's theorem in *Modern Geometry* by Roger A. Johnson (Houghton Mifflin, 1929, pp. 200–205). The proof that Feuerbach actually used consists of computing the distances between the center of the nine-point circle and the centers of the inscribed circle (ρ), the circumscribed circle (R), and the circle inscribed in the orthic (or pedal) triangle *DEF*, and showing that they equal the sum and difference of the corresponding radii. This can be expressed symbolically in the following equations, where *I* is the center of the inscribed circle, *H* is the orthocenter, and *O* is the center of the circumscribed circle:

$$(OI)^2 = R^2 - 2R\rho$$

$$(IH)^2 = 2\rho^2 - 2Rr$$

$$(OH)^2 = R^2 - 4Rr$$

$$(NI)^2 = \frac{1}{2}\left[(OI)^2 + (HI)^2\right] - (NH)^2 = \frac{1}{4}R^2 - R\rho + \rho^2 = \left(\frac{1}{2}R - \rho\right)^2.$$

The Gergonne Point

A fascinating point of concurrency in a triangle was first established by Joseph-Diaz Gergonne (1771–1859), a French mathematician. Gergonne earned a distinct place in the history of mathematics as the founder of the first purely mathematical journal, *Annales des mathématiques pures et appliqués*, which was published monthly from 1810 until 1832 and known as *Annales del Gergonne. Annales* published about 200 of Gergonne's papers, mostly on geometry, and served as a forum for the establishment of projective and algebraic geometry. We now consider a rather simple theorem established by Gergonne, as it exhibits concurrency and is easily proved using Ceva's theorem in the following way.

Gergonne's Theorem: The lines containing a vertex of a triangle and the point of tangency of the opposite side with the inscribed circle are concurrent. This point of concurrency is known as the *Gergonne point* of the triangle.

Proof of Gergonne's Theorem: In Figure 2-45, circle O is tangent to sides AB, AC, and BC at points N, M, and L, respectively. It follows that $AN = AM$, $BL = BN$, and $CM = CL$. Each of these equalities may be written as $\frac{AN}{AM} = 1$, $\frac{BL}{BN} = 1$, and $\frac{CM}{CL} = 1$.

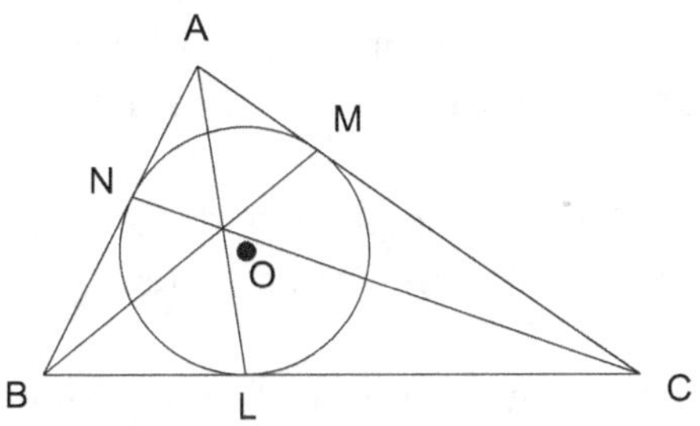

Figure 2-45

We multiply these three fractions to get $\frac{AN}{AM} \cdot \frac{BL}{BN} \cdot \frac{CM}{CL} = 1$. Therefore, $\frac{AN}{BN} \cdot \frac{BL}{CL} \cdot \frac{CM}{AM} = 1$, which as a result of Ceva's theorem implies that AL, BM, and CN are concurrent. This point of concurrency is the Gergonne point of $\triangle ABC$.

The Gergonne point is also applicable to exterior circles, as shown in Figure 2-46. The same properties as above all hold true for the exterior circle as for the interior circle.

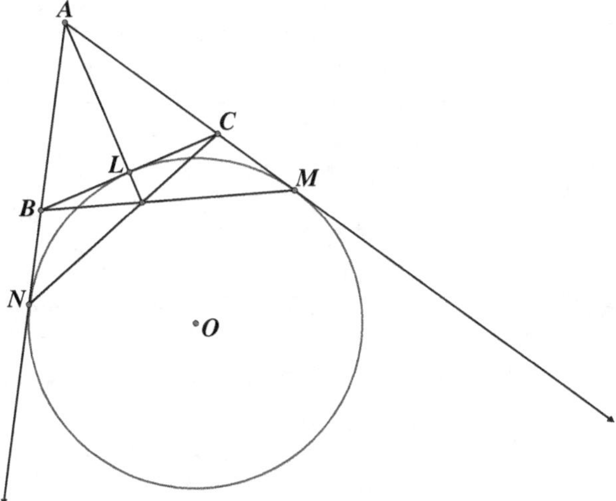

Figure 2-46

Morley's Theorem on Angle Trisectors of a Triangle

We will close this chapter on named theorems with a construction first published in 1900 by the English mathematician Frank Morley (1860–1937). It is one of the more difficult theorems in geometry to prove, but its beauty lies in the simplicity of the statement.

Morley's Theorem: The intersections of the adjacent angle trisectors of any triangle always meet in three points determining an equilateral triangle.

Figure 2-47 shows a number of differently-shaped triangles with their angles trisected. In each case, the intersections of the adjacent trisectors determine an equilateral triangle. We invite the reader to try this with a variety of differently shaped triangles to see that it holds true for all triangles.

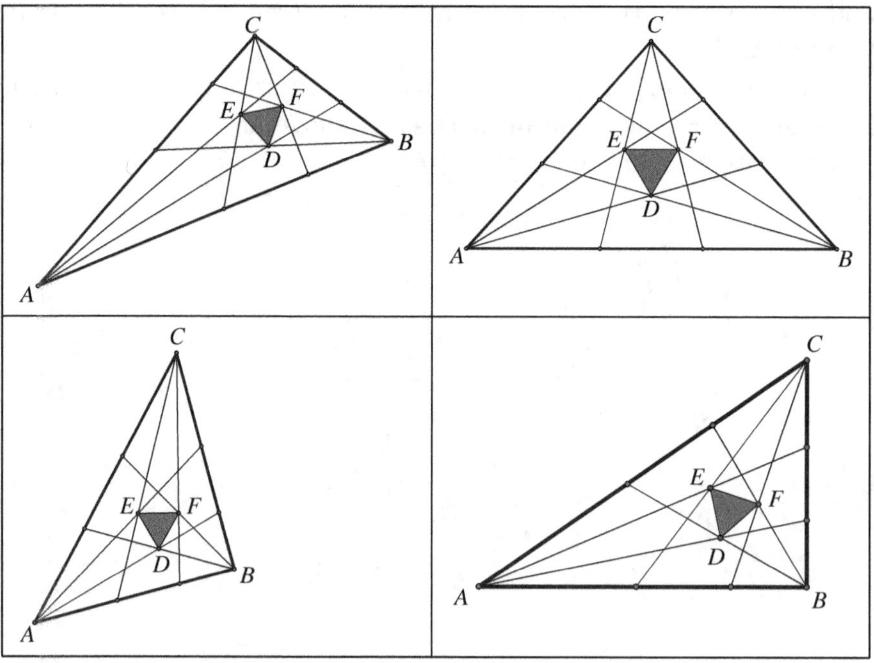

Figure 2-47

We also find that there is a concurrency in this configuration as well. In Figure 2-48, we notice that *CD*, *AF*, and *BE* are concurrent.

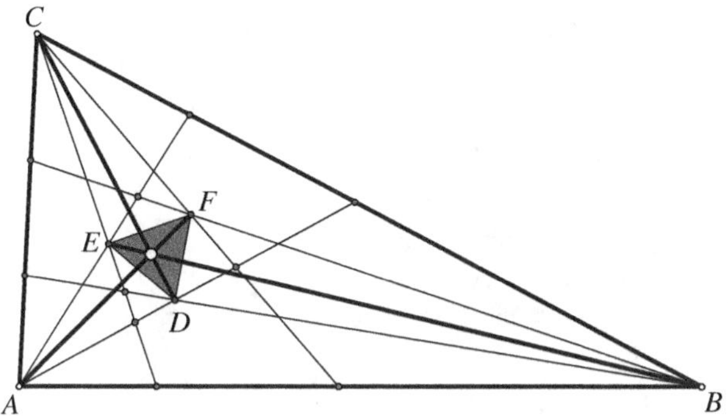

Figure 2-48

Morley's Theorem Proof 1:[15] The mathematician Yoshinori Hashimoto's original proof is short and without a figure. Let α, β, and γ be arbitrary positive angles with $\alpha + \beta + \gamma = 60°$. For any angle η we put $\eta' := \eta + 60°$. Let triangle *DEF* be an equilateral triangle, and *A* [resp. *B*, *C*] be the point lying opposite to *D* [resp. *E*, *F*] with respect to *EF* [resp. *FD*, *DE*] and satisfying $\angle AFE = \beta'$ and $\angle AEF = \gamma'$ [resp. $\angle BDF = \gamma'$, $\angle BFD = \alpha'$; $\angle CED = \alpha'$, $\angle CDE = \beta'$]. Then $\angle EAF = 180° - (\beta' + \gamma') = \alpha$, and similarly, $\angle FBD = \beta$ and $\angle DCE = \gamma$. By symmetry it is enough to show that $\angle BAF = \alpha$ and $\angle ABF = \beta$.

The perpendiculars from *F* to *AE* and *BD* have the same length *s*. If the perpendicular from *F* to *AB* has length $h < s$, then $\angle BAF < \alpha$ and $\angle ABF < \beta$. If, on the other hand, $h > s$, then $\angle BAF > \alpha$ and $\angle ABF > \beta$. Since $\angle BAF + \angle ABF = \alpha' + \beta' + 60° - 180° = \alpha + \beta$, we see that $h = s$, $\angle BAF = \alpha$, and $\angle ABF = \beta$.

[15] Yoshinori Hashimoto. "A short proof of Morley's theorem". *Elemente der Mathematik* 62 (2007), p. 121.

Morley's Theorem Proof 2: We now present Hashimoto's proof with Figure 2-49 and some additional explanations.

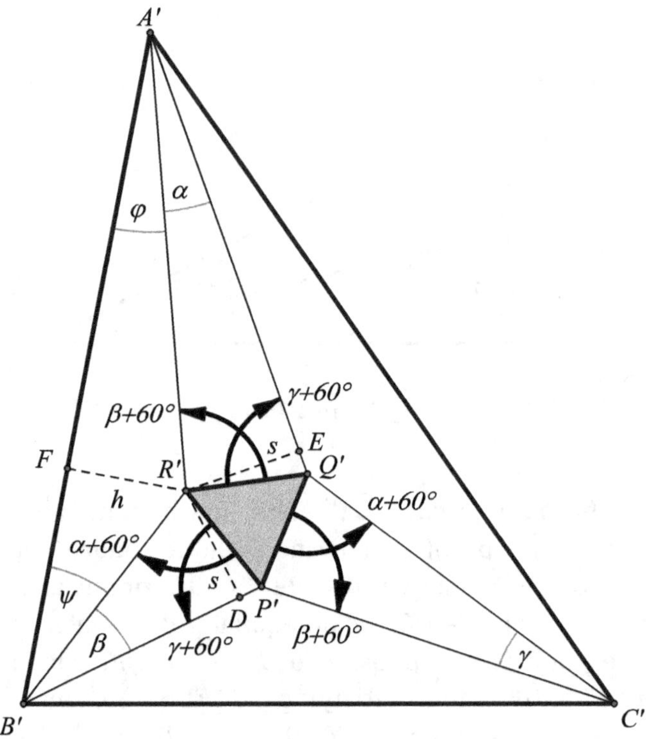

Figure 2-49

Let α, β, and γ be arbitrary angle measurements with $\alpha + \beta + \gamma = 60°$. For each of these angles we set the following: $\alpha' := \alpha + 60°$, $\beta' := \beta + 60°$, and $\gamma' := \gamma + 60°$. We have the triangle $P'Q'R'$ as equilateral. On each of the sides of the triangle $P'Q'R'$ we draw the triangles $Q'A'R'$, $R'B'P'$ and $P'C'Q'$ so that the following angles are produced:

$$\angle Q'A'R' = \alpha, \ \angle Q'R'A' = \beta' = \beta + 60°, \ \angle A'Q'R' = \gamma' = \gamma + 60°,$$

$$\angle P'B'R' = \beta, \ \angle R'P'B' = \gamma' = \gamma + 60°, \ \angle B'R'P' = \alpha' = \alpha + 60°,$$

$\angle Q'C'P' = \gamma$, $\angle P'Q'C' = \alpha' = \alpha + 60°$, and $\angle Q'P'C' = \beta' = \beta + 60°$.

It then follows that

$$\angle Q'A'R' = 180° - (\beta' + \gamma') = 180° - (\beta + 60° + \gamma + 60°) = 60° - (\beta + \gamma) = \alpha,$$

and analogously, $\angle P'B'R' = \beta$, and $\angle Q'C'P' = \gamma$. As a result of the symmetry, it suffices to show that $\varphi = \angle R'A'B' = \alpha$ and $\psi = \angle R'B'A' = \beta$. The perpendicular from R' to $A'Q'$ and $B'P'$ have the same length, s. If the perpendicular distance from R' to $A'B'$ is h, where $h < s$, then the following is true: $\varphi = \angle R'A'B' < \alpha$ and $\psi = \angle R'B'A' < \beta$.

On the other hand, if $h > s$, it then follows that $\varphi = \angle R'A'B' > \alpha$ and $\psi = \angle R'B'A' > \beta$.

However, then $\angle A'R'B' = 360° - \angle A'R'Q' - \angle Q'R'P' - \angle P'R'B' = 360° - (\beta + 60°) - 60° - (\alpha + 60°) = 180° - (\alpha + \beta)$.

Thus, $\varphi + \psi = 180° - \angle A'R'B' = 180° - 180° - (\alpha + \beta) = \alpha + \beta$.

Therefore, it follows that $h = s$, $\varphi = \angle R'A'B' = \alpha$, and $\psi = \angle R'B'A' = \beta$.

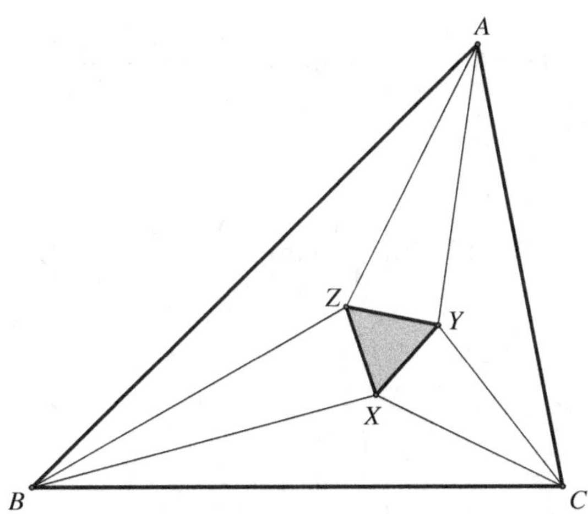

Figure 2-50

Morley's theorem Proof 3:[16] In Figure 2-50,

$$\sin \angle AYC = \sin\left(\pi - \frac{\alpha + \gamma}{3}\right) = \sin\left(\frac{\alpha + \gamma}{3}\right) = \sin\left(\frac{\pi - \beta}{3}\right) = \sin\left(\frac{2\pi + \beta}{3}\right).$$

$$\sin 3\theta = \sin\theta \cos 2\theta + \sin 2\theta \cos\theta = \sin\theta \, (3\cos^2\theta - \sin^2\theta)$$

$$= 4\sin\theta\left(\frac{\sqrt{3}\cos\theta + \sin\theta}{2}\right)\cdot\left(\frac{\sqrt{3}\cos\theta - \sin\theta}{2}\right)$$

$$= 4\sin\theta \cdot \sin\left(\frac{\pi}{3} + \theta\right)\cdot\sin\left(\frac{\pi}{3} - \theta\right). \tag{I}$$

From the law of sines for triangle AYC,

$$AY \cdot \sin\left(\frac{\pi - \beta}{3}\right) = AC \cdot \sin\left(\frac{\gamma}{3}\right) = 2R \cdot \sin\beta \cdot \sin\left(\frac{\gamma}{3}\right),$$

where R is the circumradius of $\triangle ABC$.

Therefore, by equation (I) we have

$$AY = 8R \cdot \sin\left(\frac{\beta}{3}\right)\cdot\sin\left(\frac{\gamma}{3}\right)\cdot\sin\left(\frac{\pi + \beta}{3}\right).$$

Similarly,

$$AZ = 8R \cdot \sin\left(\frac{\gamma}{3}\right)\cdot\sin\left(\frac{\beta}{3}\right)\cdot\sin\left(\frac{\pi + \gamma}{3}\right).$$

Therefore,

$$\frac{AZ}{AY} = \frac{\sin\left(\dfrac{\pi + \gamma}{3}\right)}{\sin\left(\dfrac{\pi + \beta}{3}\right)}.$$

---

[16] L. Bankoff. "A Simple Proof of the Morley Theorem". *Mathematics Magazine*, 35 (1962) no. 4, pp. 223–224.

$$\text{But } \angle AZY + \angle AYZ = \pi - \frac{\alpha}{3} = \frac{2\pi + \beta + \gamma}{3} = \frac{\pi + \beta}{3} + \frac{\pi + \gamma}{3}.$$

From here, $\angle AZY = \frac{\pi + \beta}{3}$ and $\angle AYZ = \frac{\pi + \gamma}{3}$, and similarly for triangles *BXZ* and *CXY*. It thus follows that the sum of angles around *X*, excluding $\angle YXZ$, is 300°, or $\angle YXZ = 60°$. The other two angles are similarly shown to be 60°.

Concurrency, collinearity, tangency, parallelism, and perpendicularity, among countless other relationships of the various parts of a triangle, provide a seemingly endless array of properties secretly nested within geometry. One can pursue other triangle relationships *ad infinitum*; however, we feel that we have captured the delightful essences of these named-relationship here in this chapter. After all, we must leave some of these geometric gems for the reader to discover!

At this point we have achieved a strong skill set based on geometric theorems that were developed by, named for, and (occasionally) misattributed to famous mathematicians from across history. In the next chapter, we will explore some applications of these powerful tools as we investigate most unusual geometric configurations.

Chapter 3

Applications of Some Named Geometric Relationships

As we explore the following geometrical relationships, bear in mind that some require only cleverness and elementary Euclidean geometry to prove, while others will demonstrate the power of the theorems presented in the previous chapter. Those theorems expanded our knowledge beyond the high school geometry course, and some might consider them to be in the category of advanced Euclidean geometry. This chapter applies those theorems to further the power and beauty of Euclidean geometry.

An Unusual Application of the Pythagorean Theorem

The Lunes and the Triangle

The area of a circle is not typically commensurate with the areas of rectilinear figures, such as triangles, quadrilaterals, and other polygons. That is, it is quite unusual for a circle to be equal in area to a rectangle, or a parallelogram, or any other figure comprised of straight lines. This is the result of the nature of π, or *pi*, an irrational number that can rarely be compared to rational numbers. The inclusion of π in the circle-area formula usually causes a problem of equating circular areas to non-circular areas, which do not include π. However, we will do just that here. With the help of the Pythagorean theorem, we can construct a figure comprised of circular arcs that has an area equal to a triangle.

Let us consider a rather odd-shaped figure: the lune. The lune is a crescent shaped figure (such as that in which the moon often appears) formed by two circular arcs. Recall that the Pythagorean theorem states that the sum of the areas of the squares on the legs of a right triangle is equal to the area of the square on the hypotenuse. As a matter of fact, we can easily show that the "square" can be replaced by any similar figures (drawn appropriately) on the sides of a right triangle. The sum of the areas of the similar polygons on the legs of a right triangle is equal to the area of the similar polygon on the hypotenuse, such as are shown in Figure 3-1.

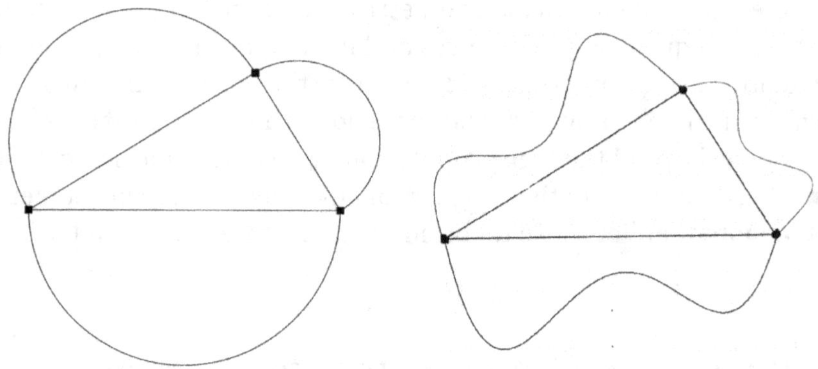

Figure 3-1

The Pythagorean theorem can then be restated for the specific case of semicircles (which are, of course, similar) to read: The sum of the areas of the semicircles on the legs of a right triangle is equal to the area of the semicircle on the hypotenuse.

Thus, for Figure 3-2, we can say that the area of the semicircles relate as follows:

$$\text{Area } P = \text{Area } Q + \text{Area } R.$$

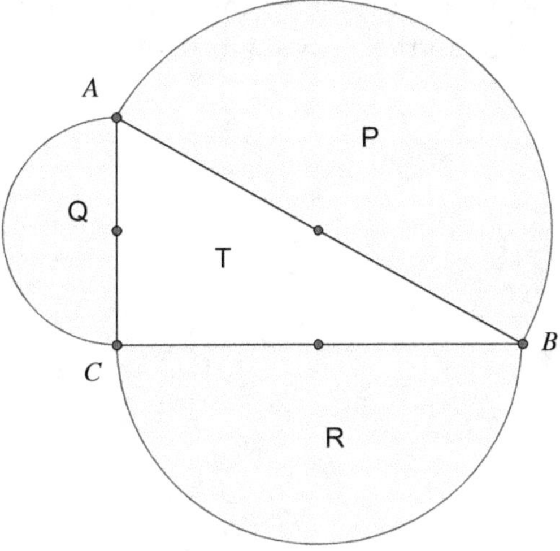

Figure 3-2

Suppose we now flip semicircle *P* over the rest of the figure (using as its axis), as shown in Figure 3-3.

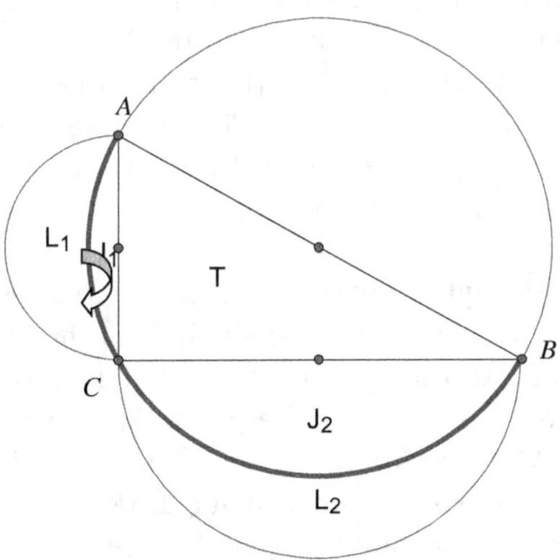

Figure 3-3

Let us now focus on the lunes formed by the two semicircles. We mark the lunes L_1 and L_2, as in Figure 3-4.

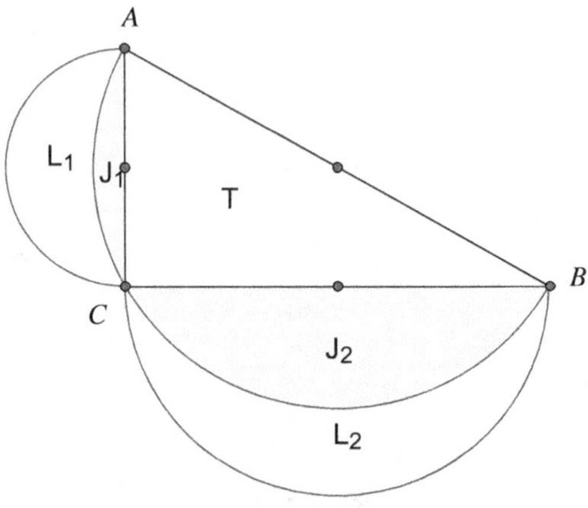

Figure 3-4

In Figure 3-2, we established that *Area P = Area Q + Area R.* In Figure 3-4, that same relationship can be written as follows: $Area\ J_1 + Area\ J_2 + Area\ T = (Area\ L_1 + Area\ J_1) + (Area\ L_2 + Area\ J_2).$

If we subtract *Area J_1 + Area J_2* from both sides of the equation, we get the astonishing result *Area T = Area L_1 + Area L_2.* That is, we have the area of a rectilinear figure (the triangle) equal to the area of non-rectilinear figures (the lunes). This is quite unusual since the measures of circular figures seem to always involve π, while rectangular (or straight line) figures do not.

Properties of the Pythagorean Figure: Squares *AKLC*, *BCNP*, and *ABRQ* are on the sides of right triangle *ABC*, as shown in Figure 3-5. There are lots of interesting geometric relationships to be found in this configuration. For example, the points *K*, *C*, and *P* are collinear, and $\triangle LCN \cong \triangle ABC$. In addition, triangles *AKQ* and *PBR* are equal in area to triangle *ABC*. Furthermore, $\triangle ACQ \cong \triangle KAB$ and $\triangle CBR \cong \triangle ABP$. There is also an unanticipated perpendicularity, since $\angle CHK = 90° = \angle CFP$. We also have concurrency because *KB* and *AP* intersect on *CD*. Amongst the many relationships that can be found in this configuration there is also some parallelism, which leads to the surprising result that $\triangle TGJ \sim \triangle ABC$. We leave the rest of these many other relationships for the reader discover.

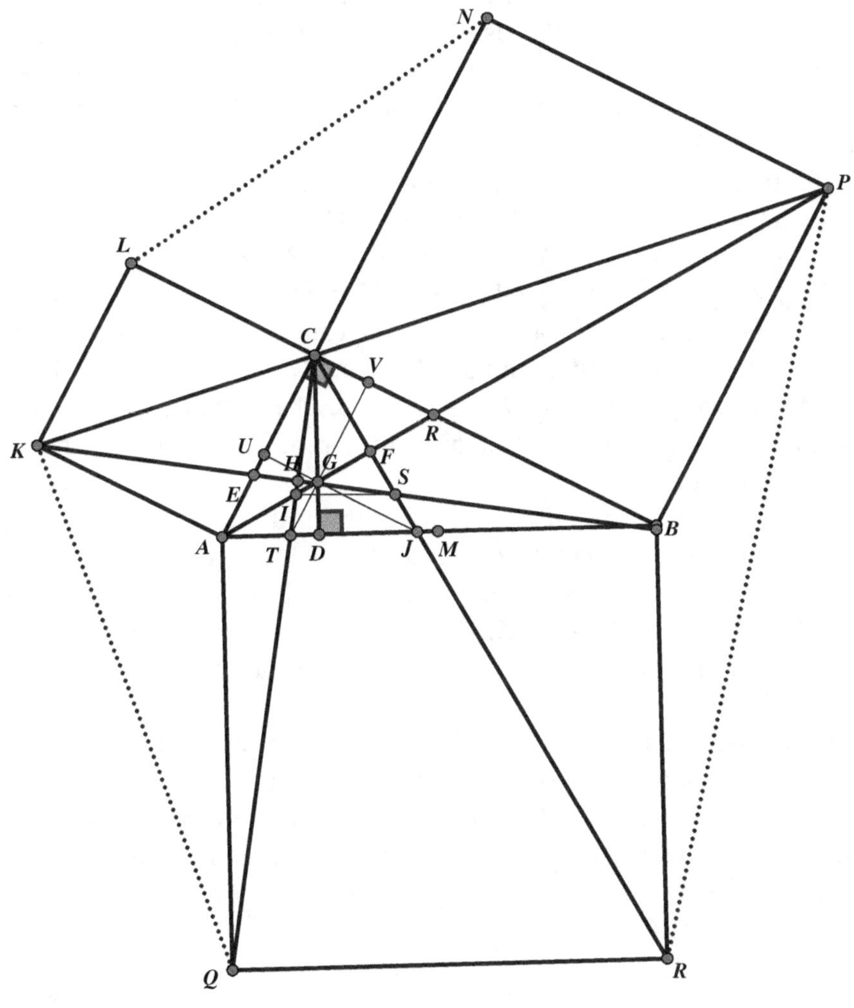

Figure 3-5

Proof: We can see that KCP in Figure 3-5 is a straight line since $\angle KCL + \angle LCN + \angle NCP = 45° + 90° + 45° = 180°$. Also, since $LC = AC$ and $NC = BC$, the right triangles LCN and ABC are congruent. On the other hand, triangles AKQ and PBR are equal in area to triangle ABC. This can be easily justified by a (rather neglected) relationship that states the two triangles are equal in area when they have two pairs of sides equal and included angles supplementary. This can also be justified by the more popular formula for area of the triangle which is $area\triangle = \frac{1}{2}ab\sin C$. In this case, angles $\angle KAQ$ and $\angle CAB$ are supplementary, as are $\angle PBR$ and $\angle ABC$, and corresponding sides $AK = AC$ and $AQ = AB$. The same is true for the triangle PBR, where $PB = BC$ and $BR = AB$.

More discovered relationships are $\triangle ACQ \cong \triangle KAB$ since $KA = CA$ and $AB = AQ$, and because $\angle KAB = 90° + \angle CAB$ and $\angle CAQ = 90° + \angle CAB$, we have that $\angle KAB = \angle CAQ$. Also, since $\angle AKE + \angle KEA = 90°$, we have $\angle CEH + \angle ECH = 90°$ and $\angle EHC = 90°$. A similar procedure establishes that $\triangle CBR \cong \triangle ABP$. The same argument can be made to show that $\angle CFG = 90°$. It follows that points C, F, D, and A are concyclic with diameter AC. Similarly, C, H, D, and B are also concyclic with diameter BC. The common chords CD, BH, and AF are concurrent; therefore, KB, and AP intersect on CD.

Careful inspection of the configuration shows that point G is the orthocenter of triangle CSI and also of triangles ACJ and ACT. From this we have the result that $GJ \perp AC$ and $GT \perp BC$, and $GJ \parallel BC$ and $GT \parallel AC$. Therefore, $\triangle TGJ \sim \triangle ABC$. There are many more curious relationships in this configuration left to the reader to discover.

Simple Application of the Pythagorean Theorem for a General Triangle: Consider an isosceles triangle ABC, where the lengths of the equal sides are s, $AD = a$, $DC = b$, and $BD = c$, as shown in Figure 3-6. The Pythagorean theorem enables us to show the following relationship: $s^2 = c^2 + ab$.

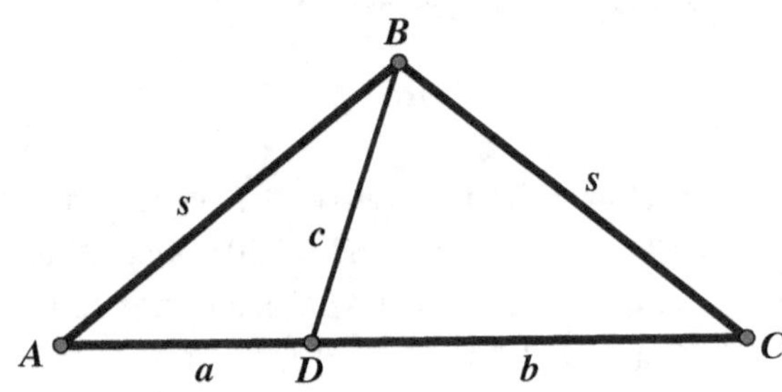

Figure 3-6

Proof: To prove this relationship, we begin by constructing the altitude $BP = h$, and let $DP = y$ and $CP = x$, as seen in Figure 3-7, so we have $b = x + y$ and $a = x - y$. We now apply the Pythagorean theorem, first to triangle BCP: $s^2 = h^2 + x^2$, and then to triangle BDP: $c^2 = h^2 + y^2$. Therefore, by subtracting these two equations, we get $s^2 = c^2 + x^2 - y^2$. However, $x^2 - y^2 = (x - y)(x + y) = ab$. Thus, $s^2 = c^2 + ab$, which is a curious application of the Pythagorean theorem for isosceles triangles.

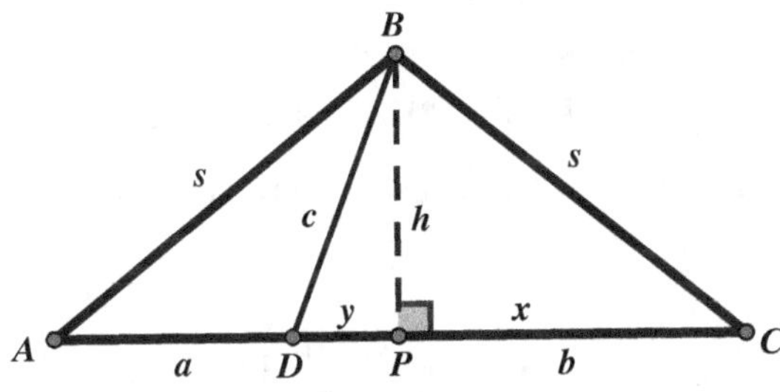

Figure 3-7

Applications of Ptolemy's Theorem

The following are some direct consequences of Ptolemy's theorem, which we encountered in the previous chapter.

Ptolemy's Theorem Application 1: If any circle passing through vertex A of parallelogram $ABCD$ intersects sides AB and AD at points P and R, respectively, and diagonal AC at point Q, as shown in Figure 3-8, the result is that $(AQ)(AC) = (AP)(AB) + (AR)(AD)$.

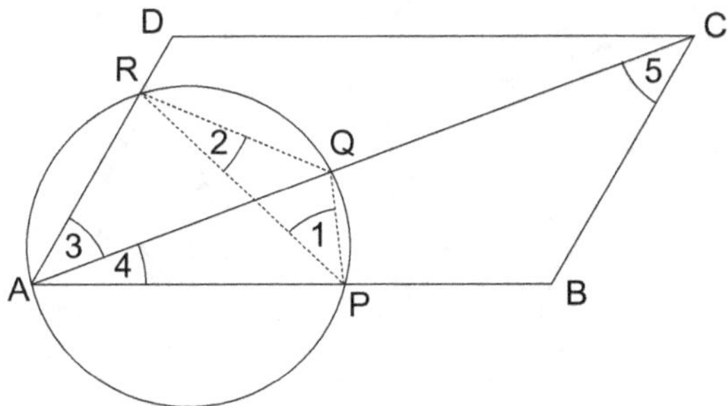

Figure 3-8

Proof: Draw *RQ*, *PQ*, and *RP*, as in Figure 3-8. Then we have $\angle 4 = \angle 2$, since both are one-half the measure of arc *PQ*. By the same reasoning, $\angle 1 = \angle 3$. Since $\angle 5 = \angle 3$, we know that $\angle 1 = \angle 5$. Therefore, $\triangle RQP \sim \triangle ABC$ (AA), and since $\triangle ABC \cong \triangle CDA$, we have $\triangle RQP \sim \triangle ABC \sim \triangle CDA$.

Then,

$$\frac{AC}{RP} = \frac{AB}{RQ} = \frac{AD}{PQ}. \tag{I}$$

By Ptolemy's theorem, in quadrilateral *RQPA*,

$$(AQ)(RP) = (RQ)(AP) + (PQ)(AR). \tag{II}$$

By multiplying each of the three equal ratios in (I) by each member of (II),

$$(AQ)(RP)\left(\frac{AC}{RP}\right) = (RQ)(AP)\left(\frac{AB}{RQ}\right) + (PQ)(AR)\left(\frac{AD}{PQ}\right).$$

Thus, $(AQ)(AC) = (AP)(AB) + (AR)(AD)$, which is our desired result.

Ptolemy's Theorem Application 2: Express the ratio of the diagonals of a cyclic quadrilateral in terms of the sides. In Figure 3-9, it is

$$\frac{AC}{BD} = \frac{(AB)(AD)+(BC)(DC)}{(DC)(AD)+(BC)(AB)}.$$

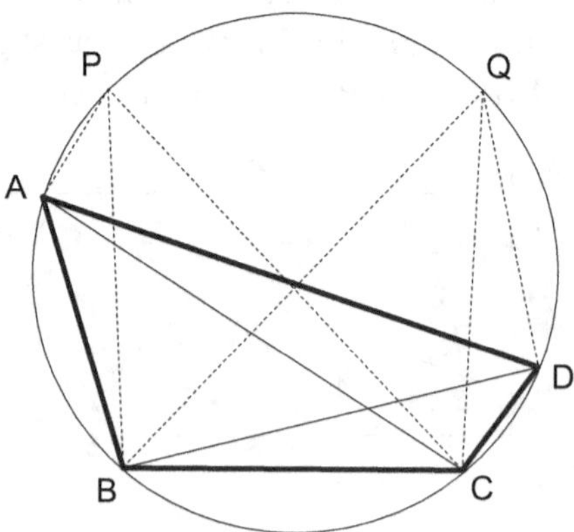

Figure 3-9

Proof: On the circumcircle of quadrilateral *ABCD*, choose points *P* and *Q* so that $PA = DC$ and $QD = AB$, as in Figure 3-9.

Applying Ptolemy's theorem to quadrilateral *ABCP*, we get

$$(AC)(PB) = (AB)(PC) + (BC)(PA). \tag{I}$$

Similarly, by applying Ptolemy's theorem to quadrilateral *BCDQ*, we get

$$(BD)(QC) = (DC)(QB) + (BC)(QD). \tag{II}$$

Since $PA + AB = DC + QD$, $\overset{\frown}{PAB} = \overset{\frown}{QDC}$, and then $PB = QC$. Similarly, since $\overset{\frown}{PBC} = \overset{\frown}{DBA}$, $PC = AD$, and since $\overset{\frown}{QCB} = \overset{\frown}{ACD}$, we have $QB = AD$. Finally, by dividing (I) by (II), and substituting for all terms containing *Q* and *P*, we arrive at

$$\frac{AC}{BD} = \frac{(AB)(AD)+(BC)(DC)}{(DC)(AD)+(BC)(AB)}.$$

Ptolemy's Theorem Application 3: A point *P* is chosen inside parallelogram *ABCD* such that $\angle APB$ is supplementary to $\angle CPD$, as shown in Figure 3-10. Then we have

$$(AB)(AD) = (BP)(DP) + (AP)(CP).$$

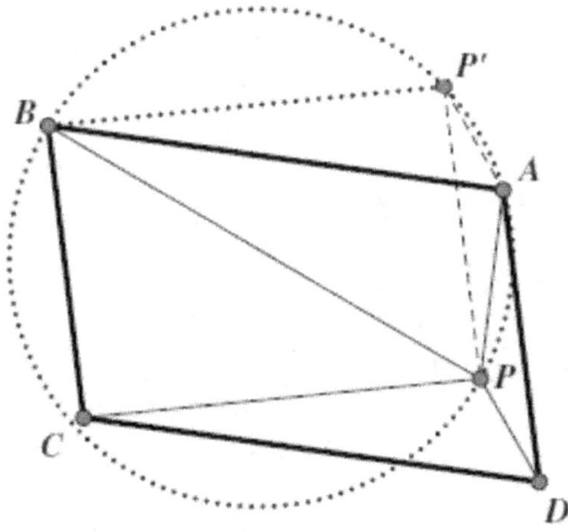

Figure 3-10

Proof: In Figure 3-10, on side of parallelogram $ABCD$, draw $\triangle AP'B \cong \triangle DPC$ so that

$$DP = AP' \text{ and } CP = BP'. \tag{I}$$

Since $\angle APB$ is supplementary to $\angle CPD$, and $\angle BP'A = \angle CPD$, we have $\angle APB$ is supplementary to $\angle BP'A$. Therefore, quadrilateral $BP'AP$ is cyclic. Now applying Ptolemy's theorem to cyclic quadrilateral $BP'AP$, we get $(AB)(P'P) = (BP)(AP') + (AP)(BP')$.

Substituting from (I), we get

$$(AB)(P'P) = (BP)(DP) + (AP)(CP). \tag{II}$$

Since $\angle BAP' = \angle CDP$, and $CD \| AB, PD \| P'A$. Therefore, $PDAP'$ is a parallelogram, and $P'P = AD$. Thus, from (II), $(AB)(AD) = (BP)(DP) + (AP)(CP)$, which is what we sought.

The next five applications develop nice pattern about regular polygons.

Ptolemy's Theorem Application 4: If isosceles $\triangle ABC$ ($AB = AC$) is inscribed in a circle, as seen in Figure 3-11, and a point P is on $\overset{\frown}{BC}$, it follows that $\frac{PA}{PB+PC} = \frac{AC}{BC}$, which is a constant for the given triangle.

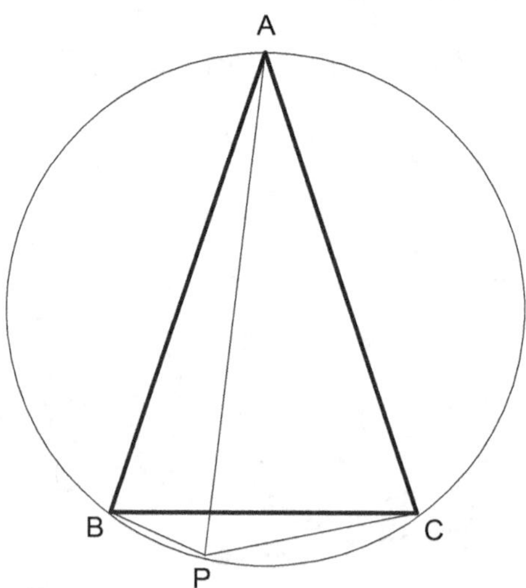

Figure 3-11

Proof: We apply Ptolemy's theorem to cyclic quadrilateral *ABPC*, shown in Figure 3-11, to get $(PA)(BC) = (PB)(AC) + (PC)(AB)$. Since $AB = AC$, $(PA)(BC) = AC(PB + PC)$, and $\frac{PA}{PB+PC} = \frac{AC}{BC}$.

Ptolemy's Theorem Application 5: In Figure 3-12, if equilateral $\triangle ABC$ is inscribed in a circle, and a point P is on $\overset{\frown}{BC}$, the following results: $PA = PB + PC$.

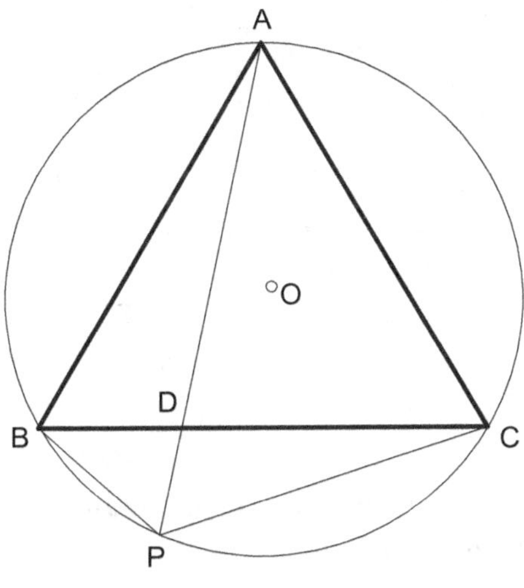

Figure 3-12

Proof: In Figure 3-12, since quadrilateral *ABPC* is cyclic, we may apply Ptolemy's theorem:

$$(PA)(BC) = (PB)(AC) + (PC)(AB). \tag{I}$$

However, since $\triangle ABC$ is equilateral, $BC = AC = AB$. Therefore, from (I), $PA = PB + PC$.

Ptolemy's Theorem Application 6: In Figure 3-13, if square $ABCD$ is inscribed in a circle, and a point P is on $\overset{\frown}{BC}$, the following holds true:

$$\frac{PA+PC}{PB+PD}=\frac{PD}{PA}.$$

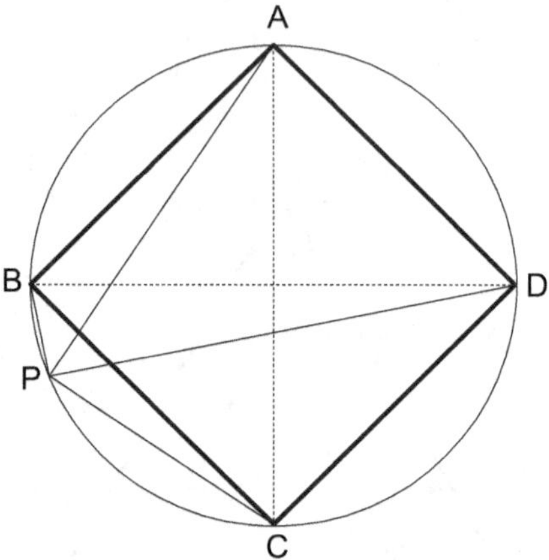

Figure 3-13

Proof: In Figure 3-13, consider isosceles $\triangle ABD$ $(AB = AD)$. Using the results of Application 4, we have

$$\frac{PA}{PB+PD}=\frac{AD}{DB}. \tag{I}$$

Similarly, in isosceles $\triangle ADC$,

$$\frac{PD}{PA+PC}=\frac{DC}{AC}. \tag{II}$$

Since $AD = DC$ and $DB = AC$,

$$\frac{AD}{DB}=\frac{DC}{AC} \tag{III}$$

From (I), (II), and (III), we get $\frac{PA}{PB+PD} = \frac{PD}{PA+PC}$,   or $\frac{PA+PC}{PB+PD} = \frac{PD}{PA}$.

Ptolemy's Theorem Application 7: If regular pentagon *ABCDE* is inscribed in a circle, as can be seen in Figure 3-14, and point *P* is on $\overarc{BC}$, we have that $PA + PD = PB + PC + PE$.

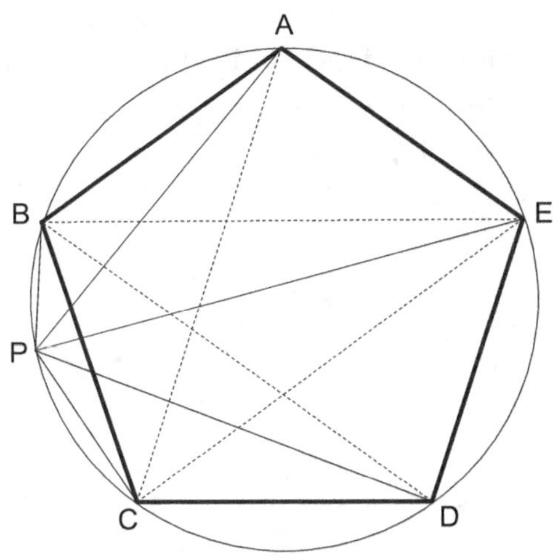

Figure 3-14

Proof: In quadrilateral *ABPC*,

$$(PA)(BC) = (BA)(PC) + (PB)(AC). \tag{I}$$

We apply Ptolemy's theorem to quadrilateral *BPCD* to get

$$(PD)(BC) = (CD)(PB) + (PC)(BD). \tag{II}$$

Since $BA = CD$ and $AC = BD$, by adding (I) and (II), we obtain

$$BC(PA + PD) = BA(PB + PC) + AC(PB + PC). \tag{III}$$

However, since $\triangle BEC$ is isosceles, based upon Application 4,

$$\frac{CE}{BC} = \frac{PE}{PB+PC}, \quad \text{or} \quad \frac{(PE)(BC)}{(PB+PC)} = CE = AC. \tag{IV}$$

Substituting (IV) into (III), we get

$$BC(PA+PD)=BA(PB+PC)+\frac{(PE)(BC)}{(PB+PC)}(PB+PC).$$

But $BC = BA$. Therefore, $PA + PD = PB + PC + PE$.

Ptolemy's Theorem Application 8: In Figure 3-15, if regular hexagon $ABCDEF$ is inscribed in a circle, and point P is on $\overset{\frown}{BC}$, prove that $PE + PF = PA + PB + PC + PD$.

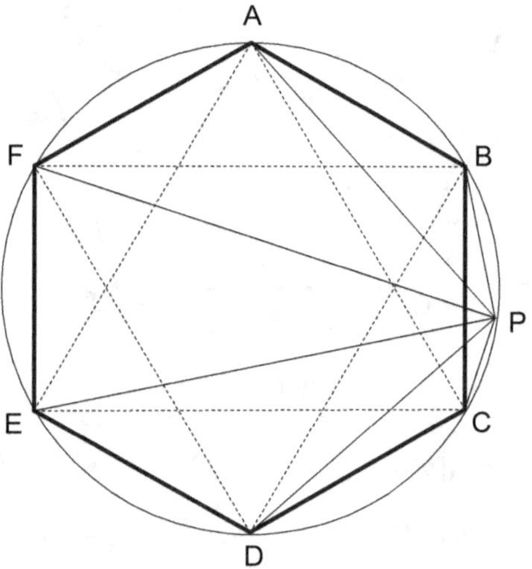

Figure 3-15

Proof: Lines are drawn between points A, E, and C to make equilateral $\triangle AEC$, as shown in Figure 3-15. Using the results of Application 5 of Ptolemy's theorem, we have

$$PE = PA + PC. \tag{I}$$

In the same way, in equilateral $\triangle BFD$,

$$PF = PB + PD. \tag{II}$$

We add (I) and (II) to get $PE + PF = PA + PB + PC + PD$.

Ptolemy's Theorem Application 9: Consider the point P inside the parallelogram $ABCD$ with two right angles, $\angle APD$ and $\angle BPC$, as shown in Figure 3-16. Prove that $AP \cdot CP + BP \cdot DP = AD \cdot CD$.

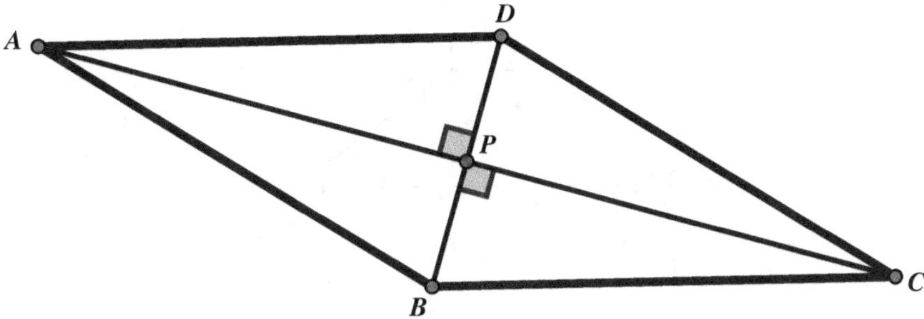

Figure 3-16

Proof: We recognize that point P is the intersection of the diagonals of a rhombus, since the diagonals are perpendicular. To employ Ptolemy's theorem, we need to have a cyclic quadrilateral. In Figure 3-17, we draw $CP \parallel BR$ and $BP \parallel CR$, thus creating rectangle $BPCR$, as all rectangles are cyclic quadrilaterals. We apply Ptolemy's theorem to this rectangle and get $BR \cdot CP + BP \cdot CR = PR \cdot BC$. Since $AP = BR$, $DP = CR$, and $PR = BC = AD = CD$, we then have the required equation: $AP \cdot CP + BP \cdot DP = AD \cdot CD$.

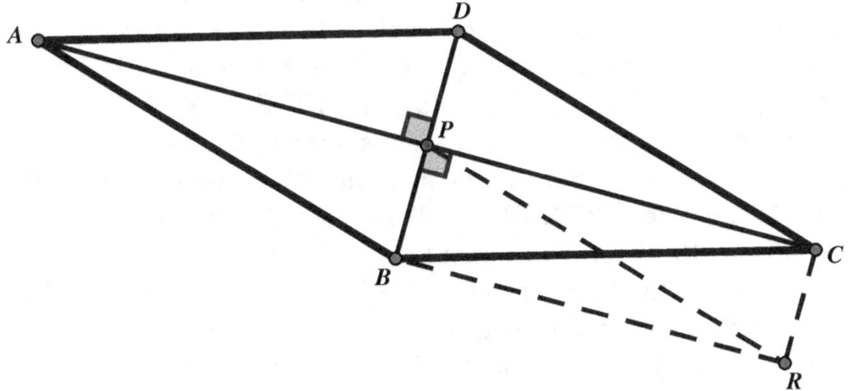

Figure 3-17

Ptolemy's Theorem Application 10: A triangle inscribed in a circle with radius 5 has two sides measuring 5 and 6. Find the measure of the third side of the triangle. Although this problem can be solved by other means, the solution using Ptolemy's theorem is rather nice.

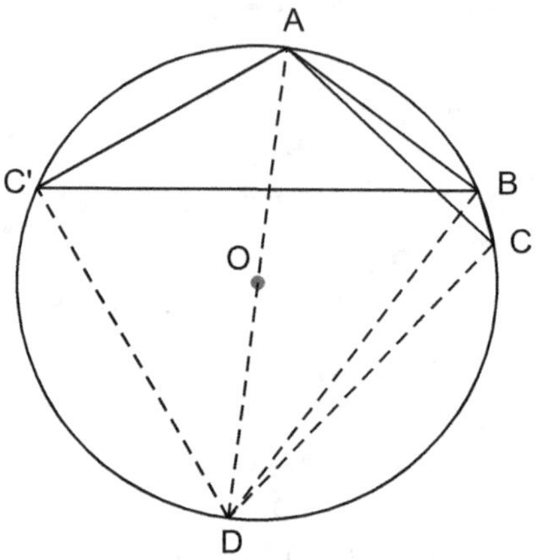

Figure 3-18

Proof: In Figure 3-18, we notice that there are two cases to consider in this problem. Both $\triangle ABC$ and $\triangle ABC'$ are inscribed in circle O, with $AB = 5$ and $AC = AC' = 6$. We are to find the lengths of BC and BC'. Draw diameter AOD, which measures 10, and draw DC, DB, and DC'. Also, $\angle AC'D = \angle ACD = \angle ABD = 90°$. Consider the case where $\angle A$ in $\triangle ABC$ is acute. In right $\triangle ACD$, $DC = 8$, and in right $\triangle ABD$, $BD = 5\sqrt{3}$. By applying Ptolemy's theorem to quadrilateral $ABCD$, we get $(AC)(BD) = (AB)(DC) + (AD)(BC)$, or $(6)(5\sqrt{3}) = (5)(8) + (10)(BC)$, and $BC = 3\sqrt{3} - 4$.

Now consider the case where $\angle A$ is obtuse, as in $\triangle ABC'$. In right $\triangle AC'D$, we have $DC' = 8$. By applying Ptolemy's theorem to quadrilateral $ABDC$, we get $(AC')(BD) + (AB)(DC') = (AD)(BC')$, and therefore $(6)(5\sqrt{3}) = (5)(8) = (10)(BC')$, and $BC' = 3\sqrt{3} + 4$.

This leads to the consideration of cyclic quadrilaterals whose areas are a maximum for the given side lengths. The cyclic quadrilaterals are rich in interesting properties, as Brahmagupta's formula and Ptolemy's theorem illustrate. We will revisit cyclic quadrilaterals throughout this chapter. One should bear in mind that while a triangle's area is defined by the lengths of its three sides, quadrilaterals with four sides of given length have variable size areas. The maximum area of a quadrilateral with clearly defined side lengths is a cyclic quadrilateral.

Applications of Ceva's Theorem

One of the most useful applications Ceva's theorem is in proving the concurrency of the various line segments encountered in elementary geometry. The simplest application of Ceva's theorem is to prove the concurrency of the medians of a triangle. To best appreciate the "power" of Ceva's theorem, first recall the conventional method of proving the medians of a triangle concurrent. Suffice it to say, it is quite long and complex. When compared to that rather cumbersome proof, the following method should provoke some excitement about Ceva's theorem.

Ceva's Theorem Application 1: Prove that the medians of a triangle are concurrent.

Proof: In $\triangle ABC$, line segments AL, BM, and CN are medians, as shown in Figure 3-19. We know that $AN = NB$, $BL = LC$, and $CM = MA$. Multiplying these equalities gives us $(AN)(BL)(CM) = (NB)(LC)(MA)$, or $\frac{AN}{NB} \cdot \frac{BL}{LC} \cdot \frac{CM}{MA} = 1$. Thus, by Ceva's theorem, AL, BM, and CN are concurrent.

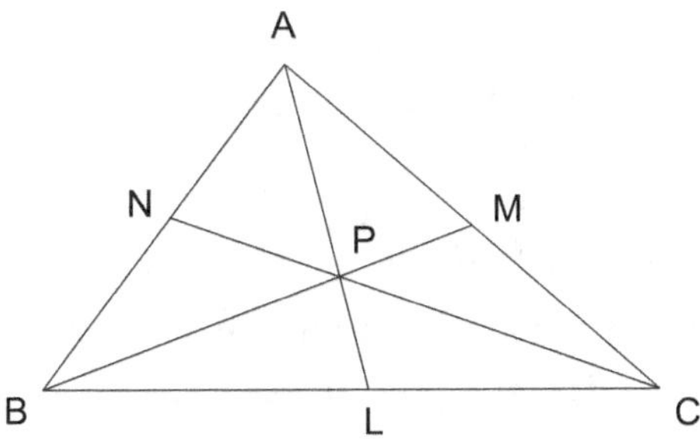

Figure 3-19

Ceva's Theorem Application 2: Prove that the altitudes of a triangle are concurrent.

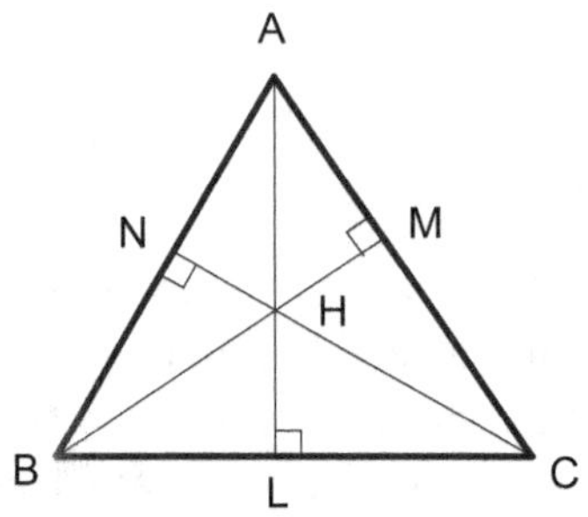 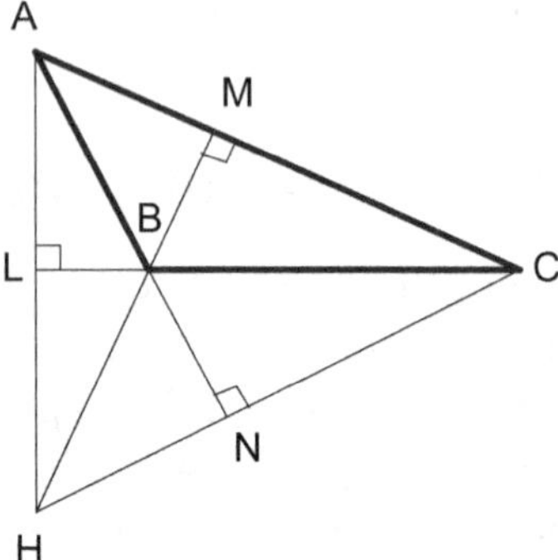

Figure 3-20

Proof: Again, it would be advisable to compare the conventional proof for the concurrency of the altitudes of a triangle to the following proof, which uses Ceva's theorem. In $\triangle ABC$, AL, BM, and CN are altitudes. You may follow this proof for both diagrams of Figure 3-20, since the same proof holds true for both an acute and an obtuse triangle.

$$\triangle ANC \sim \triangle AMB, \text{ so that } \frac{AN}{MA} = \frac{AC}{AB} \tag{I}$$

$$\triangle BLA \sim \triangle BNC, \text{ so that } \frac{BL}{NB} = \frac{AB}{BC} \tag{II}$$

$$\triangle CMB \sim \triangle CLA, \text{ so that } \frac{CM}{LC} = \frac{BC}{AC} \tag{III}$$

Multiplying (I), (II), and (III) gives us $\frac{AN}{MA} \cdot \frac{BL}{NB} \cdot \frac{CM}{LC} = \frac{AC}{AB} \cdot \frac{AB}{BC} \cdot \frac{BC}{AC} = 1$. This indicates that the altitudes are concurrent by Ceva's theorem.

To prove that the three angle bisectors of a triangle are concurrent the following proof should be helpful.

Ceva's Theorem Application 3: Prove that the bisector of any interior angle of a non-isosceles triangle and the bisectors of the two exterior angles at the other vertices are concurrent.

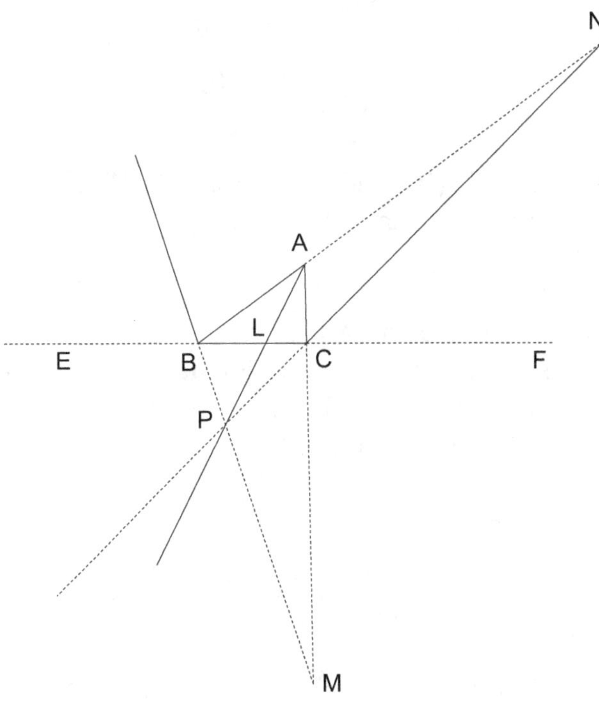

Figure 3-21

Proof: In $\triangle ABC$ in Figure 3-21, AL bisects $\angle BAC$ and meets BC at L, BM bisects exterior $\angle ABE$ and meets AC at M, and CN bisects exterior $\angle ACF$ and meets AB at N. Since the bisector (AL) of an interior angle of a triangle partitions the opposite side proportionally to the remaining two sides of the triangle,

$$\frac{BL}{LC} = \frac{AB}{AC}. \tag{I}$$

An exterior angle bisector partitions the side that it intersects proportionally to the remaining sides of the triangle. Here, this produces the following proportions:

$$\text{For } BM: \frac{CM}{MA} = \frac{BC}{AB} \tag{II}$$

$$\text{For } CN: \frac{AN}{NB} = \frac{AC}{BC} \tag{III}$$

By multiplying (I), (II), and (III) we get $\frac{BL}{LC}\cdot\frac{CM}{MA}\cdot\frac{AN}{NB} = \frac{AB}{AC}\cdot\frac{BC}{AB}\cdot\frac{AC}{BC} = 1$. By Ceva's theorem, we may conclude that AL, BM, and CN are concurrent.

Ceva's Theorem Application 4: Sometimes the question of concurrency is a bit disguised, as in the situation depicted in Figure 3-22. In $\triangle ABC$, $PQ \parallel BC$ and intersects AB and AC at points P and Q, respectively. Prove that PC and QB intersect at a point on median AM.

Proof: Since

$$PQ \parallel BC, \quad \frac{AP}{PB} = \frac{AQ}{QC} \text{ or } \frac{AP}{PB}\cdot\frac{QC}{AQ} = 1. \tag{I}$$

Since AM is a median, $BM = MC$. Therefore, $\frac{BM}{MC} = 1$. $\qquad$ (II)

By multiplying (I) and (II), we get $\frac{AP}{PB}\cdot\frac{QC}{AQ}\cdot\frac{BM}{MC} = 1$. Thus, by Ceva's theorem AM, QB, and PC are concurrent, or QB and PC intersect at a point on AM.

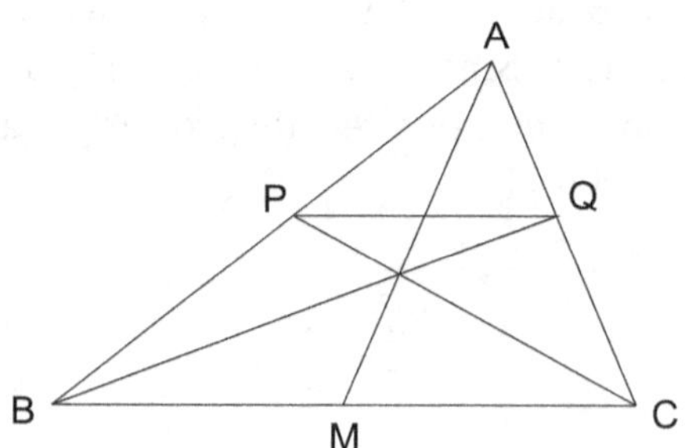

Figure 3-22

Up to this point, all our applications have been used to prove obvious concurrency. The following applications will demonstrate different uses of Ceva's theorem.

Ceva's Theorem Application 5: One concurrency generates another concurrency.

The curiosity we show in Figure 3-23 is surely one of the most remarkable surprises regarding the concurrency of lines within triangles. We begin with triangle *ABC*, which has three concurrent line segments, *AD*, *BE*, and *CF*, intersecting at point *P*. When we draw a circle that contains points *D*, *E*, and *F*, we have the following amazing result: This circle intersects the sides of the triangle in three additional points, namely, points *X*, *Y*, and *Z*. The three segments joining these points with the opposite vertex, that is, *AX*, *BY*, and *CZ*, will always be concurrent as well. Bear in mind that our initial lines that intersected at point *P* could have been *any* three concurrent lines, and they still would have determined a circle that generated another concurrency at point *Q*. Truly amazing!

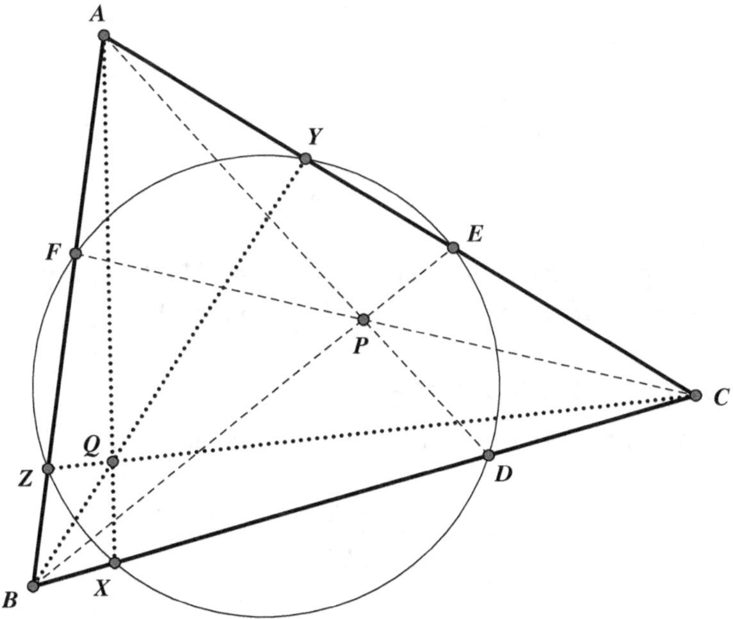

Figure 3-23

Proof: Since AD, BE, and CF intersect at point P, as shown in Figure 3-23, Ceva's theorem gives us $\frac{BD}{CD} \cdot \frac{CE}{AE} \cdot \frac{AF}{BF} = 1$. The secants AC and AB are drawn from an external point; when they intersect the circle we have $AY \cdot AE = AF \cdot AZ$, which is equivalent to $\frac{AF}{AE} = \frac{AY}{AZ}$. Analogously for vertices B and C, we obtain $\frac{BD}{BF} = \frac{BZ}{BX}$ and $\frac{CE}{CD} = \frac{CX}{CY}$. Substituting these expressions in $\frac{BD}{CD} \cdot \frac{CE}{AE} \cdot \frac{AF}{BF} = 1$, which we can also write as $\frac{BD}{BF} \cdot \frac{CE}{CD} \cdot \frac{AF}{AE} = 1$, we thus obtain $\frac{BZ}{BX} \cdot \frac{CX}{CY} \cdot \frac{AY}{AZ} = 1$, which we can write as $\frac{CX}{BX} \cdot \frac{AY}{CY} \cdot \frac{BZ}{AZ} = 1$. By the converse of Ceva's theorem, we find that AX, BY, and CZ are also concurrent, as we sought to prove.

Ceva's Theorem Application 6: In Figure 3-24, we have $\triangle ABC$ where CD is the altitude to AB, P is any point on DC, AP intersects CB at Q, and BP intersects CA at R. It follows that $\angle RDC = \angle QDC$.

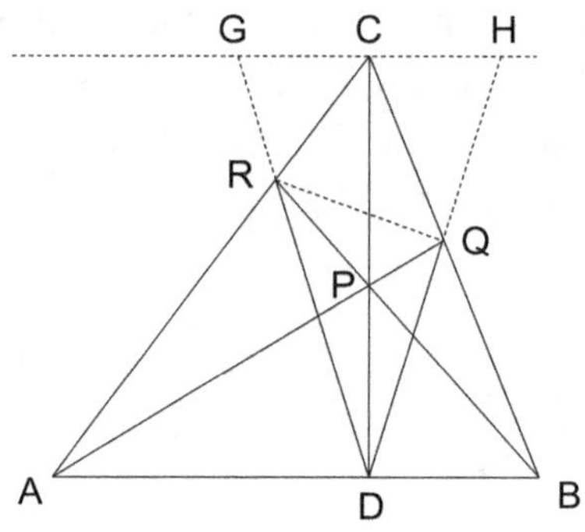

Figure 3-24

Proof: In Figure 3-24 let DR and DQ intersect the line containing C and parallel to AB at points G and H, respectively. $\triangle CGR \sim \triangle ADR$, so that $\frac{CR}{RA} = \frac{GC}{AD}$. (I)

Also $\triangle BDQ \sim \triangle CHQ$, so that $\frac{BQ}{QC} = \frac{DB}{CH}$. (II)

We now apply Ceva's theorem to $\triangle ABC$ to get

$$\frac{CR}{RA} \cdot \frac{AD}{DB} \cdot \frac{BQ}{QC} = 1$$ (III)

Substituting (I) and (II) into (III) gives us $\frac{GC}{AD} \cdot \frac{AD}{DB} \cdot \frac{DB}{CH} = 1$, or $\frac{GC}{CH} = 1$.

This implies that $GC = CH$. Thus, CD is the perpendicular bisector of GH.

Hence, $\triangle GCD \cong \triangle HCD$, and therefore, $\angle RDC = \angle QDC$.

Applications of Menelaus' Theorem

Chapter 2 familiarized us with Menelaus' theorem, and we shall now apply it to several geometric conundra. Each of the following unnamed theorems produce interesting results, and are easily proved using Menelaus' theorem.

Menelaus' Theorem Application 1: Prove that the interior angle bisectors of two angles of a non-isosceles triangle and the exterior angle bisector of the third angle meet the opposite sides in three collinear points.

Proof: In $\triangle ABC$, BM and CN are the interior angle bisectors, while AL bisects the exterior angle at A, as shown in Figure 3-25. Since the bisector of an angle (interior or exterior) of a triangle partitions the opposite side proportionally to the two remaining sides, $\frac{AM}{MC} = \frac{AB}{BC}$, $\frac{BN}{NA} = \frac{BC}{AC}$, and $\frac{CL}{BL} = \frac{AC}{AB}$.

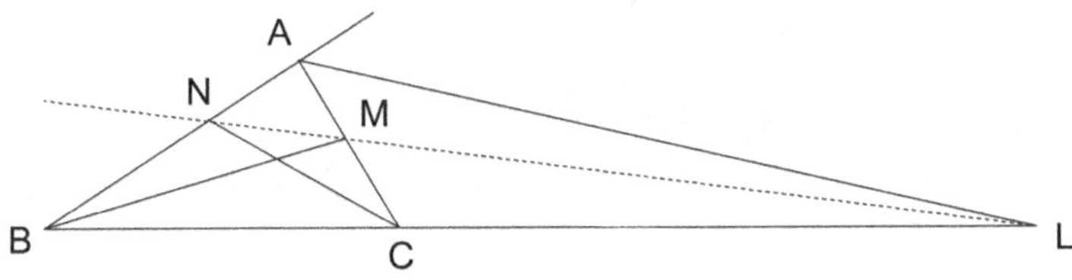

Figure 3-25

Therefore, by multiplication: $\frac{AM}{MC} \cdot \frac{BN}{NA} \cdot \frac{CL}{BL} = \frac{AB}{BC} \cdot \frac{BC}{AC} \cdot \frac{AC}{AB} = 1$. However, $\frac{CL}{BL} = \frac{-CL}{LB}$, therefore, $\frac{AM}{MC} \cdot \frac{BN}{NA} \cdot \frac{CL}{LB} = -1$. Thus, by Menelaus' theorem, N, M, and L must be collinear.

Menelaus' Theorem Application 2: Prove that the exterior angle bisectors of any non-isosceles triangle meet the opposite sides in three collinear points.

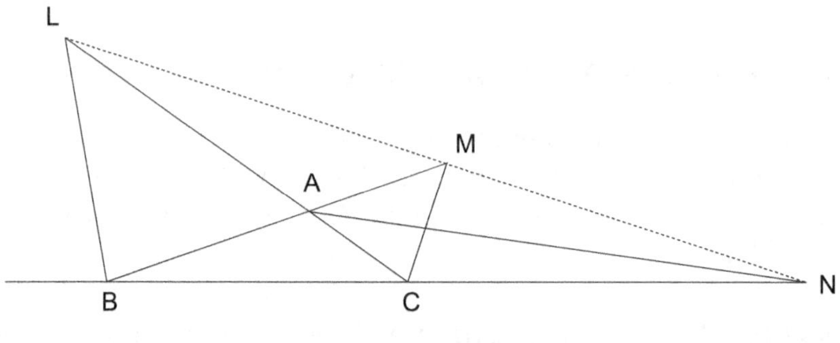

Figure 3-26

Proof: In $\triangle ABC$, the bisectors of the exterior angles at A, B, and C meet the opposite sides (extended) at points N, L, and M, respectively, as shown in Figure 3-26. Recall that the bisector of an angle (interior or exterior) of the triangle partitions the opposite side proportionally to the two remaining sides, therefore, we have the following: $\frac{CL}{AL} = \frac{BC}{AB}$, $\frac{AM}{BM} = \frac{AC}{BC}$, and $\frac{BN}{CN} = \frac{AB}{AC}$.

This generates the quality of the products $\frac{CL}{AL} \cdot \frac{AM}{BM} \cdot \frac{BN}{CN} = \frac{BC}{AB} \cdot \frac{AC}{BC} \cdot \frac{AB}{AC} = -1$, since all three ratios are negative. Thus, by Menelaus' theorem, L, M, and N are collinear.

Menelaus' Theorem Application 3: A circle through vertices B and C of $\triangle ABC$ intersects AB at P and AC at R. If PR meets BC at Q, prove that $\frac{QC}{QB} = \frac{(RC)(AC)}{(PB)(AB)}$.

Proof: Consider $\triangle ABC$ in Figure 3-27 with transversal QPR, where we get $\frac{RC}{AR} \cdot \frac{AP}{PB} \cdot \frac{QB}{CQ} = -1$ (Menelaus' theorem).

Then, considering absolute values,

$$\frac{QC}{QB} = \frac{RC}{AR} \cdot \frac{AP}{PB}. \tag{I}$$

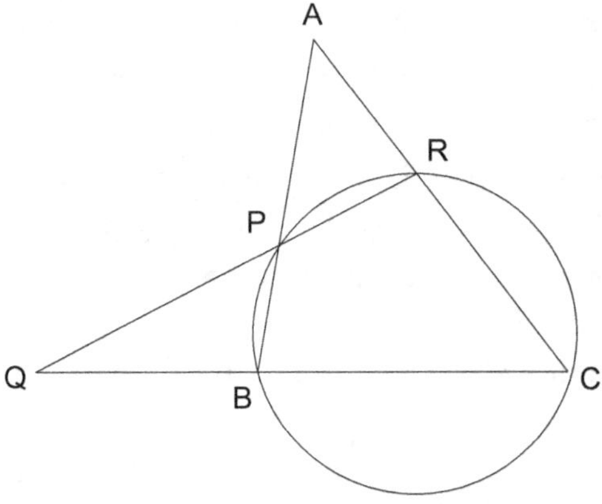

Figure 3-27

However, $(AP)(AB) = (AR)(AC)$, since if two secant segments intersect outside the circle, then the product of the lengths of one secant segment and its external segment equals the product of the lengths of the other secant segment and its external segment, or

$$\frac{AP}{AR} = \frac{AC}{AB}.$$
(II)

By substituting (II) into (I), we get $\frac{QC}{QB} = \frac{(RC)(AC)}{(PB)(AB)}$, which is our desired result.

Menelaus' Theorem Application 4: As shown in Figure 3-28, in right $\triangle ABC$, P and Q are on BC and AC, respectively, such that $CP = CQ = 2$. Through R, the point of intersection of AP and BQ, a line is drawn passing through C and meeting AB at S. Also, PQ meets AB at T. If hypotenuse $AB = 10$ and $AC = 8$, find TS.

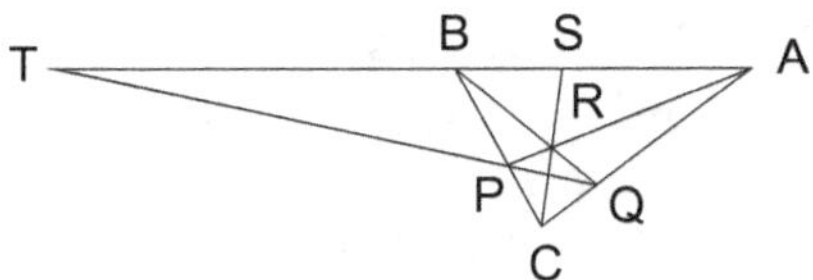

Figure 3-28

Solution: In right $\triangle ABC$, hypotenuse $AB = 10$, and $AC = 8$, so $BC = 6$ (by the Pythagorean theorem). In $\triangle ABC$, since AP, BQ and CS are concurrent, $\frac{AQ}{QC} \cdot \frac{CP}{PB} \cdot \frac{BS}{SA} = 1$, by Ceva's theorem. We substitute to get $\frac{6}{2} \cdot \frac{2}{4} \cdot \frac{BS}{10-BS} = 1$, and therefore, $BS = 4$. Now consider $\triangle ABC$ with transversal QPT. $\frac{AQ}{QC} \cdot \frac{CP}{PB} \cdot \frac{BT}{TA} = -1$ (Menelaus' theorem). Since we a re not dealing with directed line segments, this may be restated as $(AQ)(CP)(BT) = (QC)(PB)(AT)$.

By substitution, $(6)(2)(BT) = (2)(4)(BT + 10)$. Then $BT = 20$, and $TS = 24$.

Menelaus' Theorem Application 5: Figure 3-29 depicts quadrilateral $ABCD$, where AB and CD meet at P, and AD and BC meet at Q. Diagonals AC and BD meet PQ at X and Y, respectively. Prove that $\frac{PX}{XQ} = -\frac{PY}{YQ}$.

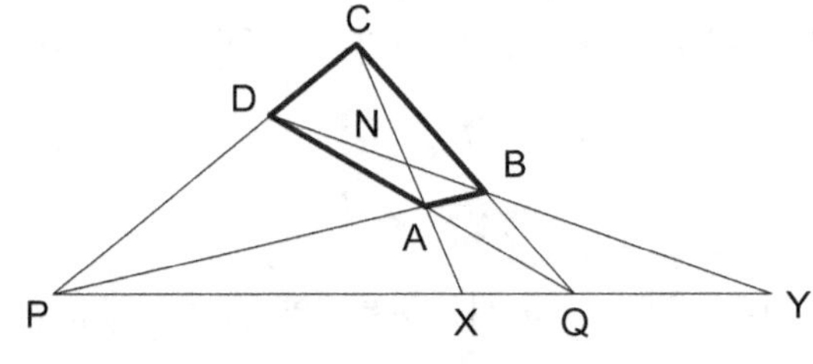

Figure 3-29

Proof: Consider $\triangle PQC$ with PB, QD, and $CX\square$ concurrent.
By Ceva's theorem:

$$\frac{PX}{XQ} \cdot \frac{QB}{BC} \cdot \frac{CD}{DP} = 1. \tag{I}$$

Now consider $\triangle PQC$ with DBY as a transversal.
By Menelaus' theorem:

$$\frac{PY}{YQ} \cdot \frac{QB}{BC} \cdot \frac{CD}{DP} = -1. \tag{II}$$

Therefore, from (I) and (II): $\frac{PX}{XQ} = -\frac{PY}{YQ}$.

Applications of Simson's Theorem

Simson's Theorem Application 1: The Simson line has many interesting properties; a few will be presented here. In Figure 3-30, if the altitude AD of $\triangle ABC$ meets the circumcircle at P, then the Simson line of P with respect to $\triangle ABC$ is parallel to the line tangent to the circle at A.

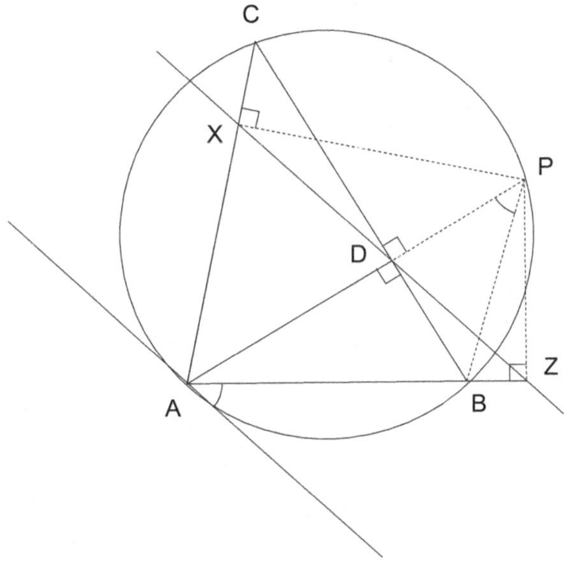

Figure 3-30

Proof: Since PX and PZ are perpendicular respectively to sides AC and AB of $\triangle ABC$, points X, D, and Z determine the Simson line of P with respect to $\triangle ABC$. Draw PB as seen in Figure 3-30. Consider quadrilateral $PDBZ$, where $\angle PDB = \angle PZB = 90°$, indicating that $PDBZ$ is a cyclic quadrilateral. In $PDBZ$, we have

$$\angle DZB = \angle DPB. \tag{I}$$

In the circumcircle of $\triangle ABC$, we have $\angle GAB = \frac{1}{2}(\widehat{AB})$, and $\angle DPB$ $(\angle APB) = \frac{1}{2}(\widehat{AB})$. Therefore,

$$\angle GAB = \angle DPB. \tag{II}$$

From (I) and (II), by transitivity, $\angle DZB = \angle GAB$, and thus, Simson line XDZ is parallel to tangent GA.

Simson's Theorem Application 2: In Figure 3-31, perpendiculars *PX*, *PY*, and *PZ* are drawn from point *P* on the circumcircle of △*ABC* to sides *AC*, *AB*, and *BC*, respectively. Then, $(PA)(PZ) = (PB)(PX)$.

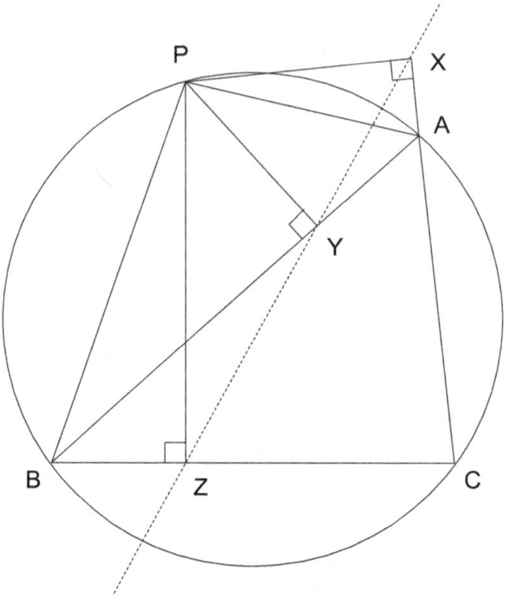

Figure 3-31

Proof: Since $\angle PYB = \angle PZB = 90°$, quadrilateral *PYZB* is cyclic, and one side subtends congruent angles at the two opposite vertices

$$\angle PBY = \angle PZY. \tag{I}$$

Since $\angle PXA = \angle PYA = 90°$, quadrilateral *PXAY* is cyclic and

$$\angle PXY = \angle PAY. \tag{II}$$

Since *X*, *Y*, and *Z* are collinear (the Simson line), then △*PAB* ~ △*PXZ*, and $\frac{PA}{PX} = \frac{PB}{PZ}$, or $(PA)(PZ) = (PB)(PX)$.

Simson's Theorem Application 3: The measure of the angle determined by the Simson lines of two given points on the circumcircle of a given triangle is equal to one-half the measure of the arc determined by the two points.

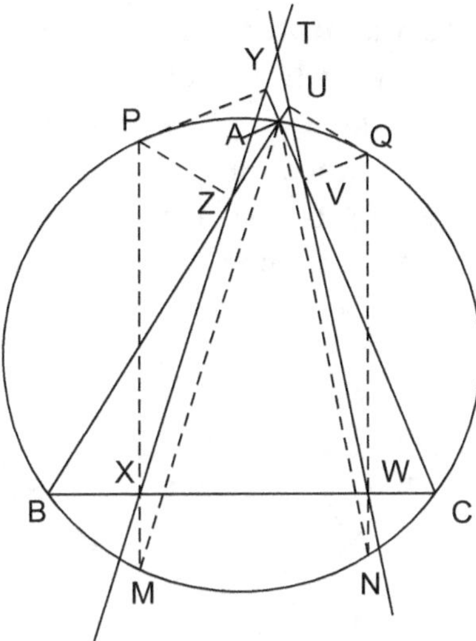

Figure 3-32

Proof: In Figure 3-32, *XYZ* is the Simson line for point *P*, and *UVW* is the Simson line for point *Q*. Extend *PX* and *QW* to meet the circle at *M* and *N*, respectively. Then draw *AM* and *AN*. Since $\angle PZB = \angle PXB = 90°$, quadrilateral *PZXB* is cyclic, since a quadrilateral is cyclic if one side subtends congruent angles at the two opposite vertices, and therefore,

$$\angle ZXP = \angle ZBP. \tag{I}$$

Also,

$$\angle ABP = \angle AMP, \text{ or } \angle ZBP = \angle AMP. \tag{II}$$

From (I) and (II), $\angle ZXP = \angle AMP$, and $XYZ \parallel AM$. (III)

In a similar fashion, it may be shown that $UVW \parallel AN$. Hence, if *T* is the point of intersection of the two Simson lines, then $\angle XTW = \angle MAN$ because their corresponding sides are parallel. Now, $\angle MAN = \frac{1}{2}(\overset{\frown}{MN})$, but since $PM \parallel QN$, $\overset{\frown}{MN} = \overset{\frown}{PQ}$, and therefore $\angle MAN = \frac{1}{2}(\overset{\frown}{PQ})$. Thus, $\angle XTW = \frac{1}{2}(\overset{\frown}{PQ})$.

Simson's Theorem Application 4: The previous application can be applied in the following way. Consider triangle *ABC* in Figure 3-33 with the Simson line *GHD* generated by point *P*, and the Simson line *LEF* generated by point *R*. Since *PR* is the diameter of triangle *ABC*'s circumcircle, we can use the strategy from the previous application to justify the fact that the Simson lines are perpendicular.

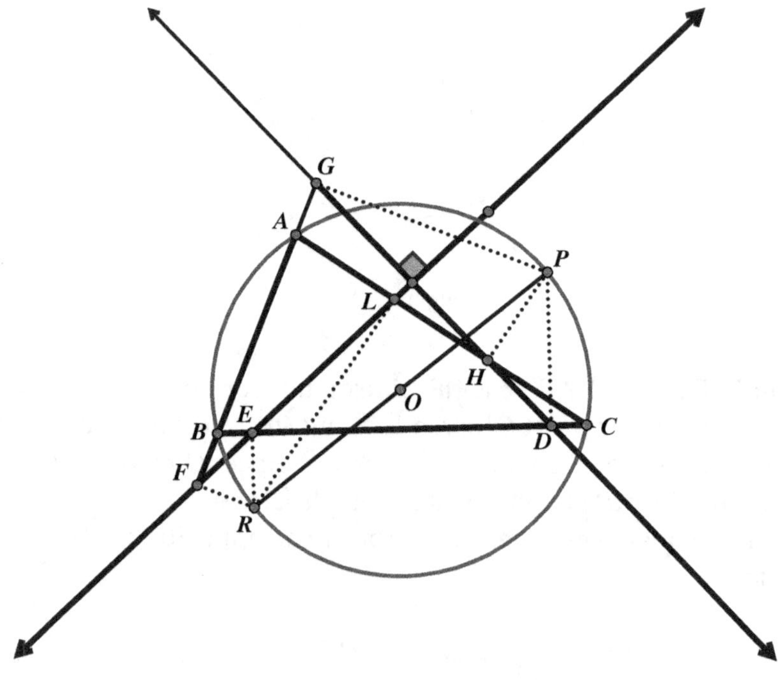

Figure 3-33

Proof: Apply the result of the previous application, which indicates that the measure of the angle determined by the Simson lines of two given points on the circumcircle of a given triangle is equal to one-half the measure of the arc determined by the two points. Here, the arc is 180°. Therefore, the angle formed is 90°.

Stewart's Theorem

Stewart's Theorem Application 1: In an isosceles triangle with two congruent sides of measure 17, a line measuring 16 is drawn from the vertex to the base. If one segment of the base, as cut by this line, exceeds the other by 8, find the measures of the two segments.

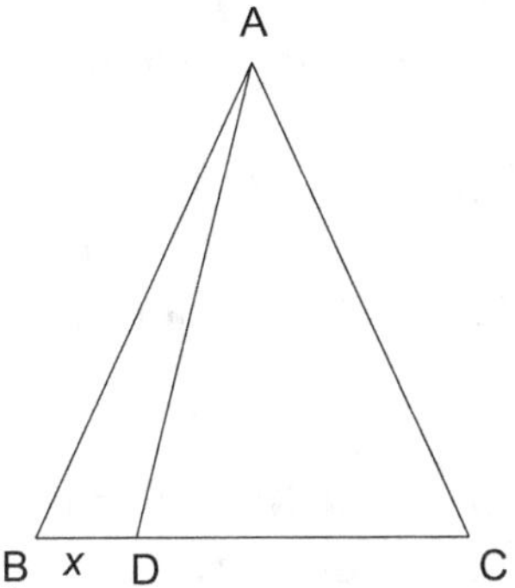

Figure 3-34

Solution: In Figure 3-34, $AB = AC = 17$, and $AD = 16$. Let $BD = x$, so $DC = x + 8$. By Stewart's theorem, we get $(AB)^2(DC) + (AC)^2(BD) = BC[(AD)^2 + (BD)(DC)]$.

Therefore, $(17)^2(x + 8) + (17)^2(x) = (2x + 8)[(16)^2 + x(x + 8)]$, and $x = 3$.

Therefore, $BD = 3$ and $DC = 11$.

Stewart's Theorem Application 2: In a right triangle, the sum of the squares of the distances from the vertex of the right angle to the trisection points along the hypotenuse is equal to $\frac{5}{9}$ the square of the hypotenuse.

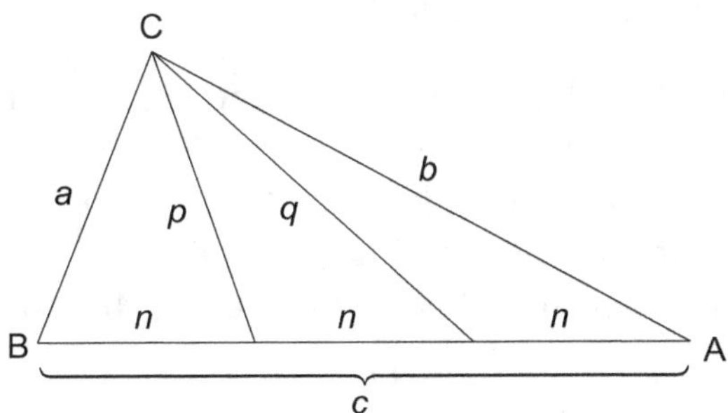

Figure 3-35

Proof: Referring to Figure 3-35, the application asks that we prove $p^2 + q^2 = 5/9(c^2)$ We apply Stewart's theorem to Figure 3-35, using p as the internal line segment, and find that $2a^2n + b^2n = c(p^2 + 2n^2)$. (I)

We repeat step (I), but use q as the internal line segment: $a^2n + 2b^2n = c(q^2 + 2n^2)$. (II)

By adding (I) and (II), we get $3a^2n + 3b^2n = c(4n^2 + p^2 + q^2)$.

Since $a^2 + b^2 = c^2$, $3n(c^2) = c(4n^2 + p^2 + q^2)$.

Since $3n = c$, $c^2 = (2n)^2 + p^2 + q^2$, but $2n = \frac{2}{3}c$; therefore, $p^2 + q^2 = c^2 - \left(\frac{2}{3}c\right)^2 = \frac{5}{9}c^2$.

Stewart's Theorem Application 3: To illustrate the power of Stewart's theorem, we offer another proof: If two angle bisectors of the triangle are congruent, then the triangles is isosceles. This straight-forward method takes this "elementary" theorem and places it (temporarily) at a more advanced point in the development of Euclidean geometry.

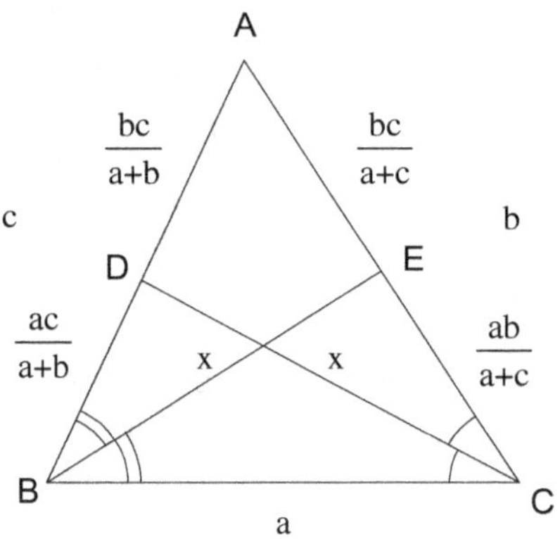

Figure 3-36

Proof: In Figure 3-36, let BE and CD be the angle bisectors in $\triangle ABC$, with $BE = CD = x$. We need to show that $b = c$. An angle bisector divides the side it is drawn to into segments of lengths proportional to the two other sides of the triangle. Thus, $BD = \frac{ac}{a+b}$, $AD = \frac{bc}{a+b}$, $AE = \frac{bc}{a+c}$, and $CE = \frac{ab}{a+c}$. We apply Stewart's theorem twice to $\triangle ABC$ to obtain

$$a^2 \frac{bc}{a+c} + c^2 \frac{ab}{a+c} = b\left(x^2 + \frac{bc}{a+c}\frac{ab}{a+c}\right) \text{ and}$$

$$a^2 \frac{bc}{a+b} + b^2 \frac{ac}{a+b} = c\left(x^2 + \frac{bc}{a+b}\frac{ac}{a+b}\right).$$

Solving for x^2, we obtain

$$x^2 = ac - \frac{ab^2c}{(a+c)^2} = ab - \frac{abc^2}{(a+b)^2}, \text{ thus, } c + \frac{bc^2}{(a+b)^2} = b + \frac{b^2c}{(a+c)^2},$$

or simply

$$c\left(1 + \frac{bc}{(a+b)^2}\right) = b\left(1 + \frac{bc}{(a+c)^2}\right).$$

If $b > c$, since $a, b, c > 0$, we have $\left(1 + \frac{bc}{(a+b)^2}\right) < \left(1 + \frac{bc}{(a+c)^2}\right)$, and the equality does not hold. If $b < c$ we have $\left(1 + \frac{bc}{(a+b)^2}\right) > \left(1 + \frac{bc}{(a+c)^2}\right)$, and the equality again does not hold. Thus, $b = c$, which completes the proof.

Miquel's Theorem

Miquel's theorem opens the door to a variety of additional theorems. Some of these are presented here.

Miquel's Theorem Application 1: The segments joining the Miquel point of a triangle to the vertices of the Miquel triangle form congruent angles with the respective sides of the original triangle.

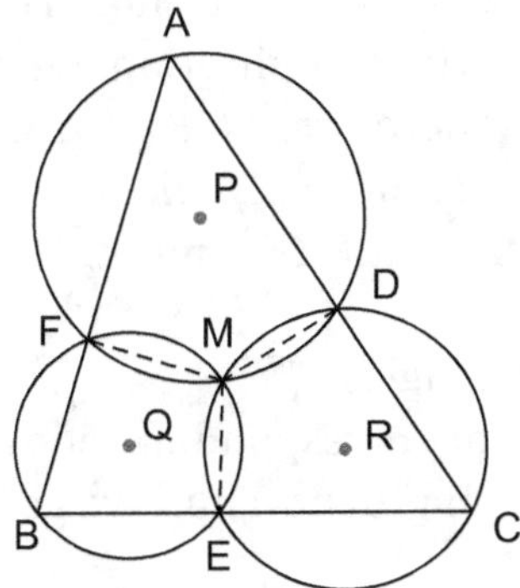

Figure 3-37

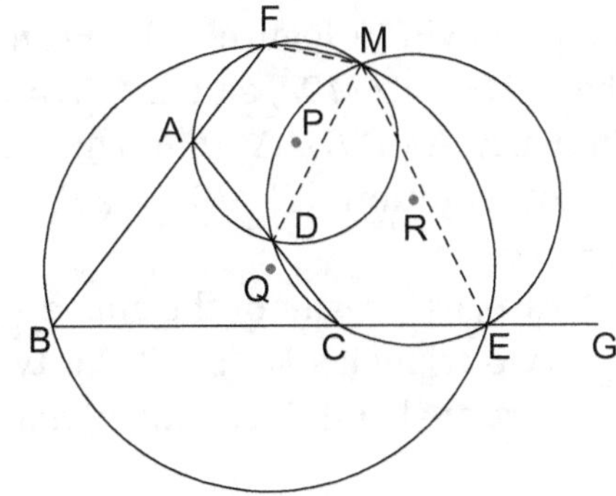

Figure 3-38

Proof: Since quadrilateral *AFMD* is cyclic, as shown in Figures 3-37 and 3-38, $\angle AFM$ is supplementary to $\angle ADM$. But $\angle ADM$ is supplementary to $\angle CDM$. Therefore, $\angle AFM = \angle CDM$, whereupon it follows that $\angle BFM = \angle ADM$. To complete the proof, merely apply the same argument to cyclic quadrilateral *CDME*.

Miquel's Theorem Application 2: We say that a triangle is inscribed in a second triangle if each of the vertices of the first triangle lies on the sides of the second triangle. Thus, we state the following theorem: Two triangles inscribed in the same triangle that share a common Miquel point are similar.

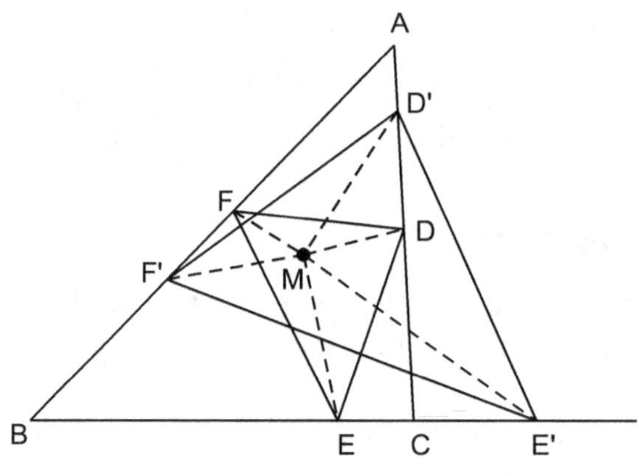

Figure 3-39

Proof: Consider $\triangle DEF$ and $\triangle D'E'F'$, which have the same Miquel point *M*, as shown in Figure 3-39. From the previous application, we find that $\angle MFB = \angle MDA$, and $\angle MF'A = \angle MD'C$. Therefore, $\triangle MF'F \sim \triangle MD'D$. Similarly, $\triangle MD'D \sim \triangle ME'E$. Thus, $\angle FMF' = \angle DMD' = \angle EME'$.

By addition, $\angle F'MD' = \angle FMD$, $\angle F'ME' = \angle FME$, and $\angle E'MD' = \angle EMD$.

Also, as a result of the above similar triangles, $\frac{MF}{MF'} = \frac{MD}{MD'} = \frac{ME}{ME'}$.

Two triangles are similar if two pairs of corresponding sides are proportional and the included angles are congruent. Thus, we get the following: $\triangle F'MD \sim \triangle FMD$, $\triangle F'ME' \sim \triangle FME$, and $\triangle E'MD' \sim \triangle EMD$. Therefore, $\frac{F'D'}{FD} = \frac{F'M}{FM}$, and $\frac{F'E'}{FE} = \frac{F'M}{FM}$. Thus, $\frac{F'D'}{FD} = \frac{F'E'}{FE}$.

Similarly, $\frac{E'D'}{ED} = \frac{F'E'}{FE}$. This proves that $\triangle DEF \sim \triangle D'E'F'$ because the corresponding sides are proportional.

Miquel's Theorem Application 3: The centers of a given triangle's Miquel circles determine a triangle similar to the given triangle.

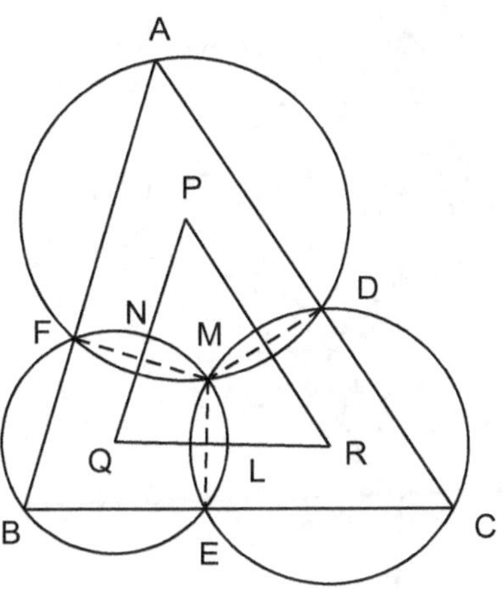

Figure 3-40

Proof: Draw common chords FM, EM, and DM. Line PQ intersects circle Q at N and line RQ intersects circle Q at L, as shown in Figure 3-40. Since the line of centers of two circles is the perpendicular bisector of their common chord, PQ is the perpendicular bisector of FM, so that $\overarc{FN} = \overarc{NM}$. Similarly, QR bisects $\overarc{EM}$, so that $\overarc{ML} = \overarc{LE}$.

Now $\angle NQL = (\overarc{NM} + \overarc{ML}) = \frac{1}{2}(\overarc{FE})$, and $\angle FBE = \frac{1}{2}(\overarc{FE})$. Therefore, $\angle NQL = \angle FBE$.

In a similar fashion it may be proved that $\angle QPR = \angle BAC$. Thus, $\triangle PQR \sim \triangle ABC$.

Miquel's Theorem Application 4: The Miquel circles provide another spectacular relationship. Consider triangle *ABC* in Figure 3-41, where three parallel lines are drawn from the triangle's three vertices to intersect each vertex's Miquel circle. Quite amazingly, these three points of intersection, and the common point of the three circles, are collinear. That is in Figure 3-41 points *E*, *D*, *P*, and *F* are collinear.

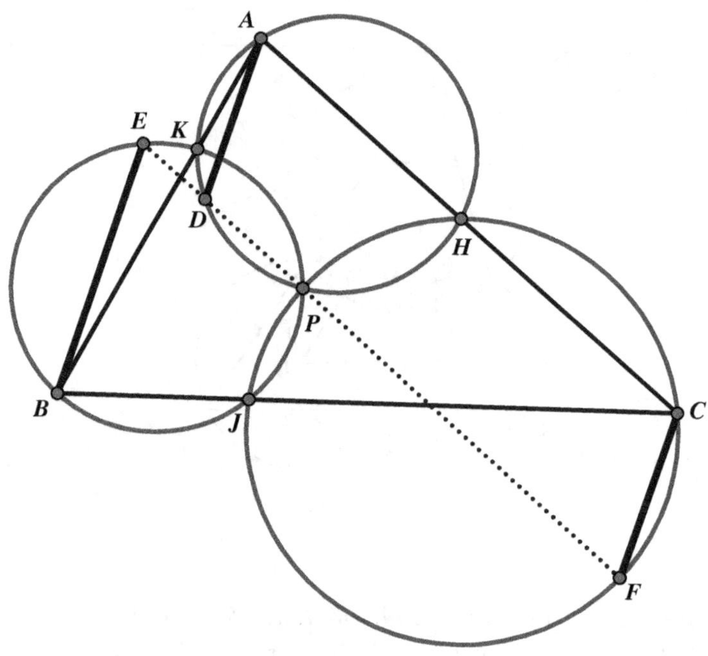

Figure 3-41

Proof: We begin by considering two cyclic quadrilaterals *APDK* and *BPEK,* as shown in Figure 3-42. In cyclic quadrilateral *APDK* we have ∠*KAD* = ∠*DPK*, and in cyclic quadrilateral *BPEK* we have ∠*EBK* = ∠*EPK*. However, since *BE* is parallel *AD*, it follows that ∠*KAD* = ∠*EBK*. Therefore, ∠*DPK* = ∠*EPK*, which implies that points *E*, *D*, and *P* are collinear, and this extends to include that the points *E*, *D*, *P*, and *F* are all collinear.

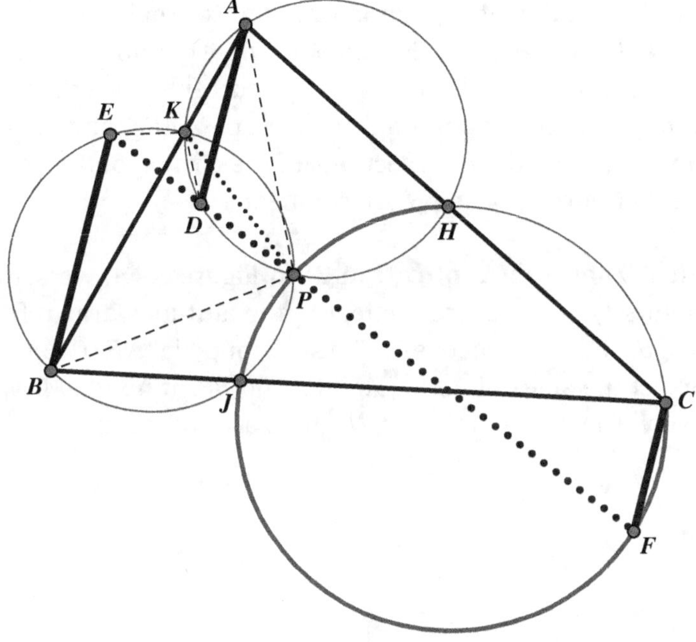

Figure 3-42

Desargues' Theorem

Desargues' Theorem Application 1: In Figure 3-43, a circle inscribed in $\triangle ABC$ is tangent to sides BC, CA, and AB at points L, M, and N, respectively. MN intersects BC at P, NL intersects AC at Q, and ML intersects AB at R. We then find that P, Q, and R are collinear.

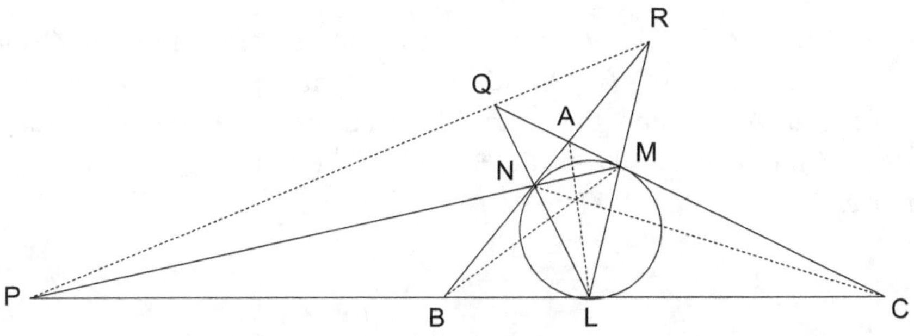

Figure 3-43

Proof: Since the tangent segments from an external point to a circle are congruent, $AN = AM$, $NB = BL$, and $MC = LC$. Therefore, $\frac{AN}{NB} \cdot \frac{BL}{LC} \cdot \frac{MC}{AM} = 1$. By Ceva's theorem, AL, BM, and CN are concurrent. Since these are the lines joining the corresponding vertices of $\triangle ABC$ and $\triangle LMN$, by Desargues' theorem, the intersections of the corresponding sides are collinear; therefore P, Q, and R are collinear.

Desargues' Theorem Application 2: In Figure 3-44, we see $\triangle ABC$, where points F, E, and D are the feet of the altitudes drawn from the vertices A, B, and C, respectively. The sides of pedal $\triangle FED$, EF, DF, and DE, intersect the sides of $\triangle ABC$, AB, AC, and BC, at points M, N, and L, respectively. The result is that M, N, and L are collinear.

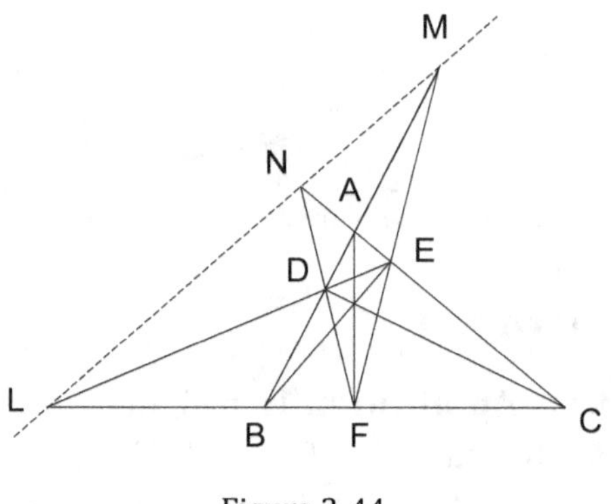

Figure 3-44

Proof: Let A, B, C and F, E, D be corresponding vertices of $\triangle ABC$ and $\triangle FED$. Line segments AF, CD, and BE are concurrent since they are the altitudes of $\triangle ABC$, and then the intersections of the corresponding sides DE and BC, FE and BA, and FD and CA are collinear by Desargues' theorem.

Pappus' Theorem

Pappus' Theorem Application 1: In Figure 3-45, sides AB, BC, and CA of $\triangle ABC$ are cut by a transversal at points Q, R, and S, respectively. The circumcircles of $\triangle ABC$ and $\triangle SCR$ intersect at P so that quadrilateral $APSQ$ is cyclic.

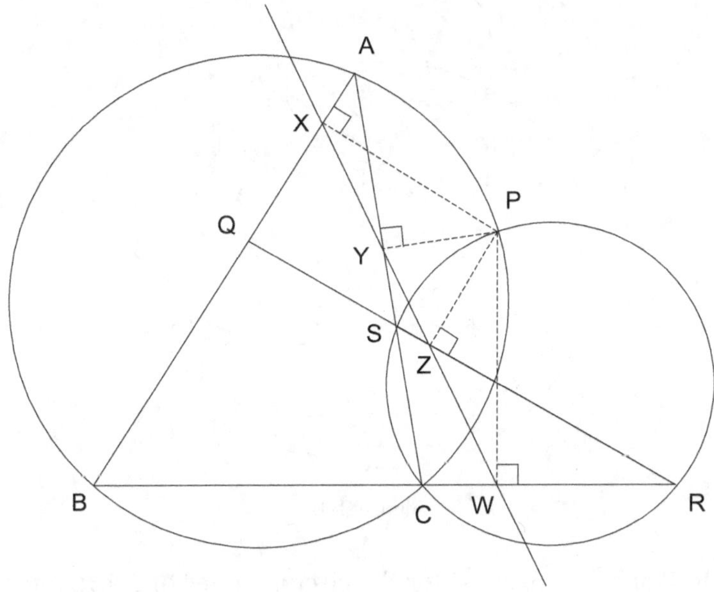

Figure 3-45

Proof: Draw perpendiculars PX, PY, PZ, and PW to AB, AC, QR, and BC, respectively, as in Figure 3-45. Since point P is on the circumcircle of $\triangle ABC$, points X, Y, and W are collinear by Simson's theorem. Similarly, since point P is on the circumcircle of $\triangle SCR$, points Y, Z, and W are collinear. It then follows that points X, Y, and Z are collinear. Thus, P must lie on the circumcircle of $\triangle AQS$ (the converse of Simson's theorem), meaning that quadrilateral $APSQ$ is cyclic.

Pappus' Theorem Application 2: In Figure 3-46, line segments *AB*, *BC*, *EC*, and *ED* form triangles *ABC*, *FBD*, *EFA*, and *EDC* so that the four circumcircles of these triangles meet at a common point.

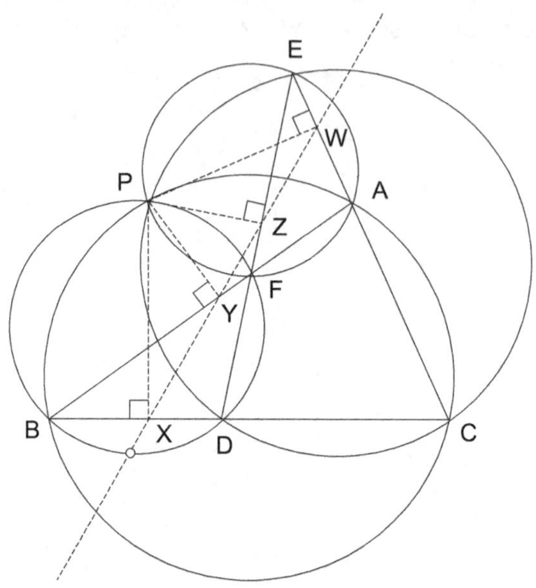

Figure 3-46

Proof: In Figure 3-46, consider the circumcircles of $\triangle ABC$ and $\triangle FBD$, which meet at *B* and *P*. From point *P*, draw perpendiculars *PX*, *PY*, *PZ*, and *PW* to *BC*, *AB*, *ED*, and *EC*, respectively. Since *P* is on the circumcircle of $\triangle ABC$, *X*, *Y*, and *W* are collinear. Also, since *P* is upon the circumcircle of $\triangle FBD$, points *X*, *Y*, and *Z* lie on its Simson line. Therefore *X*, *Y*, *Z*, and *W* are collinear. Since *Y*, *Z*, and *W* are collinear, *P* must lie on the circumcircle of $\triangle EFA$, by the converse of Simson's theorem. By the same reasoning, since *X*, *Z*, and *W* are collinear, *P* lies on the circumcircle of $\triangle EDC$. Thus, all four circles pass through point *P*.

Pascal's Theorem Variations

Variation on Pascal's Theorem 1: Pascal's theorem can be extended in the following manner: If a hexagon has its vertices on a circle in any order, then the intersections (if they exist) of the opposite sides are collinear.

As an example of this variation, you are invited to follow the proof of Pascal's theorem (see pages 52–53) using the diagram in Figure 3-47. Only one minor adjustment needs to be made, and that is the reason for steps (V) through (VII). Remember, the same pairs of "opposite sides" are used here as were used earlier.

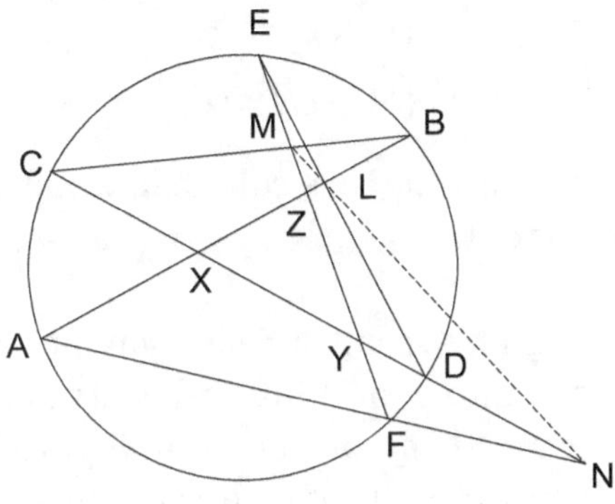

Figure 3-47

Variation on Pascal's Theorem 2: As shown in Figure 3-48, point P is any point in the interior of $\triangle ABC$. Points M and N are the feet of the perpendiculars from P to AB and AC, respectively. $AK \perp CP$ at K and $AL \perp BP$ at L. Therefore, KM, LN, and BC are concurrent.

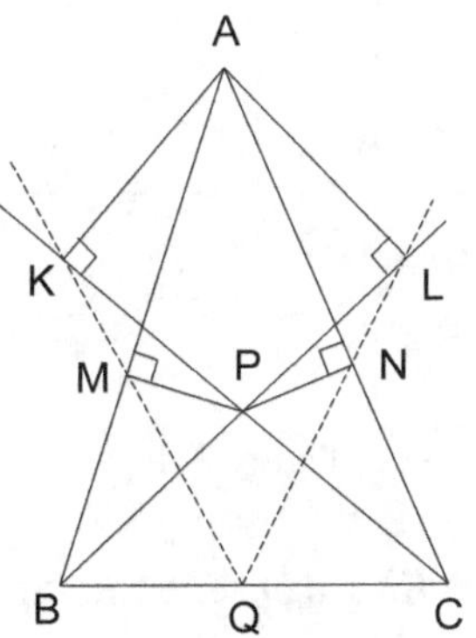

Figure 3-48

Proof: We can easily prove that points A, K, M, P, N, and L all lie on the circle with diameter AP. This can be justified by realizing that right angles *AKP* and *AMP* are inscribed in the same semicircle, as is the case for right angles *ALP* and *ANP*. Now using the variation to Pascal's theorem previously presented, we notice that for inscribed hexagon *AKMPNL*, the pairs of sides intersect as follows:

$$AM \cap LP = B$$
$$AN \cap KP = C$$
$$KM \cap LN = Q$$

By Pascal's theorem, *B*, *C*, and *Q* are collinear, which is to say that *KM*, *LN*, and *BC* are concurrent.

Variation on Pascal's Theorem 3: Select any point *P* not on △*ABC*. Draw a line ℓ that contains *P* and intersects sides *BC*, *AB*, and *AC* at points *X*, *Y*, and *Z*, respectively, as shown in Figure 3-49. Let *AP*, *BP*, and *CP* intersect the circumcircle of △*ABC* at points *R*, *S*, and *T*, respectively. It follows that *RX*, *SZ*, and *TY* are concurrent.

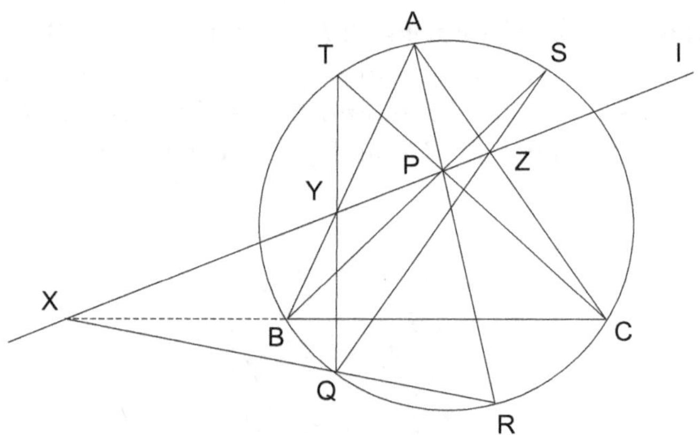

Figure 3-49

Proof: In Figure 3-49, let *RX* intersect the circumcircle at *Q*. Consider hexagon *ARQTCB* and apply Pascal's theorem to it. We notice that since *AR*∩*TC* at *P*, and *RQ*∩*CB* at *X*, then *TQ*∩*AB* at a point on ℓ, which must be *Y* (since *AB*∩ℓ at *Y*). Now consider hexagon *ARQSBC*. Similarly, since *AR*∩*SB* at *P*, and *RQ*∩*CB* at *X*, then *SQ*∩*AC* at a point on ℓ, which must be point *Z*. Thus, *RX*, *SZ*, and *TY* are concurrent.

Brianchon's Theorem

Brianchon's Theorem Application: In Figure 3-50, pentagon *ABCDE* is circumscribed about a circle, with points of tangency at *F*, *M*, *N*, *R*, and *S*. If diagonals *AD* and *BE* intersect at *P*, then *C*, *P*, and *F* are collinear.

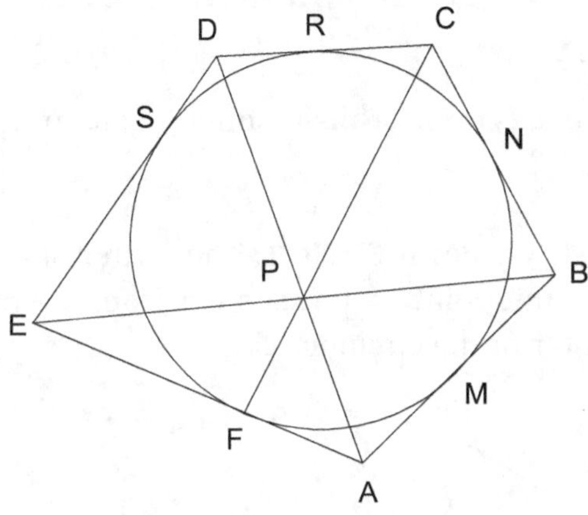

Figure 3-50

Proof: Consider the hexagon circumscribed about a circle in as shown in Figure 3-51. Merge *AF* and *EF* into one line segment, thus making *AFE* a side of a circumscribed pentagon with *F* as one point of

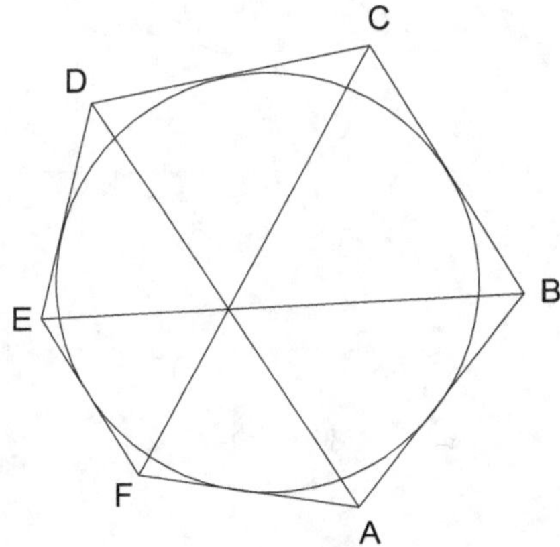

Figure 3-51

tangency, as seen in Figure 3-50. We can then view the pentagon in Figure 3-50 as a degenerate hexagon. We simply apply Brianchon's theorem to this degenerate hexagon to obtain our desired conclusion: *AD*, *BE*, and *CF* are concurrent at *P*, or points *C*, *P*, and *F* are collinear.

Feuerbach's Theorem: The Nine-Point Circle Revisited

We are now ready to establish some basic properties of the nine-point circle.

Property of the Nine-Point Circle 1: The center of a triangle's nine-point circle is the midpoint of the segment from the triangle's orthocenter to the center of the circumcircle.

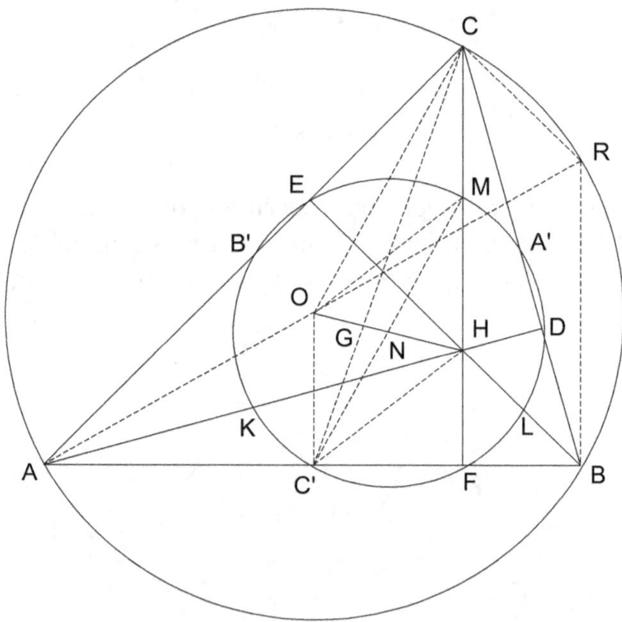

Figure 3-52

Proof: Since MC' subtends a right angle at point F, it must be the diameter of the nine-point circle, as can be seen in Figure 3-52. Therefore, the midpoint, N, of MC' is the center of the nine-point circle. Draw AO to intersect circumcircle O at point R. Then draw line segments CR and BR. OC' is a midline of $\triangle ARB$; therefore, $OC' \parallel RB$. Since $\angle ABR$ is inscribed in a semicircle, it is a right angle. Now both RB and CF are perpendicular to AB, so that $RB \parallel CF$. Similarly, $BE \parallel CR$. We therefore have parallelogram $CRBH$, so that $RB = CH$. Also, $OC' = \frac{1}{2}RB$, since OC' is a midline of $\triangle ARB$. Therefore, $OC' = \frac{1}{2}(CH) = MH$, and $OC'HM$ is a parallelogram (one pair of sides both congruent and parallel). Since the diagonals of a parallelogram bisect each other, the midpoint, N, of MC is also the midpoint of OH.

Property of the Nine-Point Circle 2: The length of the radius of the nine-point circle of a triangle is one-half the length of the radius of the circumcircle.

Proof: In Figure 3-52, we notice that MN is a midline of $\triangle OHC$. Therefore, $MN = \frac{1}{2}(OC)$, which proves that which was sought.

In a paper published in 1765, Leonhard Euler proved that the centroid, G, of a triangle trisects the segment OH, that is, $OG = \frac{1}{3}(OH)$. This line, OH, is known as the *Euler line* of a triangle.

Property of the Nine-Point Circle 3: The centroid of a triangle trisects the segment from the orthocenter to the circumcenter.

Proof: We have already proved that $OC' \parallel CH$, as seen in Figure 3-52, and that $OC' = \frac{1}{2}(CH)$. Therefore, $\triangle OGC' \sim \triangle HGC$ (AA) with a ratio of similitude of $\frac{1}{2}$. Therefore, $OG = \frac{1}{2}(GH)$, which may be stated as $OG = \frac{1}{3}(OH)$.

It now remains for us to prove that G is the centroid of $\triangle ABC$. From the triangles we just proved similar, we have $C'G = \frac{1}{2}(GC) = \frac{1}{3}(C'C)$. But since $C'C$ is a median, G must be the centroid because it appropriately trisects the median. It is interesting to note that $\frac{HN}{NG} = \frac{3}{1} = \frac{HO}{OG}$. Line *HG* is divided internally by *N* and externally by *O* in the same ratio. This is known as a *harmonic* division.

Property of the Nine-Point Circle 4: All triangles inscribed in a given circle and having a common orthocenter also have the same nine-point circle.

Proof: Since all triangles inscribed in a given circle and having a common orthocenter also must have the same Euler line, the center of the nine-point circle for all these triangles is fixed at the midpoint of the Euler line (see the first property of the nine-point circle above). Since the radius of the nine-point circle for each of these triangles is half the length of the circumradius (see the second property of the nine-point circle above), their nine-point circles have the same radius as well as a fixed center. Thus, they all must have the same nine-point circle.

Property of the Nine-Point Circle 5: The segment from the orthocenter to the intersection of the altitude (extended through the foot of the altitude) with the circumcircle is bisected by a side of the triangle.

Proof: In Figure 3-53, since both angles are inscribed in the same circle and intercept the same arc $\overset{\frown}{BC}$, $\angle CSB = \angle CAB$. In $\triangle ACF$, $\angle ACF$ is complementary to $\angle CAF$. In $\triangle CEH$, $\angle ECH$ is complementary to $\angle CHE$. But $\angle BHF = \angle CHE$. Therefore, $\angle CAF = \angle BHF$. Since both $\angle CSB$ and $\angle BHF$ are congruent to $\angle CAB$, they are congruent to each other. Thus, $\triangle HBS$ is isosceles. We can now prove $\triangle HFB \cong \triangle SFB$. It then follows that $HF = SF$, which proves the theorem for one altitude. A simple repetition may be used to verify this theorem for other altitudes.

Our next theorem shows that vertex *B* in Figure 3-53 is the midpoint of arc *TS*.

Property of the Nine-Point Circle 6: A vertex of a triangle is the midpoint of the arc of the circumcircle determined by the circumcircle's intersection with two altitudes (extended through their feet).

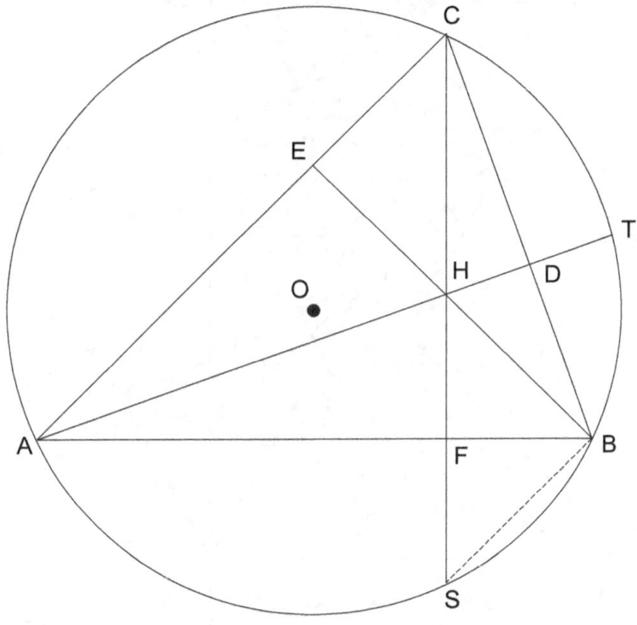

Figure 3-53

Proof: In Figure 3-53, quadrilateral *AFDC* is cyclic because $\angle AFC$ and $\angle ADC$ are congruent (right angles). In this cyclic quadrilateral, $\angle FAD = \angle DCF$ (both intercept $\overarc{DF}$). It then follows that $\overarc{SB} = \overarc{TB}$. Remember, what holds true for one pair of altitudes can also be shown to hold true for other pairs of altitudes.

This configuration leads us to another pair of similar triangles, as we will see in the following.

Property of the Nine-Point Circle 7: The triangle formed by the intersections of the altitude extensions (through the feet of the altitudes) with the circumcircle is similar to the orthic triangle, with corresponding sides parallel.

Proof: In property of the nine-point circle 5, we established that $HF = SF$ and $HD = TD$ (see Figure 3-54). Therefore, in $\triangle HST$, we see that DF is a midline and is parallel to ST. The same argument is used to prove $EF \parallel US$ and $DE \parallel TU$. It then follows that $\triangle DEF \sim \triangle TUS$.

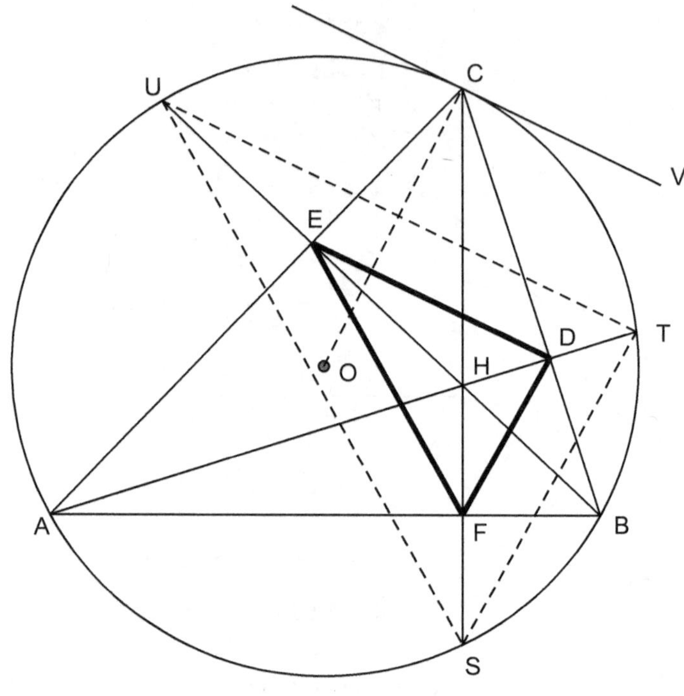

Figure 3-54

This theorem leads to a more useful relationship in our continued study of the nine-point circle.

Property of the Nine-Point Circle 8: The circumradii, which contain the vertices of the triangle, are perpendicular to the corresponding sides of the orthic triangle.

Proof: We shall prove this theorem for one of the radii in question and leave the proofs for the other two radii to the reader. In Figure 3-54, OC is a radius of the circumcircle of $\triangle ABC$ and $\triangle STU$. Property of the nine-point circle 6 proved that $\overset{\frown}{UC} = \overset{\frown}{TC}$. Therefore, OC is the perpendicular bisector of TU. Since $OC \perp TU$, OC must also be perpendicular to DE, because property of the nine-point circle 7 established $DE \parallel TU$.

We now state a theorem that follows the preceding one directly.

Property of the Nine-Point Circle 9: The tangents to the circumcircle of a triangle at the vertices of the triangle are parallel to the corresponding sides of its orthic triangle.

Proof: Again, we shall prove this theorem for only one of the sides of the orthic triangle. Radius OC is perpendicular to tangent VC, as can be seen in Figure 3-54. However, $OC \perp DE$ is a result of property of the nine-point circle 8. Therefore, $VC \parallel DE$. The same argument holds for the other sides of the orthic triangle.

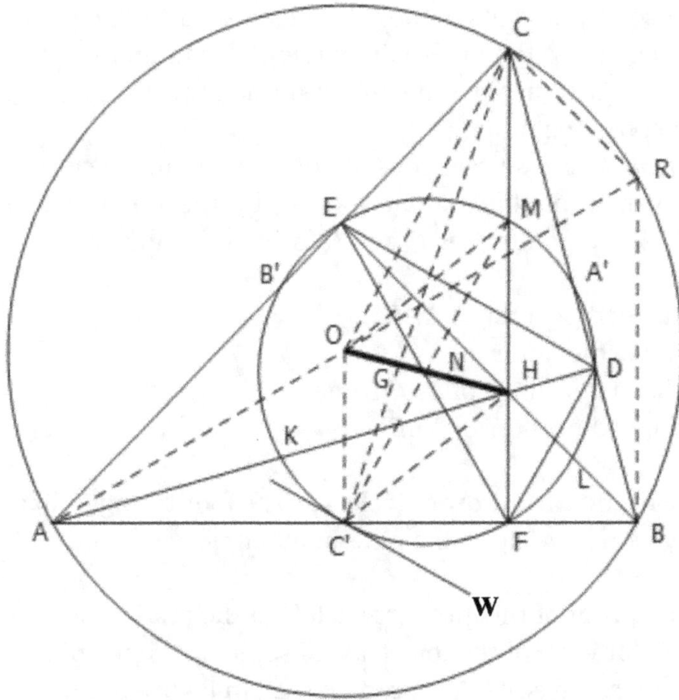

Figure 3-55

Property of the Nine-Point Circle 10: Tangents to a triangle's nine-point circle at the midpoints of the triangle's sides are parallel to the sides of the orthic triangle.

Proof: Radius NC' of the nine-point circle is perpendicular to tangent $C'W$, as can be seen in Figure 3-55. In property of the nine-point circle 8, we established that $OC \perp DE$. We showed earlier that MN was a midline of $\triangle COH$, and therefore, $MN \parallel OC$. This implies that $MNC' \parallel OC$. Thus, $C'W \parallel DE$. The proof for the remaining two sides of the orthic triangles is done in the same manner as that shown above.

Property of the Nine-Point Circle 11: Tangents to the nine-point circle at the midpoints of the given triangle's sides are parallel to the tangents to the circumcircle at the opposite vertices of the given triangle.

Proof: Since the tangents to the circumcircle at a vertex of the triangle and the tangents to the nine-point circle at the midpoints of the sides of the triangle are each parallel to the sides of the orthic triangle, they are parallel to each other.

An *orthocentric system* consists of four points, each of which is the orthocenter of the triangle formed by the remaining three. In Figure 3-55, the points A, B, C, and H form an orthocentric system:

H is the orthocenter of $\triangle ABC$,
A is the orthocenter of $\triangle BCH$,
B is the orthocenter of $\triangle ACH$, and
C is the orthocenter of $\triangle ABH$.

Property of the Nine-Point Circle 12: The four triangles of an orthocentric system have the same nine-point circle.

Proof: The proof of this property is left to the reader. All you need is to check if, for each of the four triangles, the nine determining points all lie on the same circle N, as can be seen in Figure 3-55.

One of the most famous properties of the nine-point circle was first discovered (and proved) by Karl Wilhelm Feuerbach, a German mathematician, in 1822. This property establishes a relationship between the nine-point circle and the incircle and excircles of the original triangle.

Property of the Nine-Point Circle 13: If a triangle has a fixed vertex and a fixed nine-point circle, then the locus of points that can be the circumcenter forms a circle.

Proof: Let A be the fixed point and N be the fixed nine-point circle with radius r. According to property of the nine-point circle 2, the radius of the circumcircle of ABC is $2r$. As a result, the center O of the circumcircle must lie on the circle centered at A with radius $2r$. It must be shown that all points P on this circle can be the circumcenter.

Let point P on circle A be the circumcenter. The orthocenter H is determined as the reflection of point P in N. Points B and C must also lie on the circle centered at P with radius $2r$. Let $O = P$, therefore determining $\triangle AOH$. According to property of the nine-point circle 4, all triangles inscribed in circle O with orthocenter H have N as the nine-point circle. As a result, what must be shown is that, given H, O and A, we can find B and C. Let M be the midpoint of BC. As a result, $OM = \frac{1}{2}(AH)$ and $OM \parallel AH$. Draw a perpendicular to OM at M, and let it intersect circle O at B'; then, $C' \Rightarrow B = B'$ and $C = C'$.

Property of the Nine-Point Circle 14: The nine-point circle contains two rectangles, as seen in Figure 3-56. They are rectangle $RSKL$ and rectangle $MSC'L$, where points R, S, and C' are the midpoints of the sides of triangle ABC, and points M, L, and K are the midpoints of the segments joining the orthocenter, H, to the vertices.

Proof: The line segment joining the midpoints of two sides of a triangle is parallel to the third side and one-half the length of the third side. We apply this to triangle ABC to get $SR \parallel AB$ and $SR = \frac{1}{2} AB$, and for triangle AHB we have $KL \parallel AB$ and $KL = \frac{1}{2}AB$. Therefore, $SR = KL$ and $SR \parallel KL$. Thus, $RSKL$ is a parallelogram with all the vertices on the circle, making it a rectangle. An analogous argument in triangle AHC can be made to show that $MSC'L$ is a rectangle. SM joins the midpoints of sides AC and AH; therefore, it is parallel to and half the length of AH. Similarly, in triangle ABH we have $C'L$ parallel to and half the length of AH. Therefore, $SM = C'L$ and $SR \parallel KL$. Quadrilateral $MSC'L$ is a parallelogram inscribed in a circle; thus, it is a rectangle.

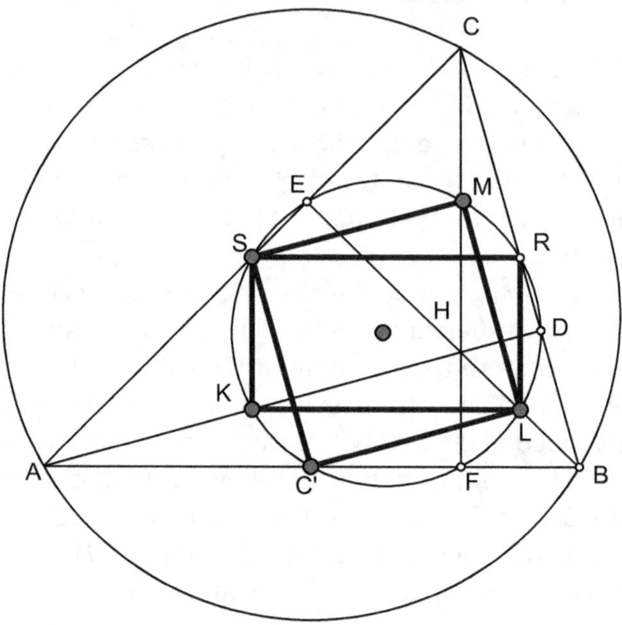

Figure 3-56

We should consider properties of lines that give another perspective on the beauty of circles and triangles. The triangle's orthocenter, centroid, incircle, circumcircle, and nine-point circle motivate further investigations.

Chapter 4

Properties of Components of Triangles

Altitudes

Property of the Altitudes 1: The lengths of a triangle's altitudes are inversely proportioned to the lengths of the triangle's sides.

Proof: In triangle ABC, let A', B', and C' be the feet of the altitudes from A, B, and C, respectively. We then have the following relation:

$$2 \cdot Area\triangle ABC = (BC)(AA') = (AC)(BB') = (AB)(CC')$$

However, in any given triangle, the lengths of a triangle's sides determine its area. According to Heron's formula, the area of any triangle with sides a, b, and c is $Area = \sqrt{s(s-a)(s-b)(s-c)}$, where s is the semiperimeter of the triangle. Thus, the area of the triangle is fixed, which in turn implies that the sides of any triangle and the corresponding altitudes are inversely proportional.

Property of the Altitudes 2: An altitude partitions a side of a triangle into two line segments. The product of the lengths of the two segments equals the product of the length of this altitude and the length of the perpendicular segment from the orthocenter to the side.

Proof: In Figure 4-1, we need to show that $(AF)(FB)=(CF)(FH)$, which means that $\frac{AF}{CF}=\frac{FH}{FB}$. Thus, we would like to show that $\triangle FBH \sim \triangle FCA$. We have $\angle BFH = 90° - \angle BAD = \angle ACF$. Also, $\angle BFH = \angle AFC = 90°$. Therefore, $\triangle FBH \sim \triangle FCA$, and therefore, $(AF)(FB)=(CF)(FH)$.

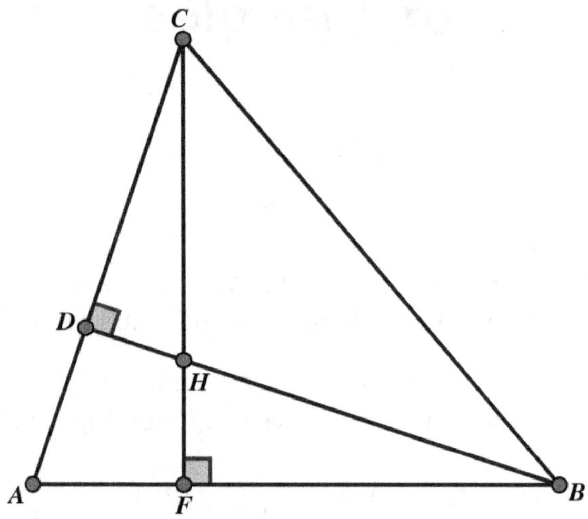

Figure 4-1

Property of the Altitudes 3: The angle formed by the circumradius containing a vertex of a given triangle and that vertex's altitude has a measure equal to the difference of the measures of the remaining two angles of the triangle.

Proof: Without the loss of generality, assume that $\angle B \geq \angle A$ and let θ denote the angle formed by the altitude and the circumradius containing vertex C. Also, let $\angle OAC' = \angle OBC' = \gamma$, $\angle OAB' = \angle OCB' = \alpha$, and $\angle A'CH = \beta$, as shown in Figure 4-2. We then have $\angle CBO = \angle BCO = \beta + \theta$. We would like to show that $\angle ABC - \angle BAC = \theta$, that is, $\beta + \theta + \gamma - (\alpha + \gamma) = \theta$, which implies $\beta = \alpha$. But $\angle A'OB = \angle A'OC =$

$90° - (\beta + \theta)$, $\angle C'OB = \angle C'OA = 90° - \gamma$, and $\angle B'OA = \angle B'OC = 90° - \alpha$. As a result, $360° = 540° - 2\alpha - 2\beta - 2\gamma - 2\theta \implies \alpha + \beta + \gamma + \theta = 90°$. But in $\triangle CBD$, we have $2\beta + \theta + \gamma = 90°$, and therefore, $\beta = \alpha$.

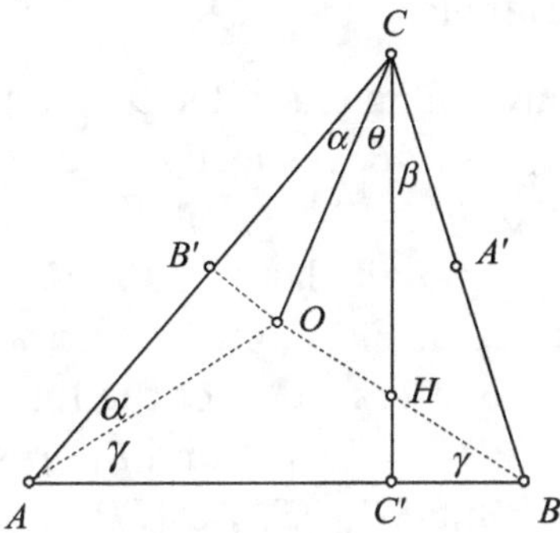

Figure 4-2

Property of the Altitudes 4: The sum of the lengths of a triangle's three altitudes is less than the perimeter of the triangle.

Proof: In Figure 4-3, since the shortest distance from a point to a line is given by the perpendicular, we have that $BE \leq BC$, $AD \leq AB$, and $CF \leq AC$. Thus, $BE + AD + CF \leq AB + BC + AC$. However, since equality causes the triangle to degenerate, the sum of the three altitudes must be strictly less than the sum of the three sides, i.e., the perimeter of the triangle.

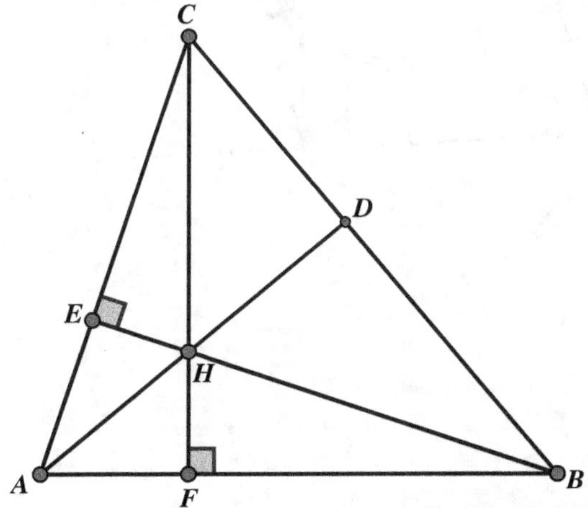

Figure 4-3

Orthocenters

Property of the Orthocenter The circumcenter of a given triangle is the orthocenter of the triangle formed by joining the midpoints of the sides of the original triangle.

Proof: Let $\triangle ABC$ be any triangle and let D, E, and F denote the midpoints of sides BC, AC, and AB, respectively, as shown in Figure 4-4. The circumcenter of $\triangle ABC$ is the point of intersection of the lines through D, E, and F perpendicular to BC, AC, and AB, respectively, which are the perpendicular bisectors. The orthocenter of $\triangle DEF$ can, in turn, be located at the intersection of the lines passing through D, E, and F and perpendicular to FE, FD, and DE, respectively, which are the altitudes of $\triangle DEF$. But $BC \parallel FE$, $AC \parallel FD$, and $AB \parallel DE$. Thus, the circumcenter of triangle ABC and the orthocenter of triangle DEF coincide.

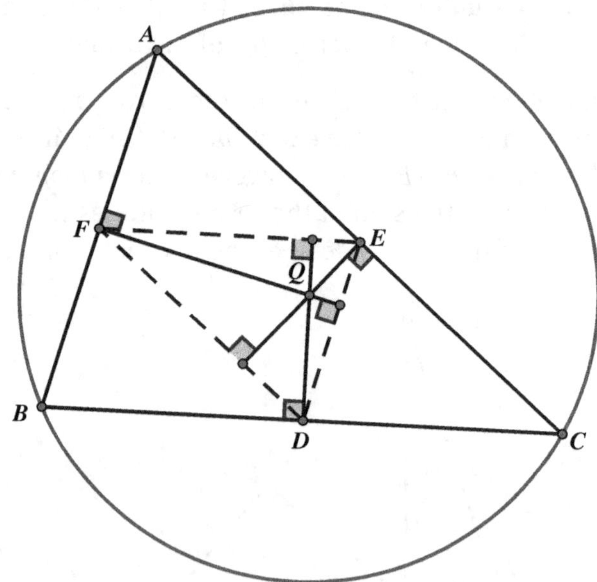

Figure 4-4

Property of the Orthocenter 2: The circumcircle of a given triangle is congruent to the circle containing two vertices and the orthocenter of the triangle.

Proof: In triangle ABC, extend AH to meet the circumcircle at A', as shown in Figure 4-5. Then, $\angle HCB = 90° - \angle ABC = \angle BAA' = \angle BCA'$. Thus, $\triangle BHC \cong \triangle BCA'$, so the circumradius of triangle ABC equals that of triangle BCA', which, in turn, equals that of triangle BHC.

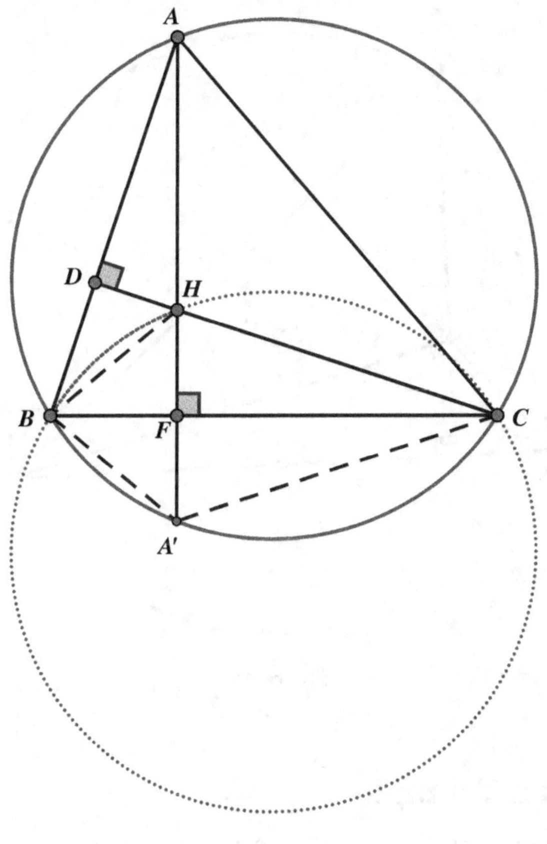

Figure 4-5

Property of the Orthocenter 3: The perpendicular segment from the circumcenter of a triangle to a side has the same length as one-half the segment from the opposite vertex to the orthocenter of the triangle.

Proof: In Figure 4-6, we let A', B' and C' denote the midpoints of sides a, b, and c, respectively. Let D, E, and F represent the feet of the altitudes drawn to a, b, and c, respectively. Let H denote the orthocenter and M the midpoint of CH. We then have MC as the diameter of the nine-point circle, since $\angle MFC' = 90°$. If AO intersects the circle O at R, then OC' is the midsegment of triangle ARB, so that $OC' \parallel CF \parallel RB$ and $OC' = \frac{1}{2} RB$. Similarly, $EB \parallel CR$. Thus, $CRBH$ is a parallelogram and $RB = CH = 2 \cdot OC'$, and therefore, $OC' = \frac{1}{2} \cdot (CH)$.

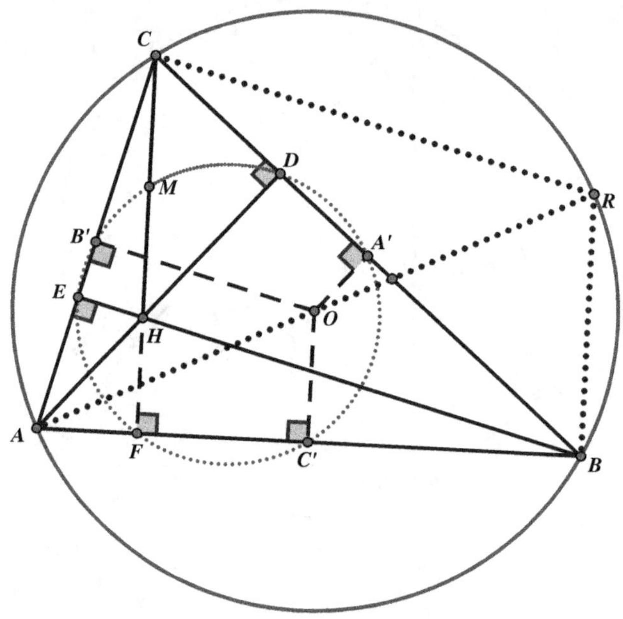

Figure 4-6

Centers of a Triangle

Property of the Centers of a Triangle: A line containing the orthocenter and the midpoint of a side of a given triangle intersects on the circumcircle the circumdiameter containing the opposite vertex.

Proof: Consider $\triangle COM$ in Figure 4-7, where M is the midpoint of CH. If we dilate $\triangle COM$ to twice its size, centering the dilation at C, we obtain $\triangle CO'H$, and $O'H$ intersects AB at its midpoint C', since $CH' = OM = \frac{O'H}{2}$, making $AO'BH$ a parallelogram.

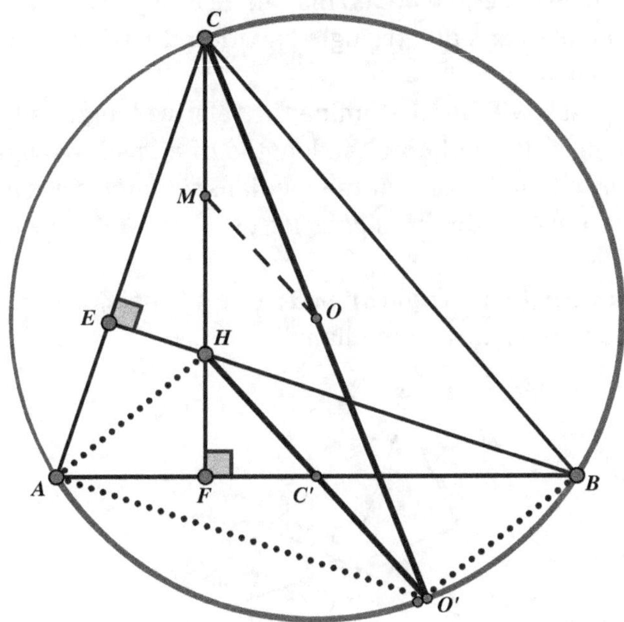

Figure 4-7

Unusual Triangle Median Properties

Properties of Triangle Medians

Although Stewart's theorem can be applied to the medians of a triangle, medians posess lots of interesting properties that are not direct results of Stewart's theorem. Some of these are certainly worthy of our attention.

When asked for a property of the medians of a triangle, the high school geometry student will probably be quick to respond that the point of intersection of the medians (the centroid, or center of gravity) is a trisection point of each median. We used Ceva's theorem to prove that the medians of a triangle are concurrent. The student may also recall that the median of a triangle partitions the triangle into two triangles of equal area. This can be easily extended to realize that the three medians of a triangle partition the triangle into six triangles of equal area.

Our first task will be to examine the relative lengths of the medians of a triangle. If you know the lengths of a given triangle's sides, you could order the lengths of the medians *without measuring them.* This is what our next theorem does for us.

Properties of Medians Application 1: In a triangle, the longest side corresponds to the shortest median.

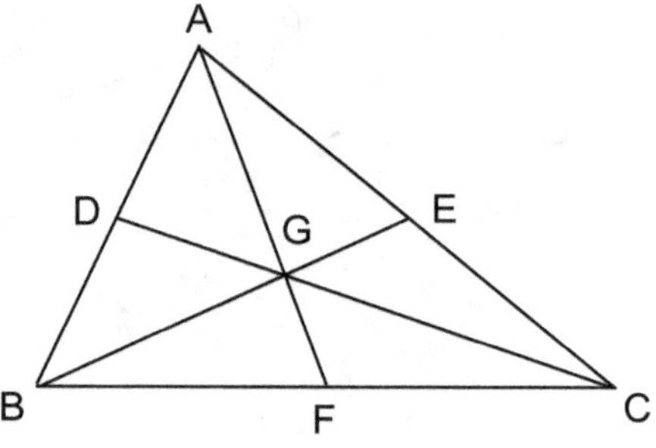

Figure 4-8

Proof: In Figure 4-8, assume $AC > AB$. We must show that $BE < CD$. Notice that $\triangle AFB$ and $\triangle AFC$ are two triangles that have two sides of the same length, that is, $BF = CF$ and $AF = AF$. Therefore, since $AC > AB$, we have $\angle AFC > \angle AFB$. Also, $\triangle GFB$ and $\triangle GFC$ are two triangles that have two sides of the same length. Therefore, since $\angle GFC > \angle GFB$, $GC > GB$. Because of the trisection property of the centroid, $CD > BE$.

We can find the length of a median with Stewart's theorem, and we know from the previous application what the relationship is between the lengths of the medians with regard to the lengths of the sides of the triangle. For the next two applications, we turn our attention to some interesting relationships about the sum of the lengths of the medians of a given triangle.

Properties of Medians Application 2: For any triangle, the sum of the lengths of the medians is less than the perimeter of the triangle.

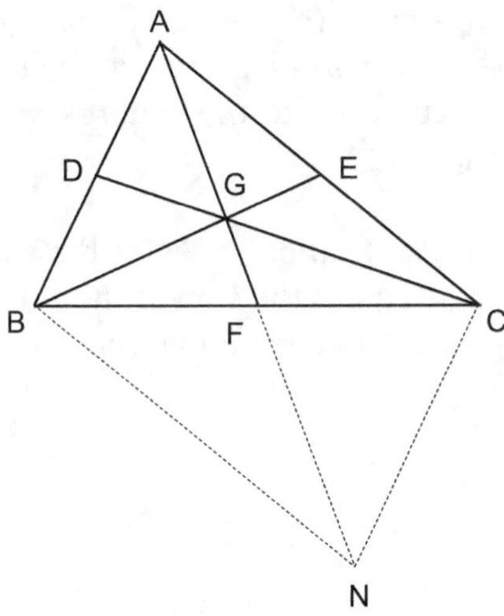

Figure 4-9

Proof: Begin the proof by choosing N on AF so that $AF = NF$, as can be seen in Figure 4-9.

Remember that AF, BE, and CD are the medians of $\triangle ABC$. Quadrilateral $ACNB$ is a parallelogram, since the diagonals bisect each other. Therefore $BN = AC$. For $\triangle ABN$, $AN < AB + BN$; using appropriate substitutions, we can express this inequality as $2(AF) < AB + AC$, or $2(m_a) < c + b$. Similarly, we can show $2(m_b) < a + c$ and $2(m_c) < a + b$. By addition: $2(m_a + m_b + m_c) < 2(a + b + c)$, or $m_a + m_b + m_c < a + b + c$.

(Note: We have simplified the notation by letting a, b, and c represent the side lengths and letting m_a, m_b, and m_c represent the median lengths to sides a, b, and c, respectively.)

Properties of Medians Application 3: For any triangle, the sum of the lengths of the medians is greater than three-fourths of the perimeter of the triangle.

Proof: In Figure 4-8, we begin by using the trisection property of the centroid, G of $\triangle ABC$.

In $\triangle BGC$, $BG + CG > BC$, or $\frac{2}{3}(m_c) + \frac{2}{3}(m_b) > a$. In a similar way we get $\frac{2}{3}(m_a) + \frac{2}{3}(m_c) > b$ and $\frac{2}{3}(m_a) + \frac{2}{3}(m_b) > c$. By addition, $\frac{4}{3}(m_a + m_b + m_c) > a + b + c$.

Therefore, $m_a + m_b + m_c > \frac{3}{4}(a + b + c)$. The previous two theorems thus tell us that $\frac{3}{4}(a + b + c) < m_a + m_b + m_c < a + b + c$.

We now turn our attention to the squares of the lengths of the medians of a given triangle.

Properties of Medians Application 4: Twice the square of the length of a median of a triangle equals the sum of the squares of the lengths of the two including sides minus one-half the square of the length of the third side.

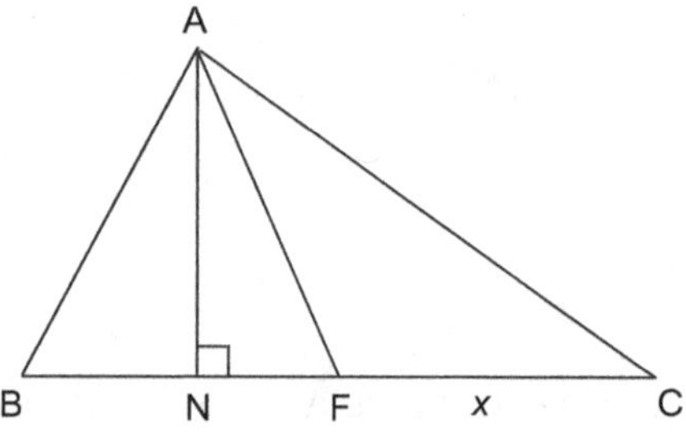

Figure 4-10

Proof: By applying Stewart's theorem to $\triangle ABC$ in Figure 4-10, we get the following:

$$(AB)^2(FC) + (AC)^2(BF) = (BF + FC)[(AF)^2 + (BF)(FC)].$$

Let $x = FC = BF$. Then the following:

$$x(AB)^2 + x(AC)^2 = 2x\left[(AF)^2 + x^2\right]$$

$$(AB)^2 + (AC)^2 = 2\left[(AF)^2 + x^2\right]$$

$$2(AF)^2 = (AB)^2 + (AC)^2 - 2x^2$$

Since $x = \frac{1}{2}(BC)$, we obtain our desired result: $2(AF)^2 = (AB)^2 + (AC)^2 - \frac{1}{2}(BC)^2$.

This theorem by itself is not too exciting, but it will help us prove some useful and interesting properties, one of which is the next application.

Properties of Medians Application 5: The sum of the squares of the lengths of a triangle's medians equals three-fourths the sum of the squares of the lengths of the sides of the triangle.

Proof: The proof of this theorem uses the result stated in Application 4, namely:

$$2m_a^2 = b^2 + c^2 - \frac{1}{2}a^2$$

$$2m_b^2 = a^2 + c^2 - \frac{1}{2}b^2$$

$$2m_c^2 = a^2 + b^2 - \frac{1}{2}c^2$$

Adding these three equations gives us:

$$2(m_a^2 + m_b^2 + m_c^2) = 2(a^2 + b^2 + c^2) - \frac{1}{2}(a^2 + b^2 + c^2)$$

$$2(m_a^2 + m_b^2 + m_c^2) = \frac{3}{2}(a^2 + b^2 + c^2)$$

$$m_a^2 + m_b^2 + m_c^2 = \frac{3}{4}(a^2 + b^2 + c^2),$$

which was our desired result.

We can use this result to establish a relationship between the sum of the squares of the lengths of the segments joining the centroid with the vertices and the sum of the squares of the lengths of the sides.

Properties of Medians Application 6: The sum of the squares of the lengths of the segments joining the centroid with the vertices is one-third the sum of the squares of the lengths of the sides.

Proof: The length of a segment joining the centroid with a vertex is two-thirds the length of its respective median. This leads to $\left(\frac{2}{3}m_a\right)^2 +\left(\frac{2}{3}m_b\right)^2 +\left(\frac{2}{3}m_c\right)^2 =\frac{4}{9}(m_a^2 +m_b^2 +m_c^2)$.

However, from medians application 5, we have $m_a^2 +m_b^2 +m_c^2 = \frac{3}{4}(a^2 +b^2 +c^2)$. Therefore, $\frac{4}{9}(m_a^2 +m_b^2 +m_c^2) =\frac{1}{3}(a^2 +b^2 +c^2)$, which is what was to be demonstrated.

The next theorem is more general and relates *any* point in the plane of a triangle to segments of the triangle.

Properties of Medians Application 7: In Figure 4-11, P is any point in the plane of $\triangle ABC$ with centroid G, then $(AP)^2 +(BP)^2 +(CP)^2 = (AG)^2 +(BG)^2 +(CG)^2 +3(PG)^2$.

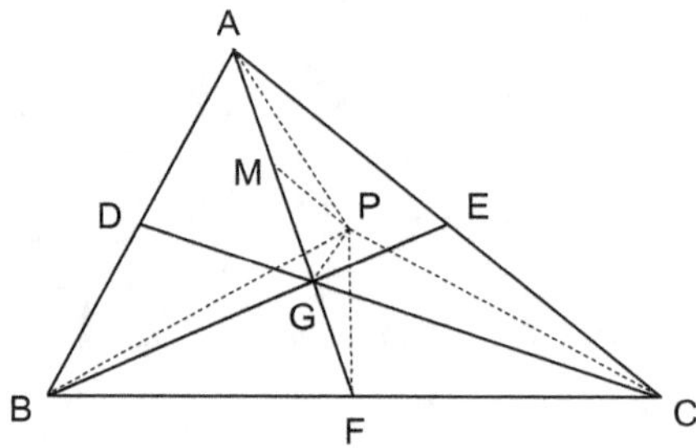

Figure 4-11

Proof: Begin by letting M be the midpoint of AG, as seen in Figure 4-11. We now apply the result of Properties of Medians Application 4 to the following triangles:

To $\triangle PBC$: $2(PF)^2 =(BP)^2 +(CP)^2 -\frac{1}{2}(BC)^2$ (I)

To $\triangle PAG$: $2(PM)^2 =(AP)^2 +(PG)^2 -\frac{1}{2}(AG)^2$ (II)

To $\triangle PMF$: $2(PG)^2 =(PM)^2 +(PF)^2 -\frac{1}{2}(MF)^2$ (III)

Since $MF=\frac{2}{3}(AF)$ and $AG=\frac{2}{3}(AF), MF = AG$. We substitute this into equation (III) and multiply by 2 to get $4(PG)^2 = 2(PM)^2 +2(PF)^2 -(AG)^2$.   (IV)

Now we add (I), (II), and (IV):

$$2(PF)^2 + 2(PM)^2 + 4(PG)^2 = (BP)^2 + (AP)^2 + 2(PM)^2 + (CP)^2$$

$$+ (PG)^2 + 2(PF)^2 - \frac{1}{2}(BC)^2 - \frac{1}{2}(AG)^2 - (AG)^2.$$

This leads to

$$(AP)^2 + (BP)^2 + (CP)^2 - 3(PG)^2 = \frac{3}{2}(AG)^2 + \frac{1}{2}(BC)^2. \qquad \text{(V)}$$

A similar argument made for median *BE* yields

$$(AP)^2 + (BP)^2 + (CP)^2 - 3(PG)^2 = \frac{3}{2}(BG)^2 + \frac{1}{2}(AC)^2. \qquad \text{(VI)}$$

For median *CD* we get

$$(AP)^2 + (BP)^2 + (CP)^2 - 3(PG)^2 = \frac{3}{2}(CG)^2 + \frac{1}{2}(AB)^2. \qquad \text{(VII)}$$

By adding (V), (VI), and (VII):

$$3\left[(AP)^2 + (BP)^2 + (CP)^2 - 3(PG)^2\right] =$$

$$\frac{3}{2}\left[(AG)^2 + (BG)^2 + (CG)^2\right] + \frac{1}{2}\left[(BC)^2 + (AC)^2 + (AB)^2\right] \qquad \text{(VIII)}$$

We now apply the result of Properties of Medians Application 5 to triangle *ABC*:

$$(AG)^2 + (BG)^2 + (CG)^2 = \frac{1}{3}\left[(BC)^2 + (AC)^2 + (AB)^2\right] \text{ or}$$

$$3\left[(AG)^2 + (BG)^2 + (CG)^2\right] = (BC)^2 + (AC)^2 + (AB)^2$$

Now substitute this into equation (VIII) to get our desired result:

$$3\left[(AP)^2 + (BP)^2 + (CP)^2 - 3(PG)^2\right] = \frac{3}{2}\left[(AG)^2 + (BG)^2 + (CG)^2\right]$$

$$+ \frac{1}{2}\left(3\left[(AG)^2 + (BG)^2 + (CG)^2\right]\right)(AP)^2 + (BP)^2 + (CP)^2$$

$$= (AG)^2 + (BG)^2 + (CG)^2 + 3(PG)^2$$

The medians of a triangle provide us with many more interesting relationships, such as the following.

Properties of Medians Application 8: In any triangle, a median and the midline that intersects it (in the interior of the triangle) bisect each other.

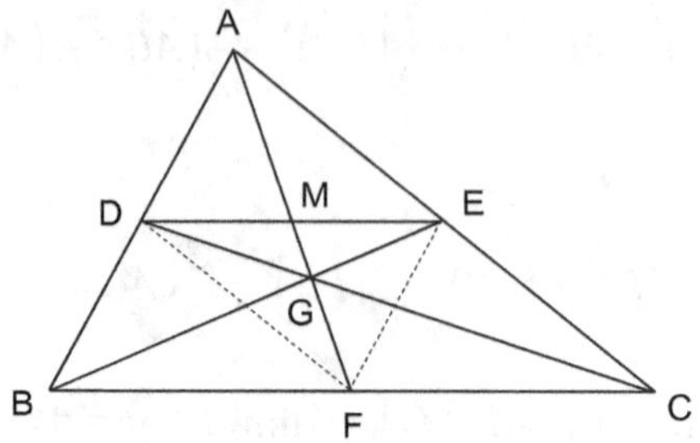

Figure 4-12

Proof: We wish to prove in Figure 4-12 that median *AF* and midline *DE* bisect each other. We draw midlines *DF* and *EF* to form quadrilateral *ADFE*, which is a parallelogram since the opposite sides are parallel. Therefore, diagonals *AF* and *DE* bisect each other.

The centroid serves as the "balancing" point of a triangle. We examine that property in our next application.

Properties of Medians Application 9: In any $\triangle ABC$, such as the one shown in Figure 4-13, we have *XYZ* as any line through the centroid *G*. Perpendiculars are drawn from each vertex of $\triangle ABC$ to this line so that $CY = AX + BZ$.

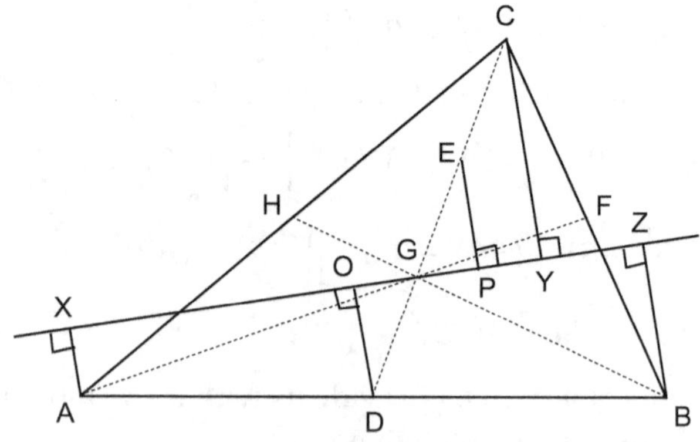

Figure 4-13

Proof: In Figure 4-13, we draw medians CD, AF, and BH. From E, the midpoint of CG, draw $EP \perp XZ$. Also draw $DQ \perp XZ$. Since $\angle CGY = \angle QGD$, and $EC = EG = DG$ (trisection property of a centroid), $\triangle QGD \cong \triangle PGE$, and $QD = EP$. Also, $AX \parallel BZ$, therefore, QD is the median of trapezoid $AXZB$, and $QD = \frac{1}{2}(AX + BZ)$ (property of the median of a trapezoid). $EP = \frac{1}{2}(CY)$, (property of a midline), therefore, $\frac{1}{2}(CY) = \frac{1}{2}(AX + BZ)$ (transitivity), and $CY = AX + BZ$.

It is interesting to note that for a given point in a given circle, an infinite number of inscribed triangles exist that have this point as a centroid. This will be proven in Application 10.

Properties of Medians Application 10: An infinite number of triangles, each having a given interior point as centroid, can be inscribed in a given circle.

Proof: This proof will be somewhat different from others so far. To show that there exists an infinite number of triangles with the necessary specifications, we will show that one such triangle, *randomly selected*, exists. This will imply that an infinite number of others, similarly constructed, also exist.

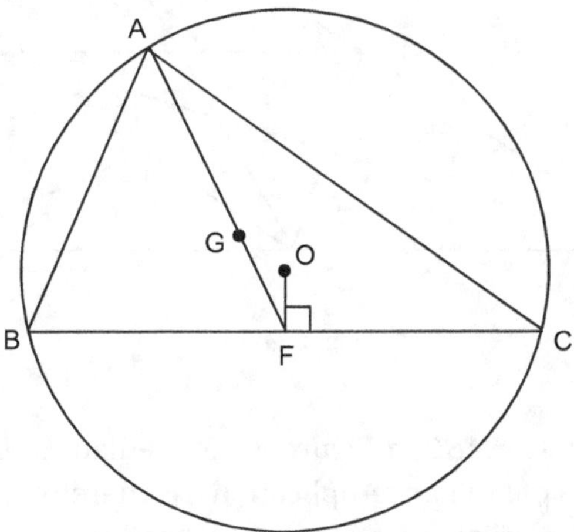

Figure 4-14

In Figure 4-14, we begin by selecting any point on circle O. This point will be at vertex A of $\triangle ABC$. Join A with given centroid G and extend AG through G to point F so that $GF = \frac{1}{2}(AG)$.

Then draw OF. At F, construct a perpendicular to OF, intersecting the circle at points B and C.

This easily justifiable construction proves that a triangle exists with the specified conditions. But since many other triangles can similarly be constructed, depending upon the selection of point A, our proof is complete.

We conclude our study of the medians of a triangle by briefly considering the *medial triangle*, a triangle formed by joining the midpoints of the sides of a triangle.

Properties of Medians Application 11: A triangle and its medial triangle have the same centroid.

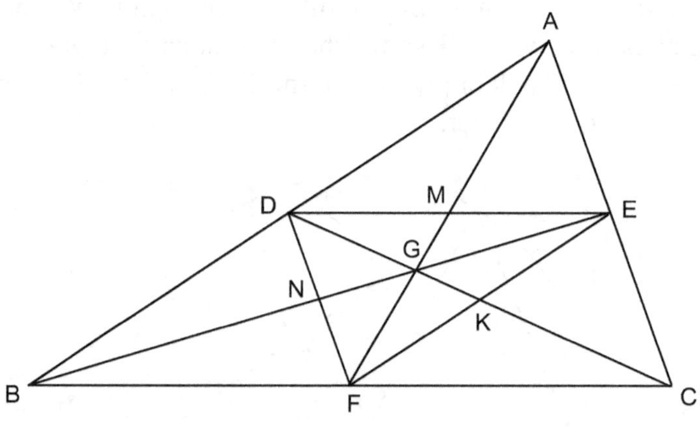

Figure 4-15

Proof: Referring to $\triangle ABC$ in Figure 4-15, median AF bisects DE at M (from Properties of Medians Application 8). Therefore, FM is a median of medial $\triangle DEF$. Similarly, DK and EN are medians of $\triangle DEF$ as well as being segments of the medians of $\triangle ABC$. Since the medians of $\triangle ABC$ meet at G, so do the medians of $\triangle DEF$.

Chapter 5

Unusual Geometric Experiences

This chapter presents a collection of 125 *geometricks*, unusual geometric problems with peculiar proofs and astounding properties. We begin with some easy "tricks" and gradually present more sophisticated illustrations. The next chapter presents the proofs of each of the geometricks. Both the statements and the proofs will open your eyes to the marvels of geometry.

Geometrick 1: Comparing Lengths in a Clever Way

Point A is outside the circle with center O. Line segment AO intersects the circle at point D, as shown in Figure 5-1. Secant AC intersects the circle at B and C. Curiously, we find that $AD < AB$.

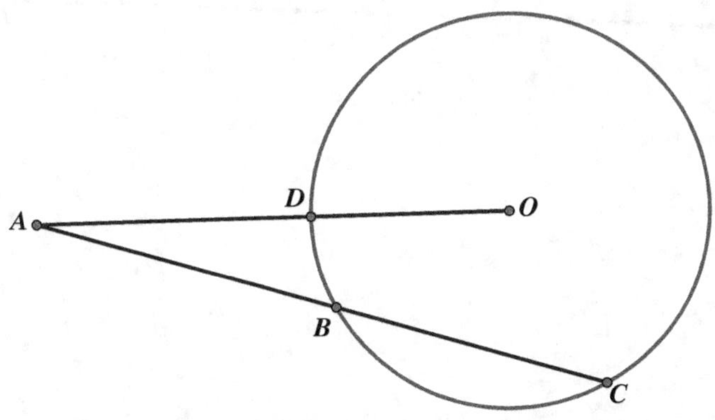

Figure 5-1

Geometrick 2: An Easily Overlooked Situation

In Figure 5-2, point *P* is outside parallelogram *ABCD*, forming triangles *APB* and *DPC*. The difference of the areas of triangle *APB* and triangle *DPC* equals half the area of parallelogram *ABCD*. This problem can be tricky because it demands a somewhat complicated solution, but can be easily solved.

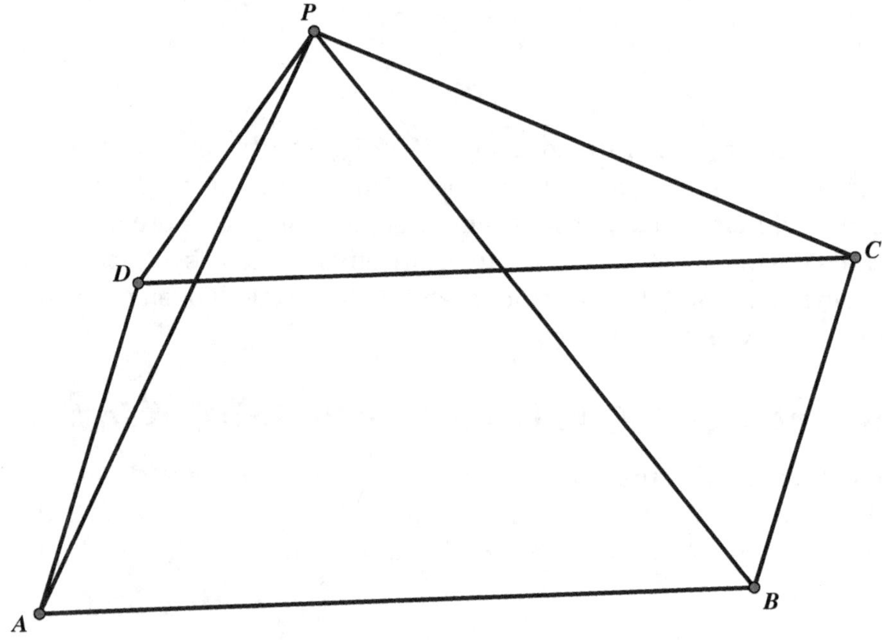

Figure 5-2

Geometrick 3: How a Circle's Diameter Relates to the Altitude of an Inscribed Triangle

Triangle *ABC* has altitude *AD* and is inscribed in circle *O* with diameter *AE*, as shown in Figure 5-3. We find that $AB \cdot AC = AD \cdot AE$.

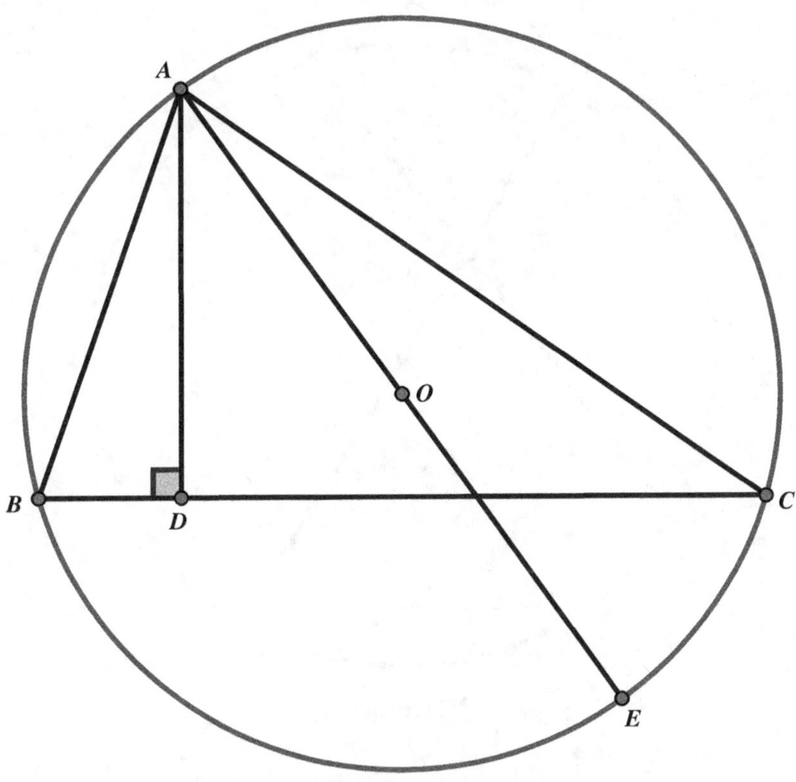

Figure 5-3

Geometrick 4: A Surprise at the Intersection of Two Chords of the Circle

In Figure 5-4, chord AB in circle O intersects diameter CD at point P. The result is $AP \cdot BP = (DO)^2 - (PO)^2$.

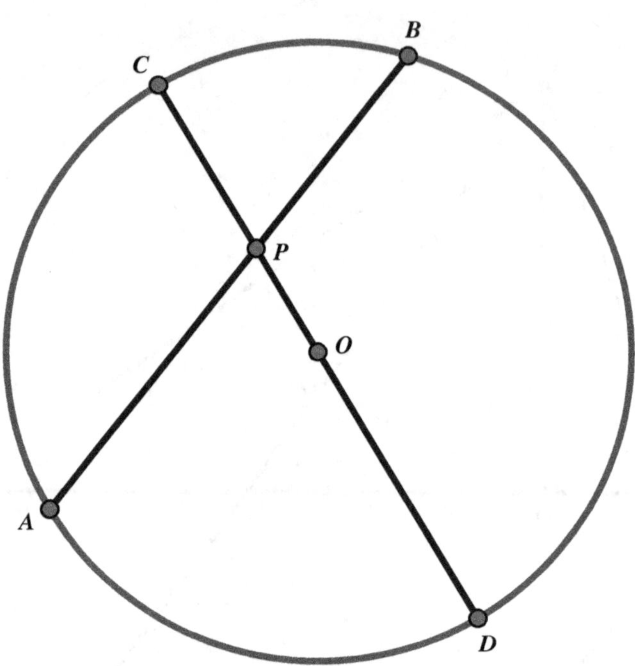

Figure 5-4

Geometrick 5: Two Tangents to a Circle Creating Parallel Lines

In Figure 5-5, we find triangles *ABC* and *DEO*. Tangents *ED* and *EB* intersect circle *O* at points *D* and *B*. Since *AB* is a diameter of circle *O*, we find that *OE* is parallel to *AC*.

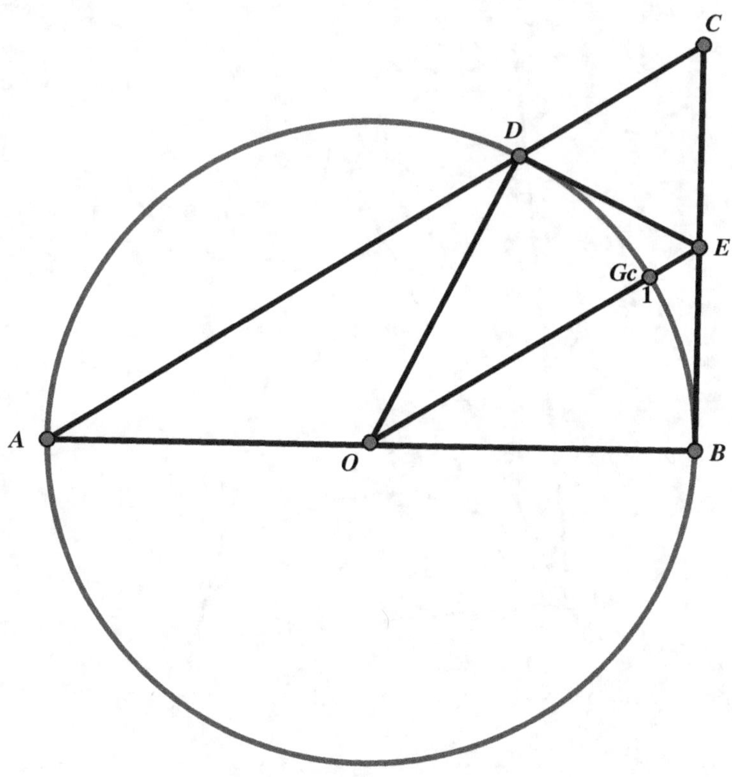

Figure 5-5

Geometrick 6: Finding the Radius of a Circle in an Unusual Fashion

As shown in Figure 5-6, lines AB and AC are tangent to a circle at points B and C. The lines AD and BD are perpendicular to AB and BC, respectively. Strangely, AD is equal to the radius of the circle.

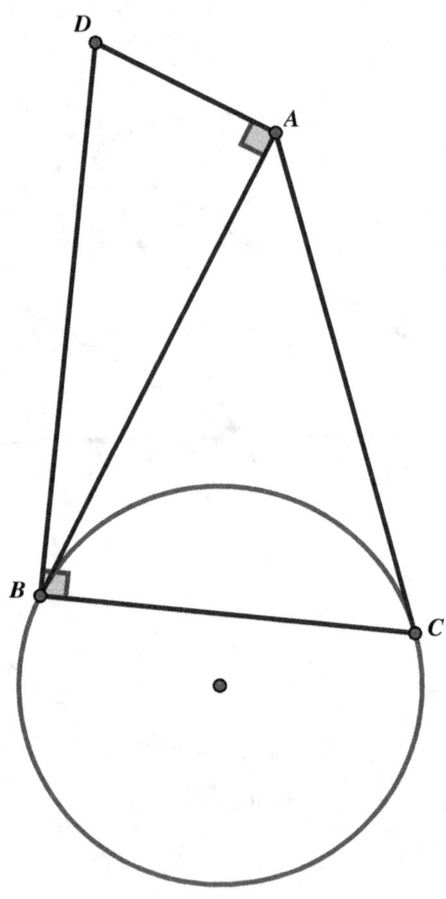

Figure 5-6

Geometrick 7: Overlapping Triangles

Isosceles triangle ABC has $AB = AC$, as shown in Figure 5-7. Points D and E lie on AB and AC, respectively, such that $AD = AE$, and $\angle ADH = \angle AEH$. We then find that $FB = GC$.

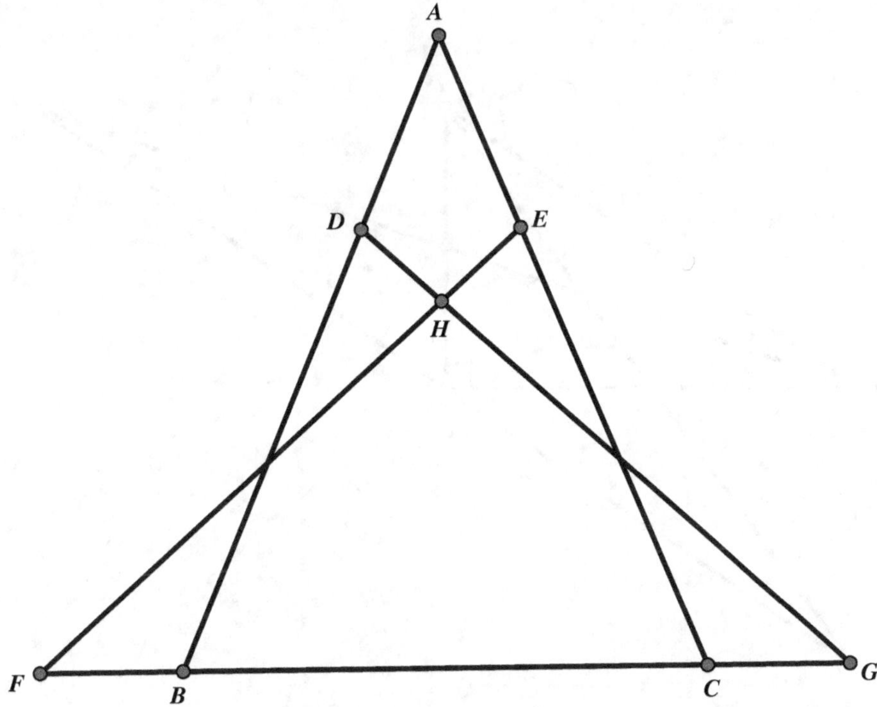

Figure 5-7

Geometrick 8: A Curious Extension of the Pythagorean Theorem

In Figure 5-8, points M and N are midpoints of sides AB and AC, respectively, of right triangle ABC. We extend BN and CM to intersect a line through A parallel to BC at points D and E, respectively. Unexpectedly, we find that $CE^2 + BD^2 = 5BC^2$.

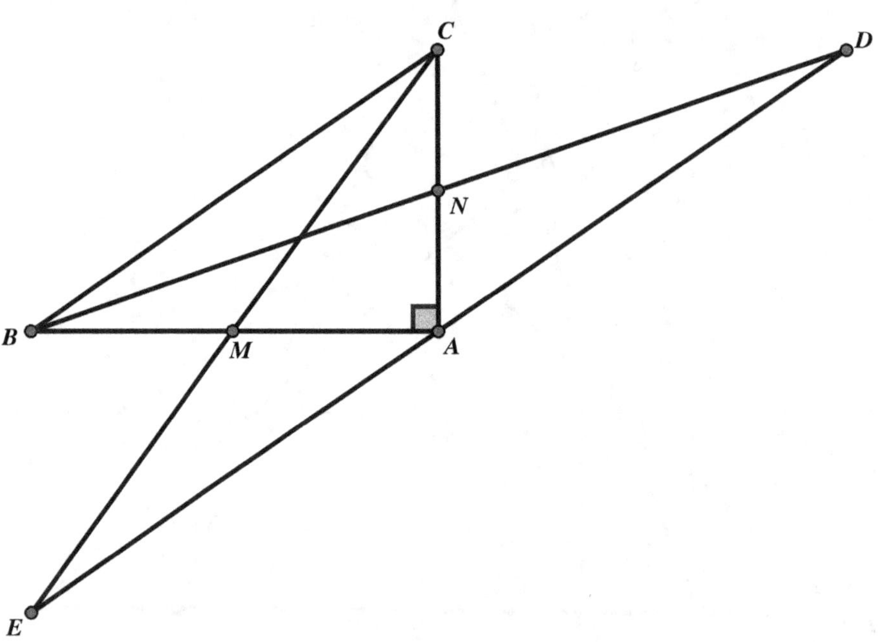

Figure 5-8

Geometrick 9: A Tangent That Bisects a Secant

As shown in Figure 5-9, the leg *AB* of right triangle *ABC* is the diameter of circle *O*, which intersects the hypotenuse at point *E*. The tangent to the circle at point *E* unexpectedly bisects the other leg *BC* of the right triangle.

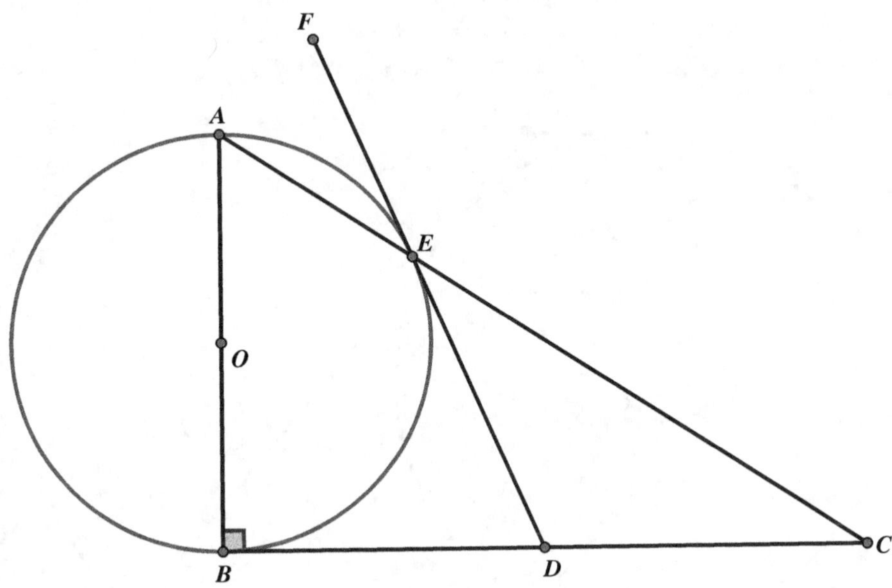

Figure 5-9

Geometrick 10: An Unexpected Placement of the Circle

Circle *O* has parallel tangents *AB* and *CD*, as shown Figure 5-10. Line segment *BC* is tangent to circle *O* at point *E*. The unexpected result is that a circle with diameter *BC* will contain the center *O* of the given circle.

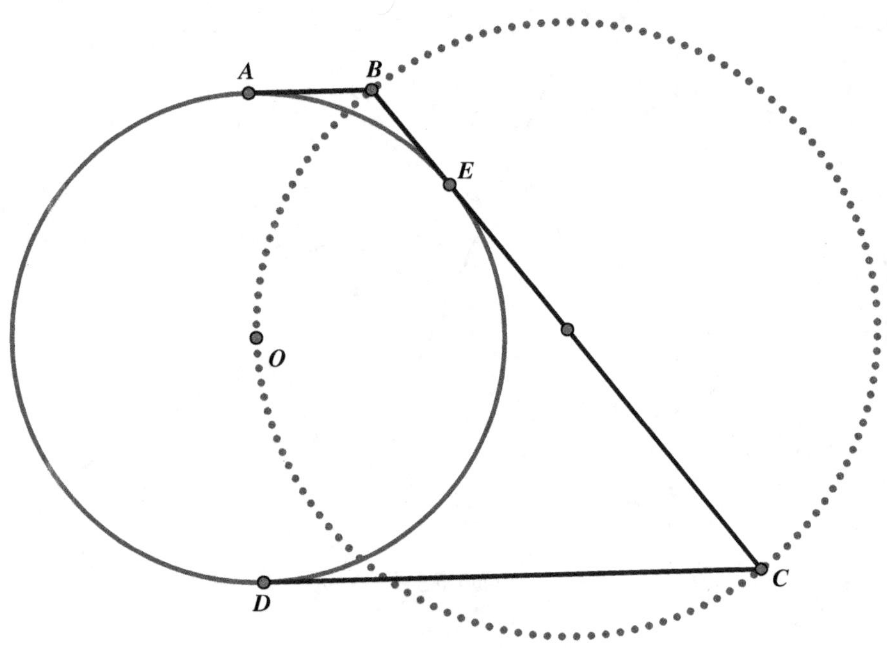

Figure 5-10

Geometrick 11: A Hidden Isosceles Triangle

Regular pentagon *ABCDE* is inscribed in circle *O*, as shown in Figure 5-11. The pentagon's diagonals *AC* and *BD* intersect at point *F*. We must now prove that triangles *FDC* and *FAB* are isosceles.

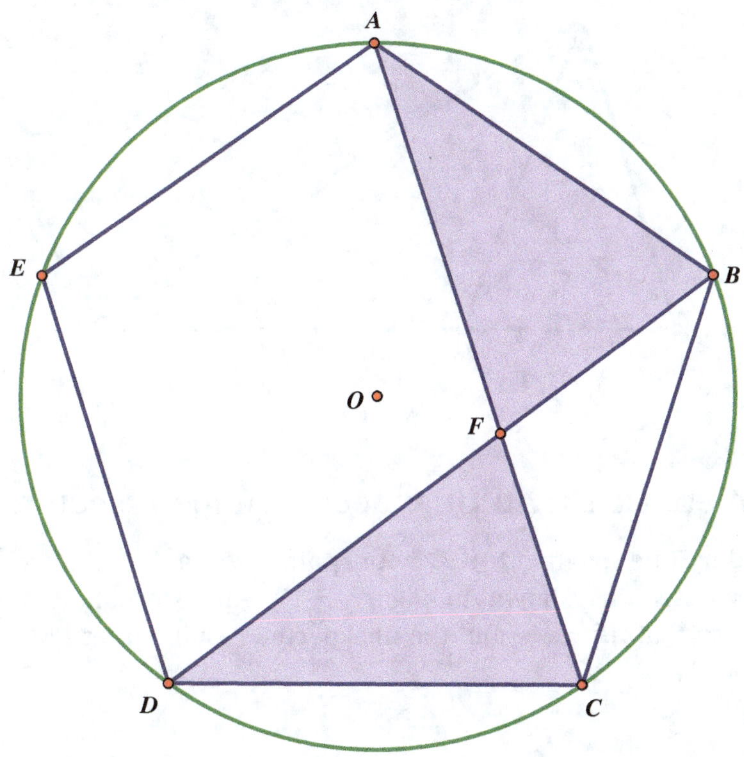

Figure 5-11

Geometrick 12: Sides of a Triangle Are Related to the Angle Bisector

The angle bisector AD in triangle ABC partitions the side to which is drawn, BC, into two sections, p and q, as shown in Figure 5-12, where side $AC = b$ and side $AB = c$. We find that $AD = \sqrt{bc - pq}$.

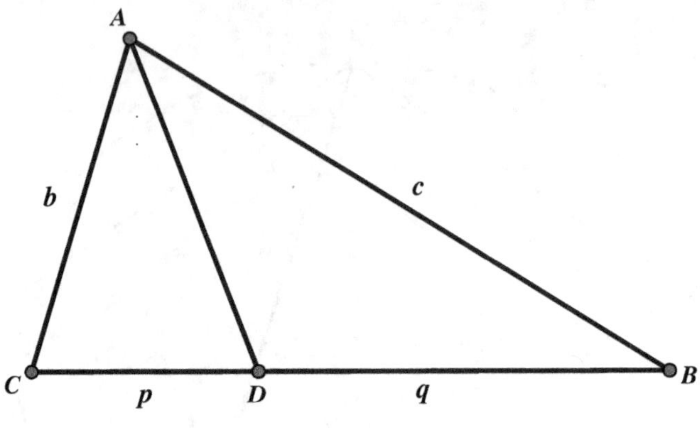

Figure 5-12

Geometrick 13: An Unexpected Angle Bisector

Consider line segment AB with points M and N, such that $AB \cdot AM = AN^2$, as shown in Figure 5-13. An external point P is placed so that $AP = AN$, and the unexpected result is that PN bisects angle MPB.

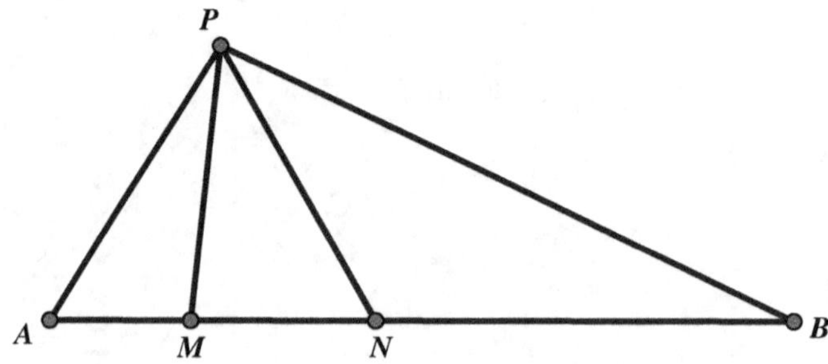

Figure 5-13

Geometrick 14: A Geometric Area Challenge

Finding the area of a given surface is often required in geometry, but the surfaces are usually common geometric figures. Here we're faced with five congruent equilateral triangles, shown in Figure 5-14, where vertices of adjacent equilateral triangles are at the midpoints of their neighbor triangle and share a common base. If the side length of each equilateral triangle is 2, the challenge is to find the area covered by the five equilateral triangles.

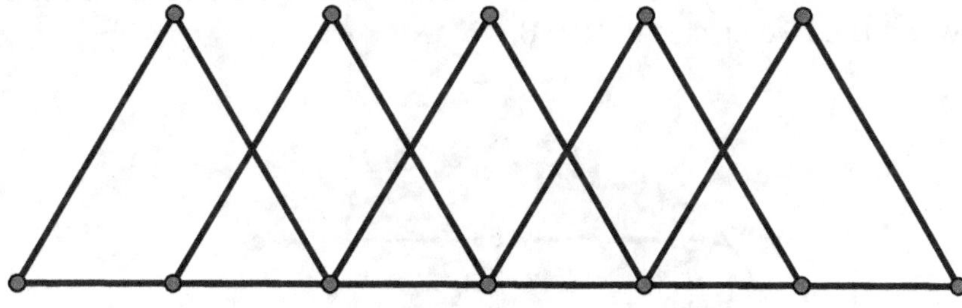

Figure 5-14

Geometrick 15: Deceptive Geometry

Figure 5-15 shows squares inscribed in congruent isosceles right triangles in two ways. Which of these shaded squares has a larger area?

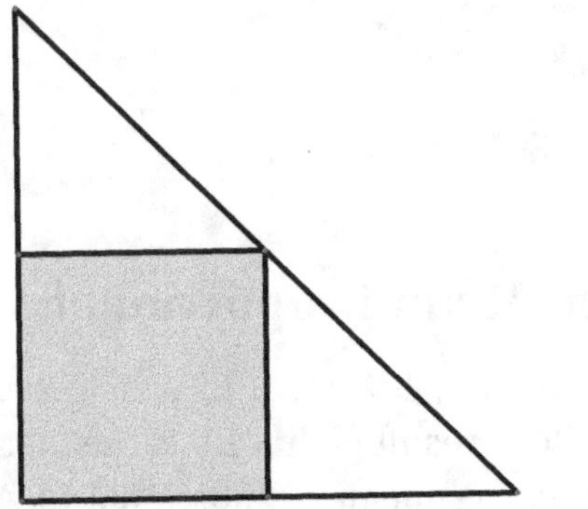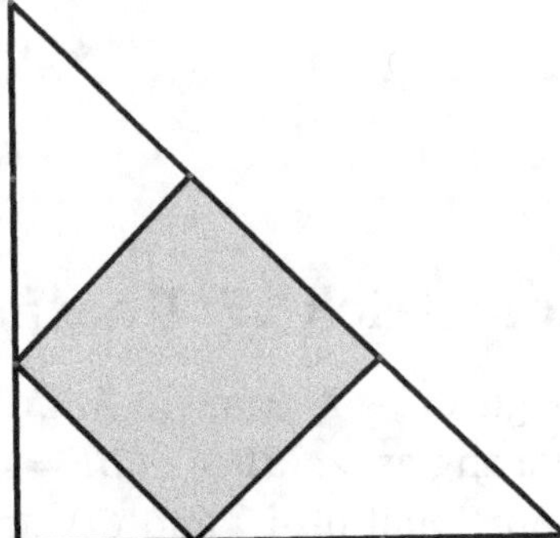

Figure 5-15

Geometrick 16: Determining the Area Between Two Concentric Circles

There are some geometric problems that look very common and can be solved in a traditional fashion. However, the beauty of such examples is that there are times when an unusually elegant method makes the solution almost trivial. Such is the case with the two concentric circles shown in Figure 5-16, where AB is a chord of the larger of the two concentric circles and is tangent to the smaller circle at point T. The challenge here is to find the area of the lighter shaded region between the two circles if $AB = 8$.

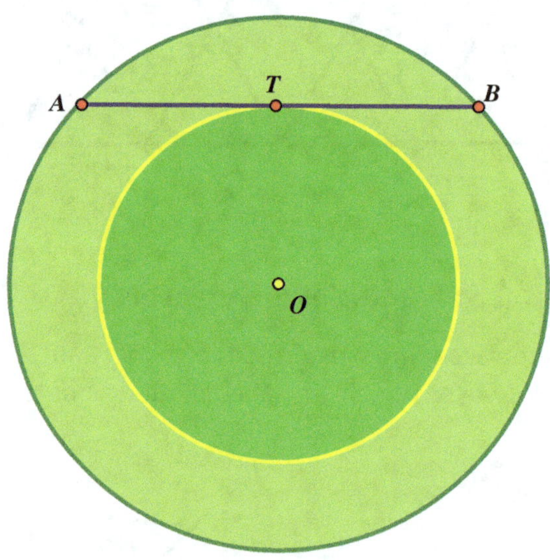

Figure 5-16

Geometrick 17: Unexpected Mean Proportional

In Figure 5-17, triangle ABC has two rays, BD and BE, which form equal angles: $\angle CBD = \angle CBE = \angle BAD$. The result is that BD is the mean proportional of AD and CD, and BE is the mean proportional of AE and CE.

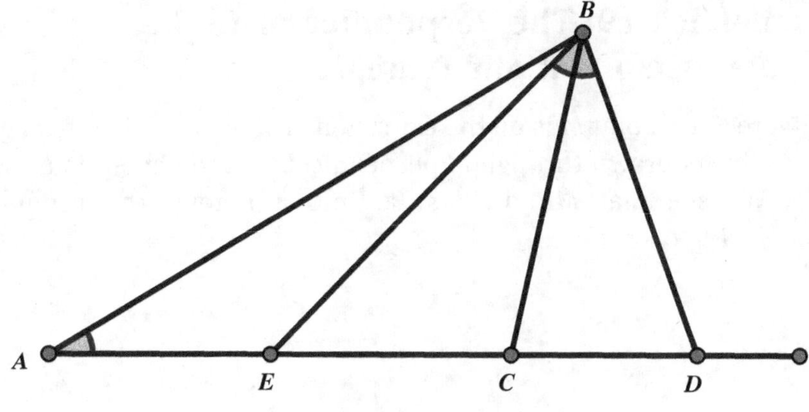

Figure 5-17

Geometrick 18: The Surprise Angle Between an Altitude and Angle Bisector

Shown in Figure 5-18, triangle ABC has altitude AD and angle bisector AE. Surprisingly, $\angle DAE = \frac{1}{2}(\angle B - \angle C)$.

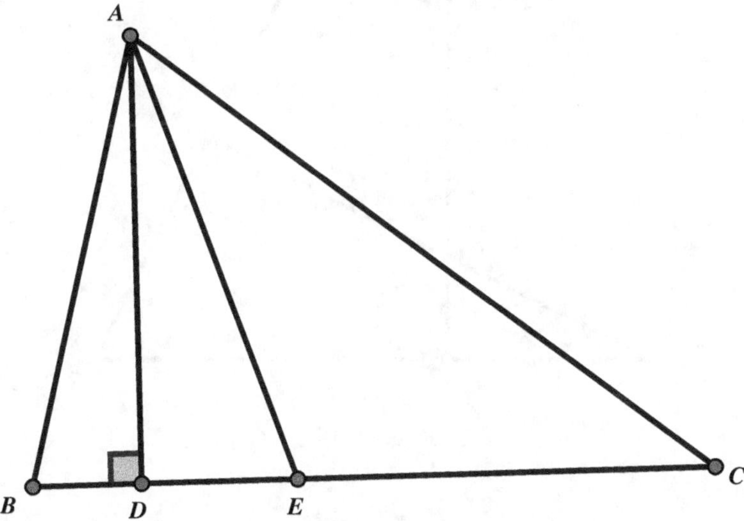

Figure 5-18

Geometrick 19: The Perpendicular to the Hypotenuse of a Right Triangle

In Figure 5-19, point M is on the extension of side BA of right triangle ABC. A line is drawn through M perpendicular to the hypotenuse BC at P. We see that MP divides BC into a noteworthy product: $PN \cdot PM = PB \cdot PC$.

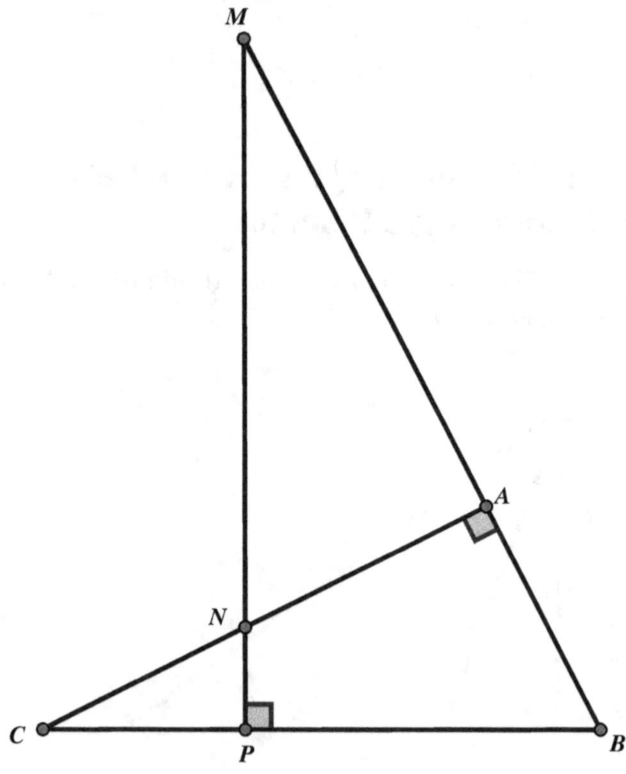

Figure 5-19

Geometrick 20: An Unexpected Chord Product

Triangle *ABC* with altitude *CD* is inscribed in a circle with center *P* and diameter *CE*, shown in Figure 5-20. The chord product that evolves is $AC \cdot BC = CD \cdot CE$.

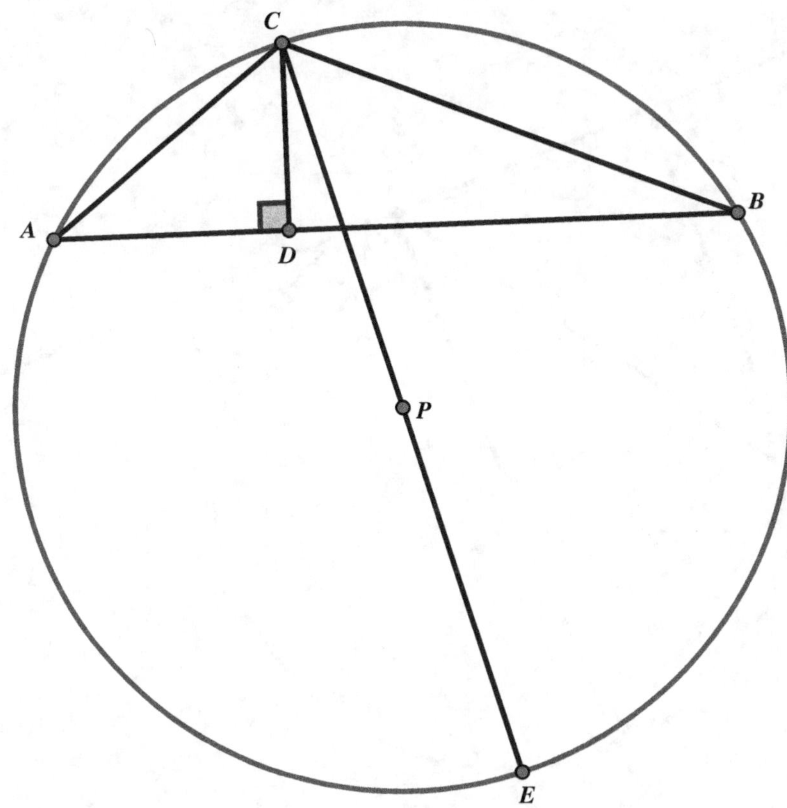

Figure 5-20

Geometrick 21: The Power of a Point on a Circle

From any point on a circle, the distance to the point of tangency is the mean proportional between the diameter of the circle and the distance from that point to the tangent line. For Figure 5-21, we then have $\frac{CD}{AD} = \frac{AD}{AB}$.

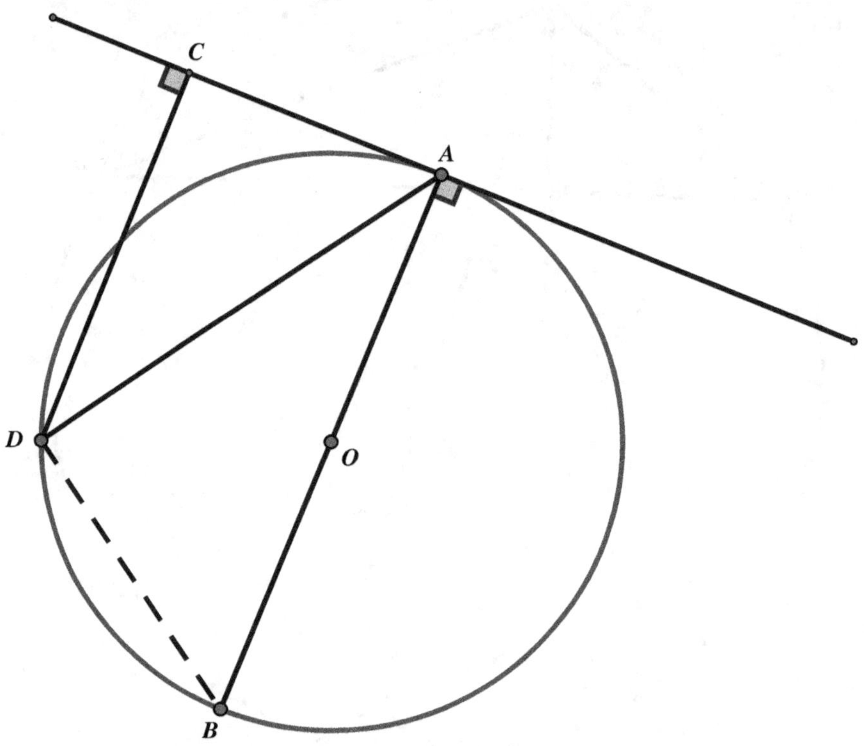

Figure 5-21

Geometrick 22: A Strange Relationship of Circle Lines

In Figure 5-22, circle P and circle O intersect at point A. APB is a diameter of the circle P and intersects circle O at point C. Curiously, we find that $AB \cdot AC = 2OC^2$.

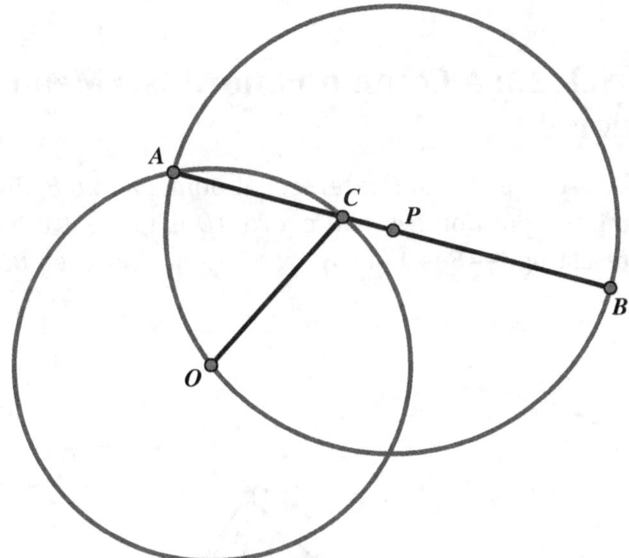

Figure 5-22

Geometrick 23: An Unusual Product of Triangle Sides

The product of two sides of a triangle is equal to the product of the altitude to the third side and the diameter of the circumscribed circle.

Geometrick 24: A Curious Point on an Inscribed Circle

The product of the distances from any point on the inscribed circle of a triangle to the three sides of the triangle is equal to the product of the distances from the same point to the sides of the inscribed triangle of the circle.

Geometrick 25: A Common Chord Is a Mean Proportional

In Figure 5-23, circles *P* and *R* intersect at points *A* and *B*. Chord *AC* is tangent to circle *P* at point *A*, and chord *AD* is tangent to circle *R* at point *A*. Interestingly, *AB* is a mean proportional between *BC* and *BD*.

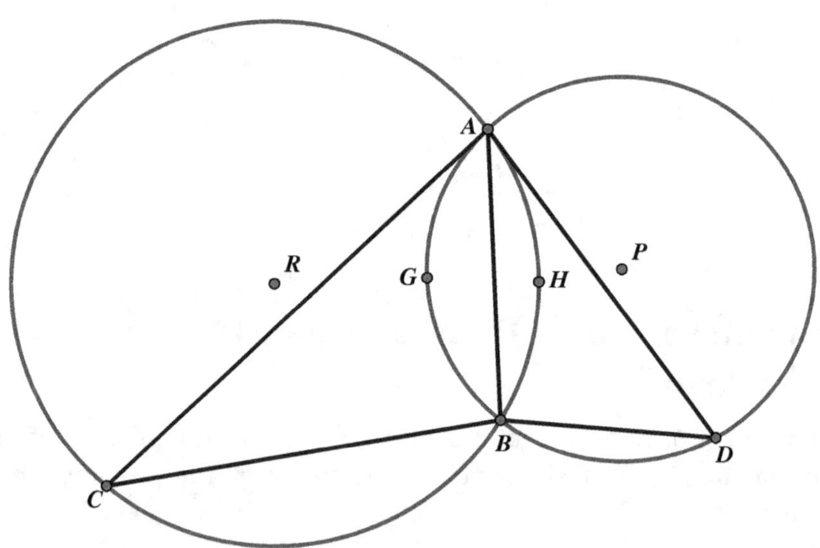

Figure 5-23

Geometrick 26: Unusual Bisection of a Right Angle

Square *ABFE* is placed on hypotenuse *AB* of triangle *ABC,* as shown in Figure 5-24. Line *CG* joins the point of intersection of the square's diagonals, *G*, and the vertex of the right angle, *C.* We find that *GC* bisects angle *ACB*.

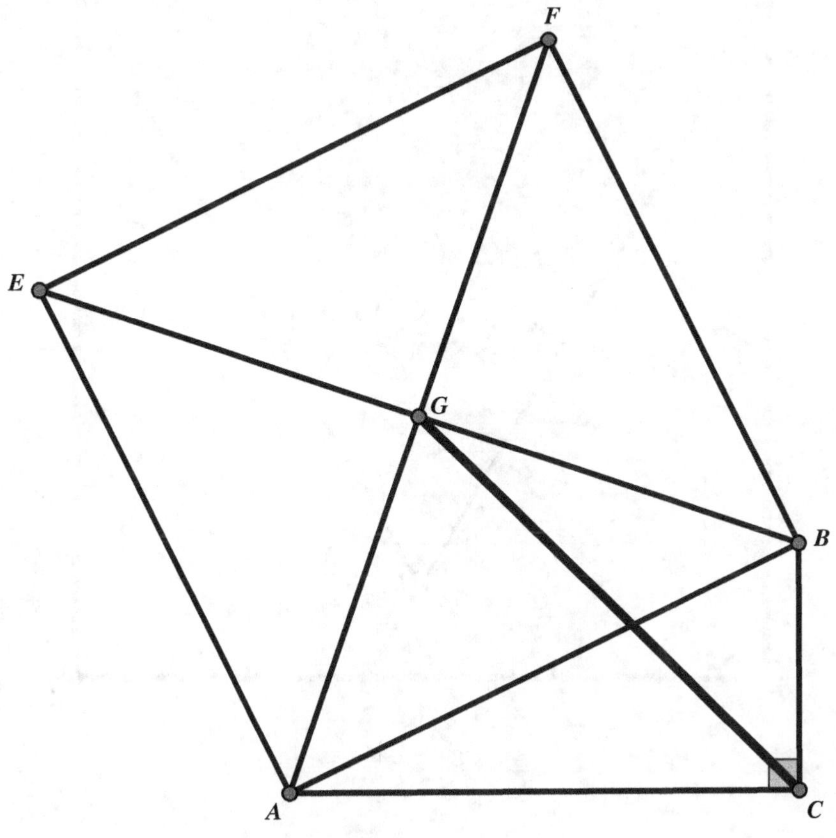

Figure 5-24

Geometrick 27: The Relationship Between an Equilateral Triangle and a Square

In Figure 5-25, equilateral triangle *CDE* lies within square *ABCD*. The challenge here is to determine the ratio of the areas of $\frac{\triangle CDE}{\triangle BCF}$.

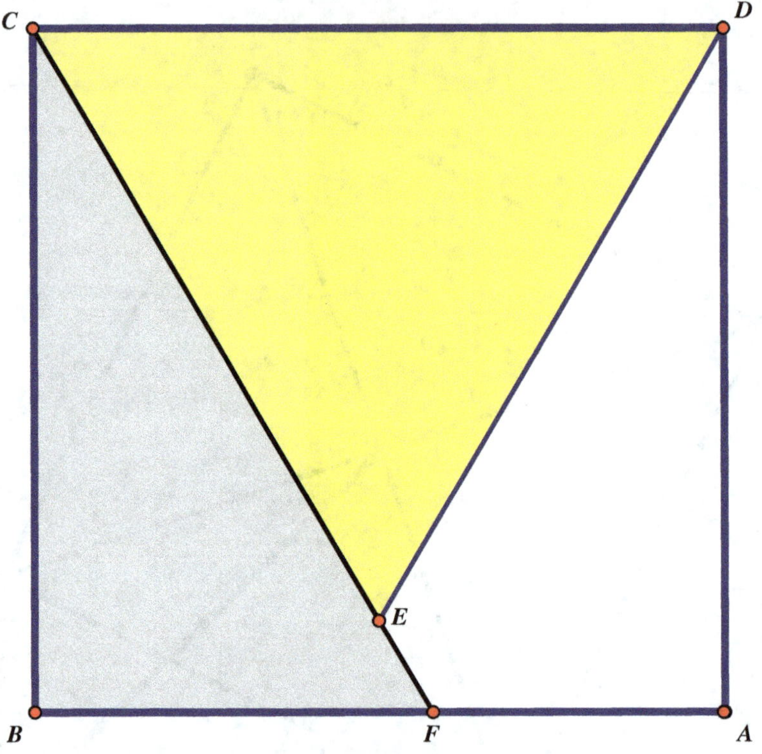

Figure 5-25

Geometrick 28: Discovering an Equilateral Triangle in a Square

In Figure 5-26, square *ABCD* contains isosceles triangle *ABF*, which has base angles of 15° each. The challenge is to show that triangle *CDF* is equilateral.

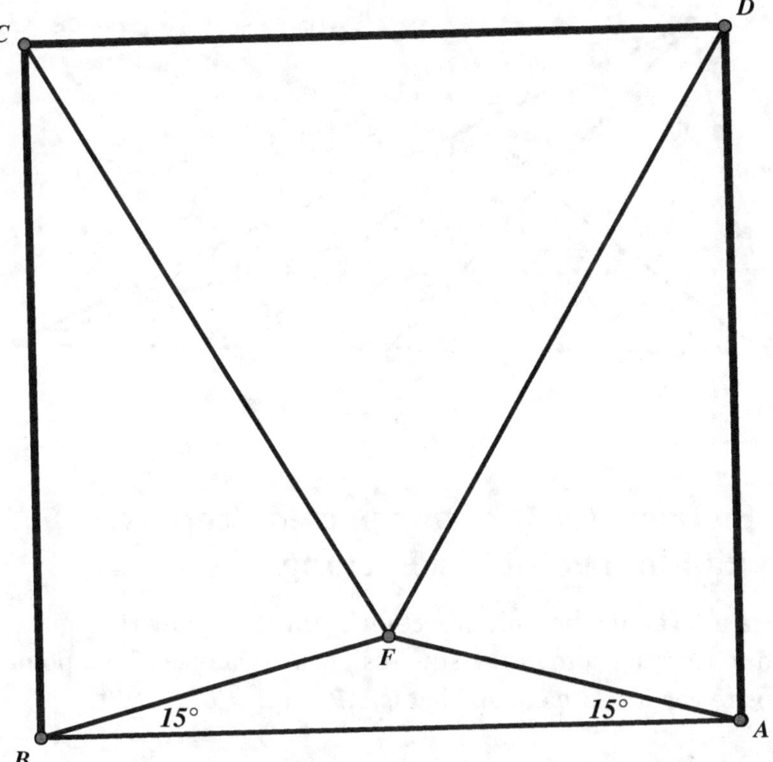

Figure 5-26

Geometrick 29: A Significant Point on the Median of a Triangle

Point *P* is the midpoint of median *AM* of triangle *ABC*, as shown in Figure 5-27. Remarkably, the area of triangle *PBC* is half the area of triangle *ABC*.

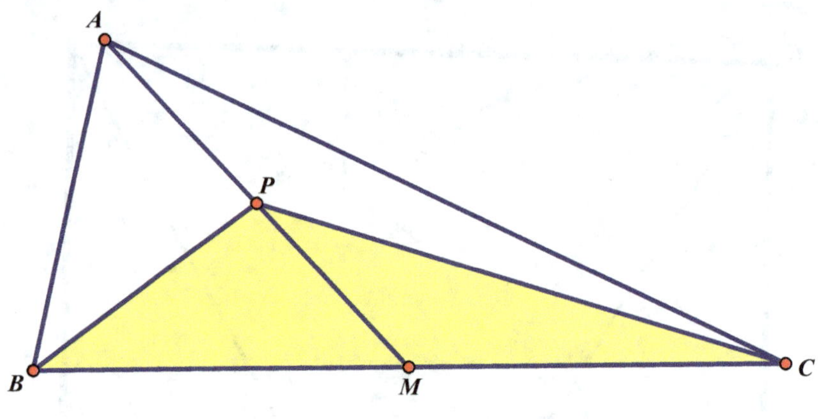

Figure 5-27

Geometrick 30: The Unexpected Property of a Random Point in a Rectangle

Point *P* is located randomly in rectangle *ABCD*, as shown in Figure 5-28. We find that the sum of the squares of the distances from point *P* to opposite vertices are equal. That is, $AP^2 + CP^2 = BP^2 + DP^2$.

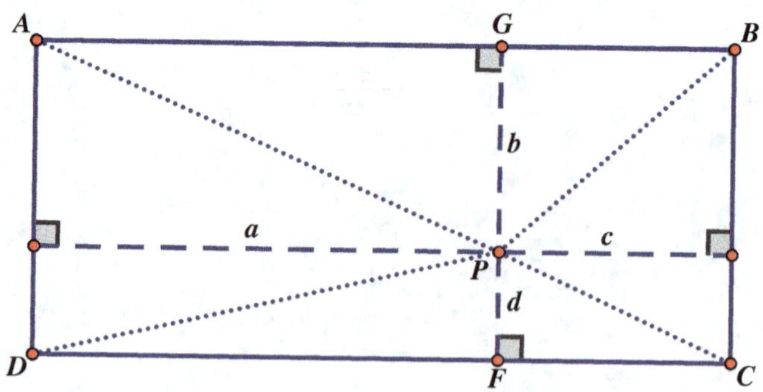

Figure 5-28

Geometrick 31: Another Peculiar Property of a Random Point in a Rectangle

The sum of the squares of the distances from random point P to the four vertices of a rectangle equals the square of the diagonal plus four times the square of the distance from point P to the point of intersection of the diagonals. In Figure 5-29, we must show that $AP^2 + CP^2 + BP^2 + DP^2 = AC^2 + 4MP^2$.

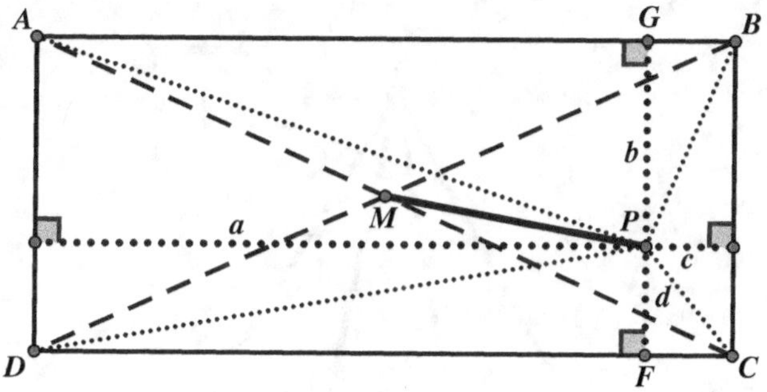

Figure 5-29

Geometrick 32: The Sum of Perpendiculars from a Random Point on the Side of a Rectangle to the Diagonals

From any point P on side AB of rectangle $ABCD$, shown in Figure 5-30, perpendiculars PE and PF are drawn to the rectangle's diagonals. The sum of the perpendiculars, $PE + PF$, is the same regardless of where point P is located on AD.

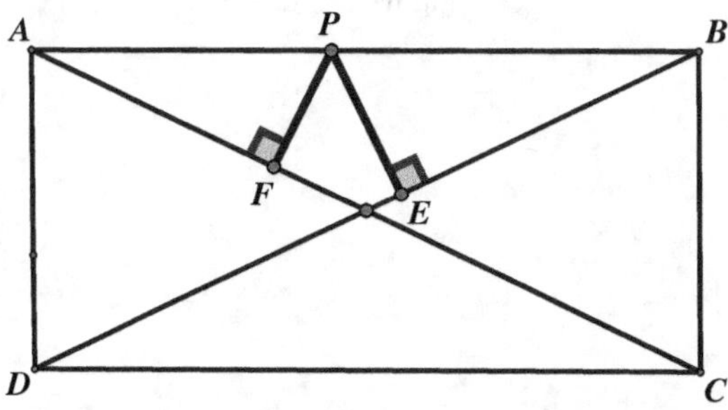

Figure 5-30

Geometrick 33: Trisection Points of a Semicircle Arc Can also Trisect the Diameter

Semicircle *DACF* has its diameter on side *DF* of equilateral triangle *EFD*, as shown in Figure 5-31. Points *A* and *C* trisect the semicircle so that $\overarc{AD} = \overarc{AC} = \overarc{FC}$, and *AE* and *CE* intersect diameter *DF* at points *G* and *H*, respectively. It turns out that *DG* = *GH* = *HF*.

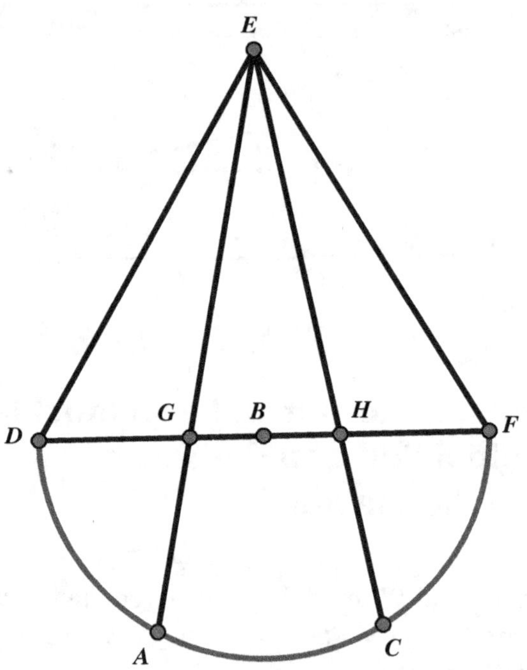

Figure 5-31

Geometrick 34: Combinations of Circular Arcs

Beyond the typical relationships among circular arcs are extraordinary combinations that lead to interesting results. Consider a semicircle with *AB* as its diameter of length 2 and *C* as the diameter's midpoint, as can be seen in Figure 5-32. The perpendicular to *AB* at point *C* intersects the semicircle at point *D*. Circular arc *BF* is drawn with center *A* and circular arc *AE* is drawn with center *B*, such that *AF* and *BE* intersect at point *D*. Circular arc *EF* is drawn with center *D*. The challenge here is to find the area between the bold arcs *ADB* and *AEFB*.

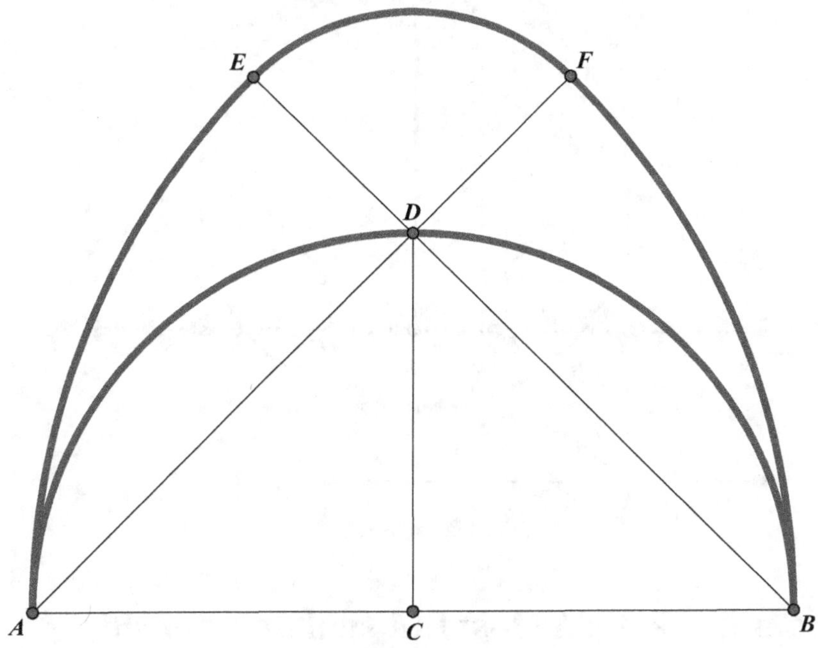

Figure 5-32

Geometrick 35: Comparing Unusual Combinations of Circular Regions

In Figure 5-33, we are asked to show that the area between bold lines, which is comprised by two concentric semicircles and two congruent semicircles whose centers are collinear, is equal to the area of the circle shown. To simplify, let $AC = BD = r - s$, and $FE = r + s$.

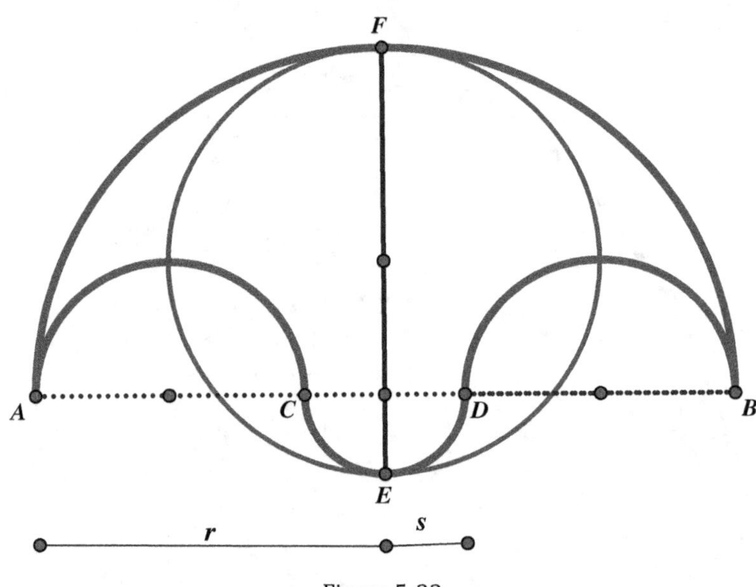

Figure 5-33

Geometrick 36: A Most Unusual Relationship

In Figure 5-34, we begin with the semicircle ABC, where we insert isosceles right triangle MON, with $MN \parallel AB$. The circumscribed circle about right triangle MON has diameter DO. Completely by surprise we find that the area of the regions $AOEM + BOFN = Area\triangle MON$.

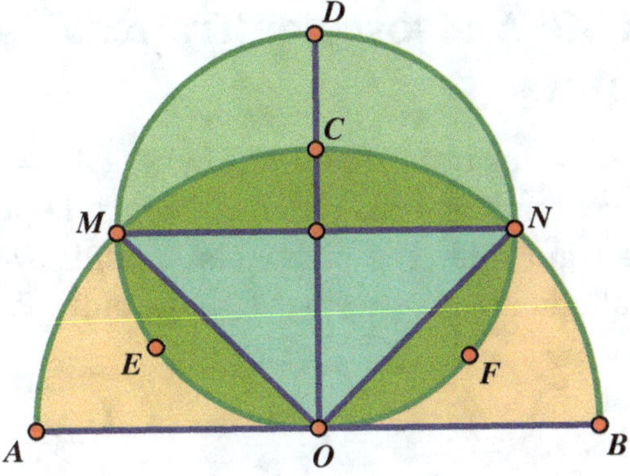

Figure 5-34

Geometrick 37: Finding the Area of a Quarter of a Quadrilateral

In Figure 5-35, the opposite sides AD and BC of quadrilateral $ABCD$ are extended to meet at point W. The midpoints of AC and BD are X and Y, respectively. We find that the relationship between the area of triangle WXY and the area of quadrilateral $ABCD$ is the following: $area\triangle WXY = \frac{1}{4} area ABCD$.

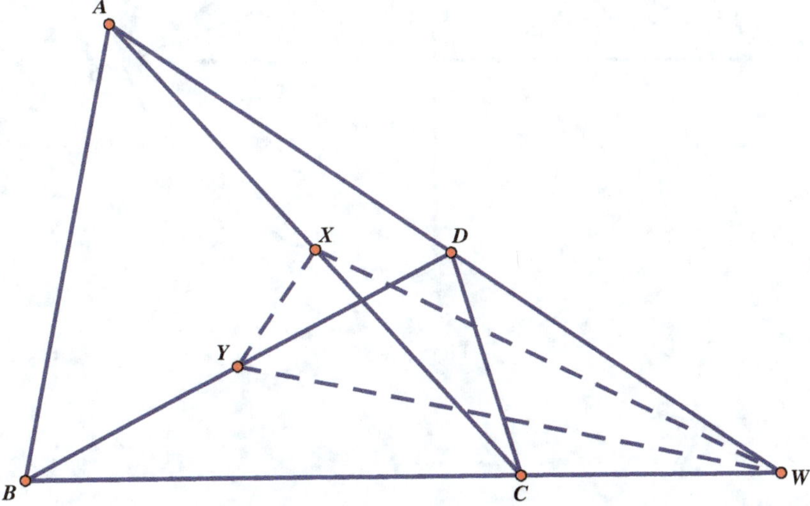

Figure 5-35

Geometrick 38: An Isosceles Trapezoid Generates Concyclic Points

In Figure 5-36, isosceles trapezoid $ABCD$, where $AD = BC$, is inscribed in a circle with center P and diameter EF, which is perpendicular to the bases of the trapezoid. The diagonals AC and BD intersect EF at point G. The challenge here is to prove that points B, P, G, and C are concyclic.

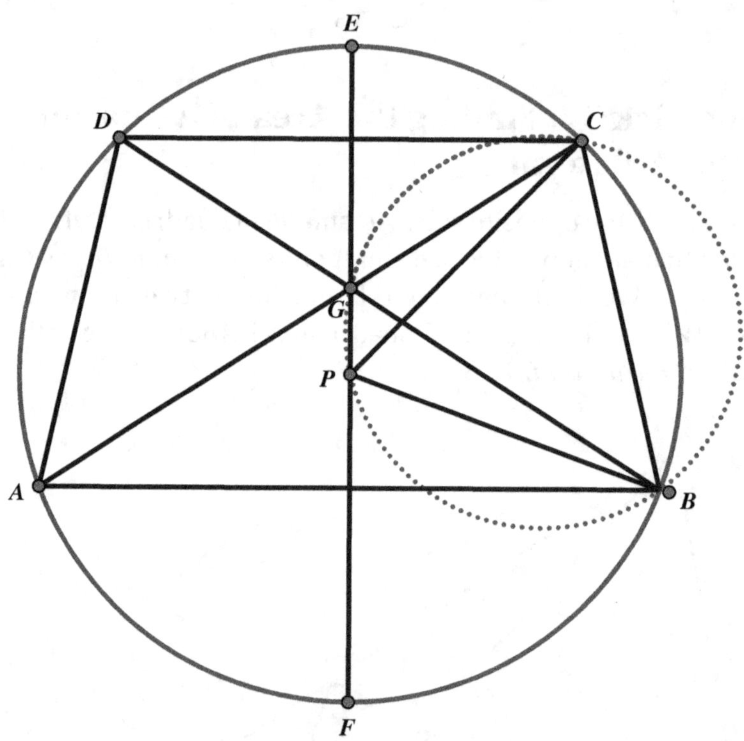

Figure 5-36

Geometrick 39: An Unanticipated Angle Bisector

Shown in Figure 5-37 are two circles that are internally tangent at point *A*. Chord *BC* of the large circle is tangent to the small circle at point *P*. Strangely, *AP* bisects angle *BAC*.

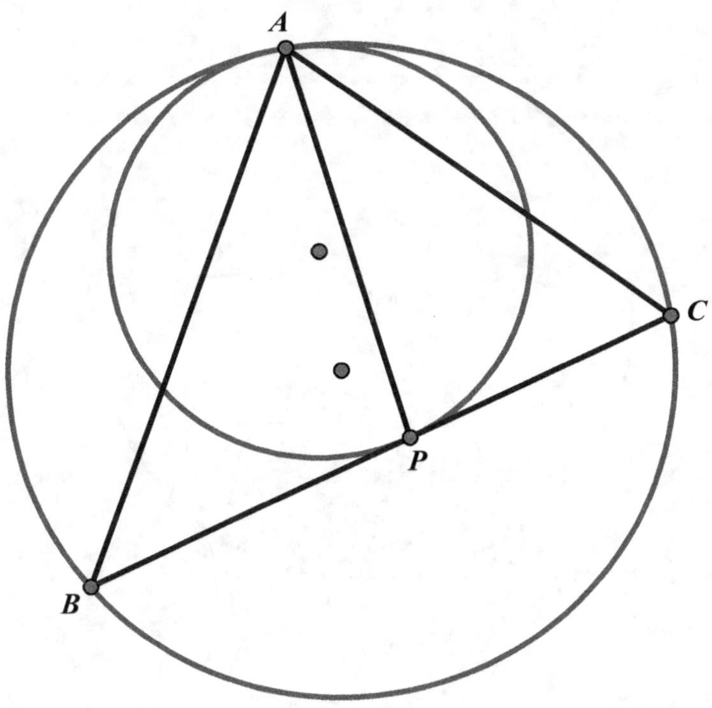

Figure 5-37

Geometrick 40: To Find the Common Area Determined by Three Intersecting Circles

Circles A, B, and C intersect symmetrically, as shown in Figure 5-38. If the radius of each circle is r, what is the common (shaded) area between the three circles?

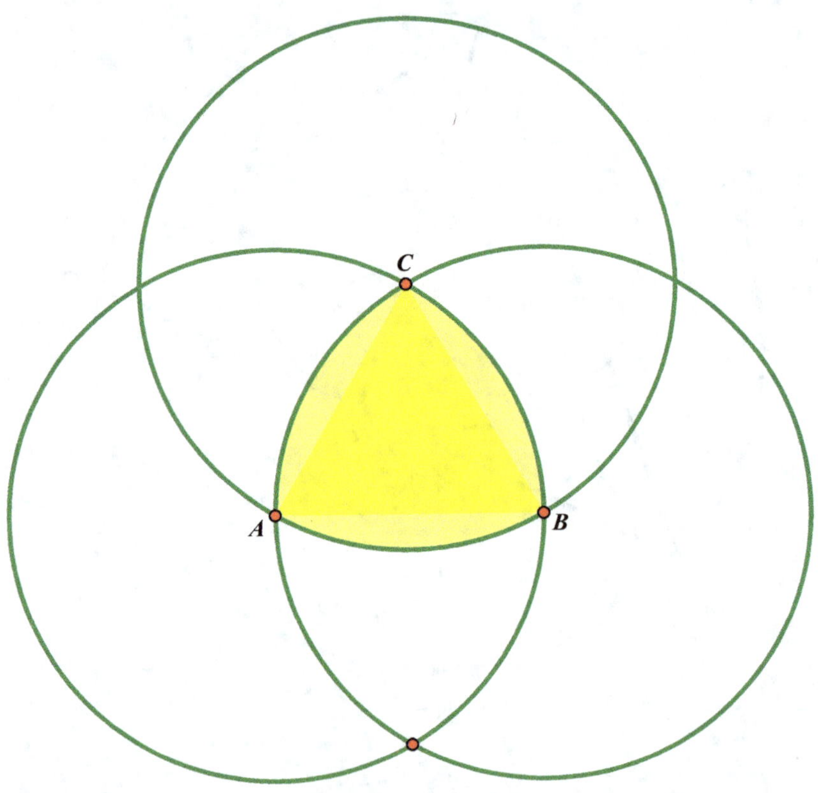

Figure 5-38

Geometrick 41: Surprise Angle Sum

When the three triangles placed on the sides of a given triangle are inscribed in circumcircles, and when the angle sum of the three remote angles is 180°, the circumcircles have a common point. In Figure 5-39, the triangles *ABC*, *CEF*, and *AFG* are placed externally on the sides of triangle *ACF*, and the sum of the remote angles $\angle B + \angle E + \angle G = 180°$. A circumcircle is drawn around each external triangle, and the circumcircles of the three triangles have a common point *H*.

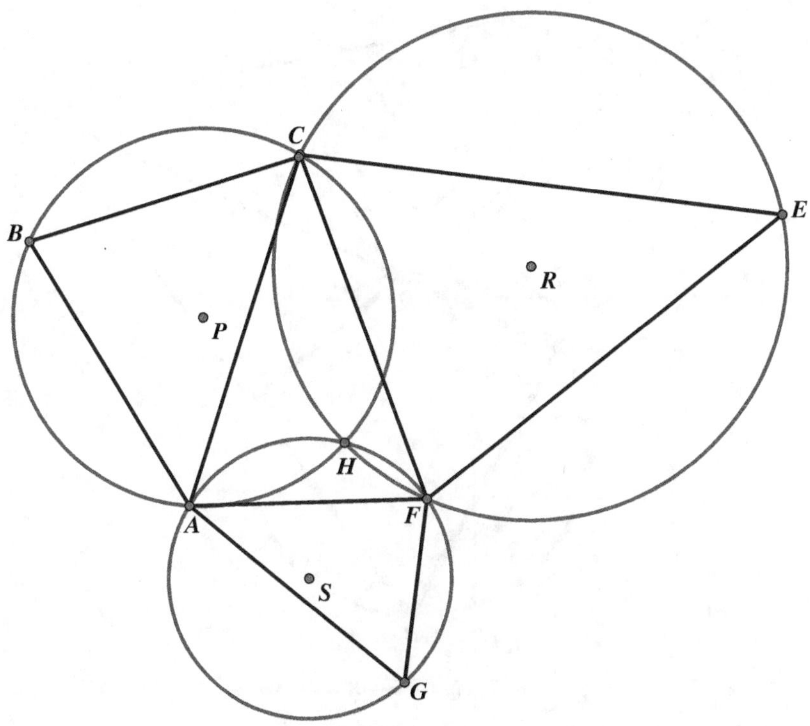

Figure 5-39

Geometrick 42: An Unexpected Collinearity

The opposite sides of the cyclic quadrilateral *ABCD* intersect at points *P* and *Q*, as shown in Figure 5-40. The circumscribed circles of triangles *ADQ* and *DCP* intersect at point *R*. Unexpectedly, we find that the points *P*, *R*, and *Q* are collinear.

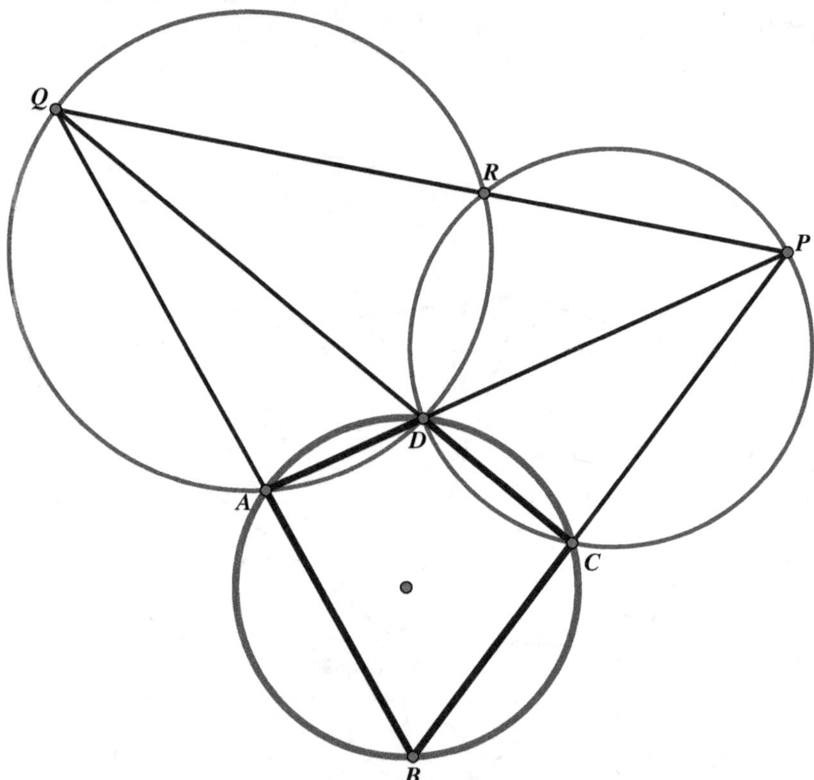

Figure 5-40

Geometrick 43: Four Congruent Concurrent Circles Define Another Circle

In Figure 5-41, four congruent circles share a common point *P*, and their common tangents create quadrilateral *ABCD*. Surprisingly, the quadrilateral *ABCD* is cyclic.

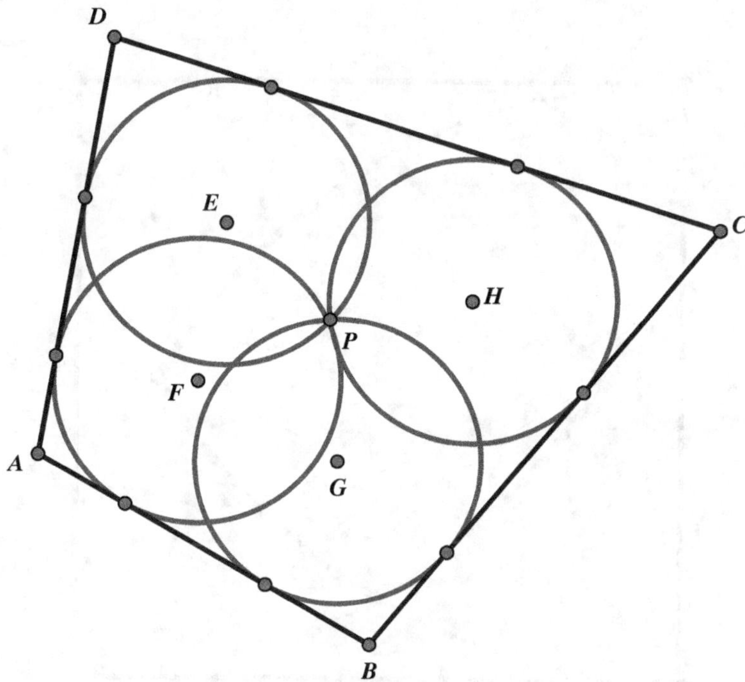

Figure 5-41

Geometrick 44: To Find the Common Area Determined by Four Intersecting Circles

In Figure 5-42, the congruent quarter circles have centers at *A*, *B*, *C*, and *D*, which are the vertices of square *ABCD*. If the radius of each circle is *r*, what is the common area *EFGH* between the four circles?

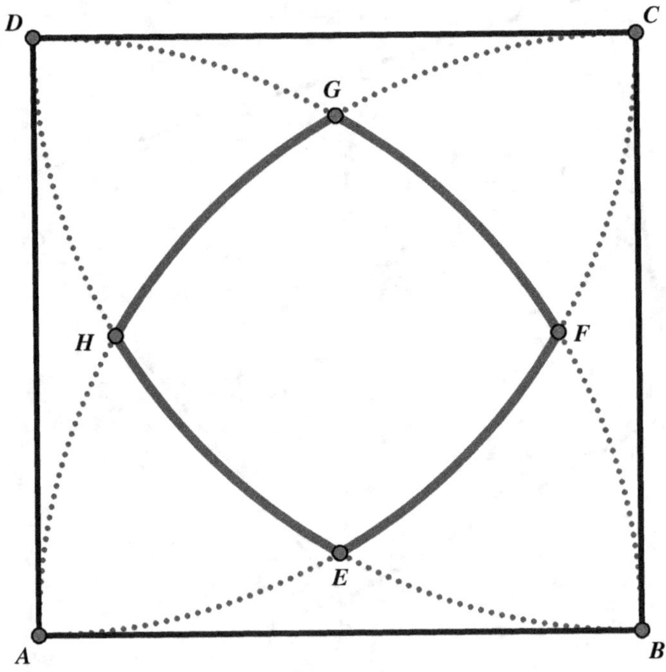

Figure 5-42

Geometrick 45: Two Connected Squares Generate Another Square

Squares *ABCD* and *CEFG* share a common vertex at point *C*, which is shown in Figure 5-43. Points *P* and *Q* are the midpoints of *BE* and *DG*, respectively. When these two points are connected with the center points of the two squares, we find that quadrilateral *PNQM* is also a square.

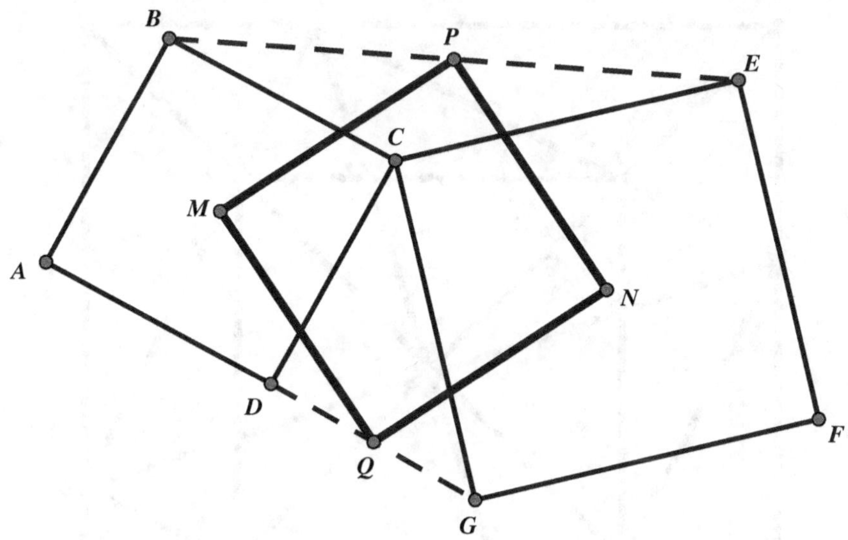

Figure 5-43

Geometrick 46: Generating Squares

The two perpendicular lines *EG* and *FH* joining opposite sides of square *ABCD* contain the point *P*, the intersection of the diagonals *AC* and *BD*. Remarkably, the quadrilateral *EFGH* is a square, and the points of intersection of square *EFGH* with the diagonals of the original square *ABCD* form another square *JKLM*, as can be seen in Figure 5-44.

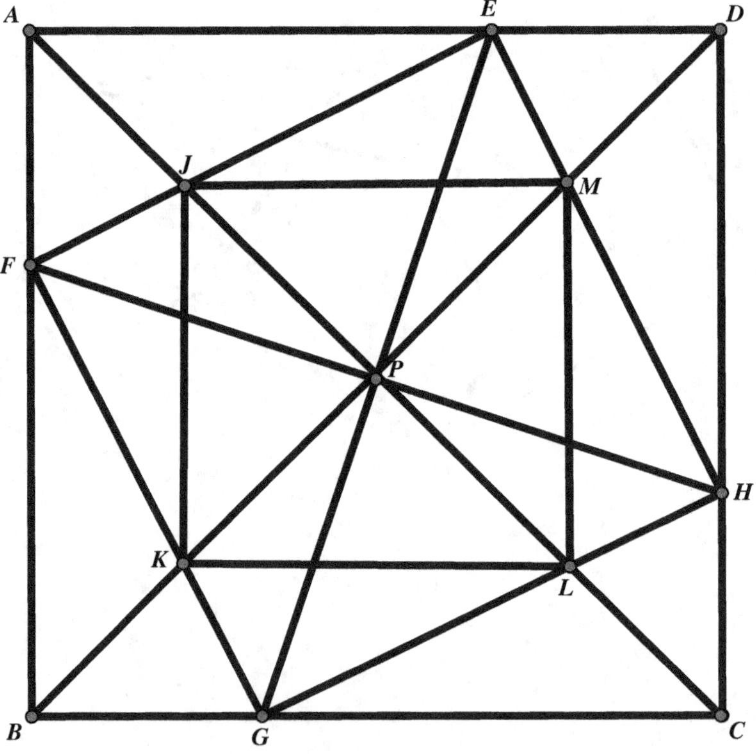

Figure 5-44

Geometrick 47: All Perpendicular Transversals in a Square Are Equal

In Figure 5-45, transversals *NJ* and *GH* are perpendicular, and their intersection point *Q* is in the square *ABCD*. Amazingly, we find that *NJ* = *GH*.

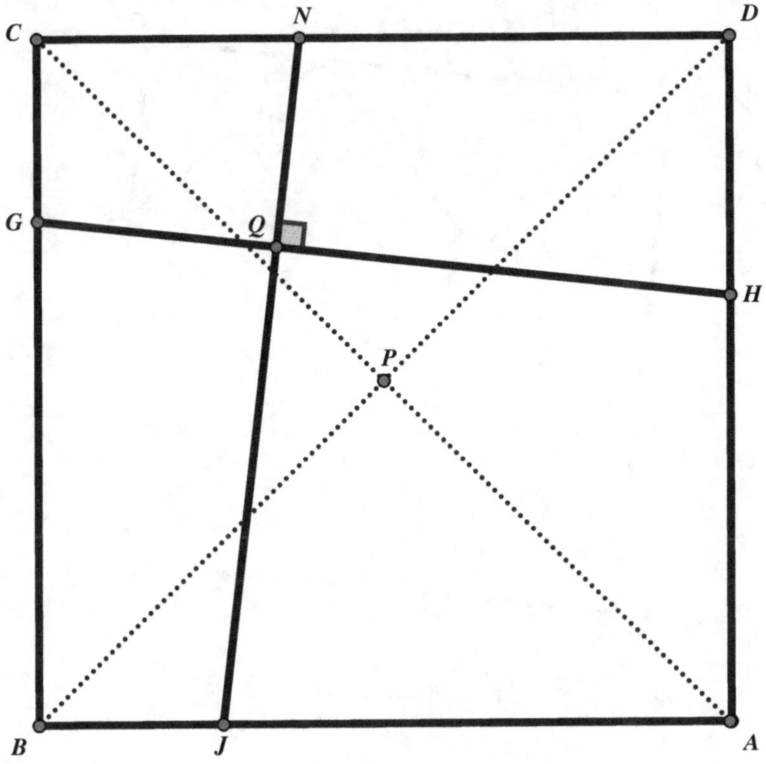

Figure 5-45

Geometrick 48: Relationship of a Square Within a Square

In Figure 5-46, square *EFHG* is on the diagonal *AC* of square *ABCD* and its vertices *E* and *F* are on sides *AD* and *CD*, respectively. We need to find the relationship between the areas of squares *EFHG* and *ABCD*.

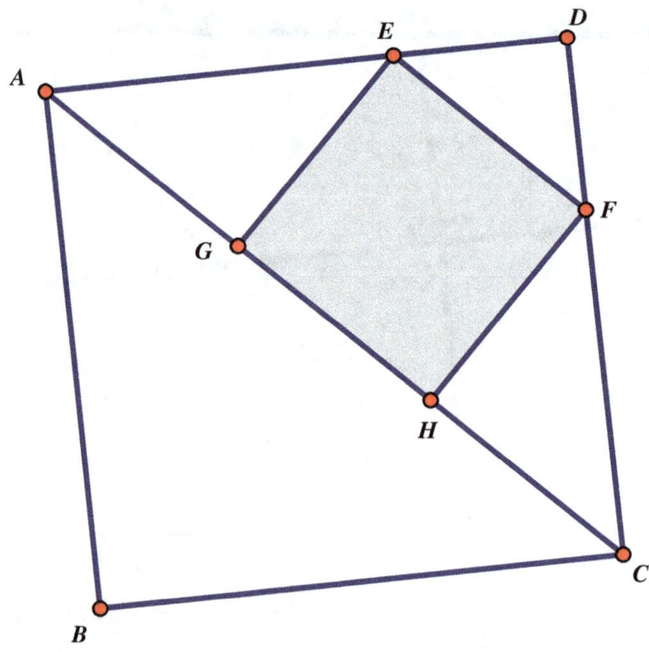

Figure 5-46

Geometrick 49: An Unexpected Area Determined

Simple problems that demonstrate unexpected solutions are quite impressive. Here we have point *M* as the midpoint of side *AB* of △*ABC*. Point *P* is any point on *AM*, as shown in Figure 5-47. The line through point *M*, parallel to *PC*, meets *BC* at *D*. What part of the area of △*ABC* is the area of △*BDP*?

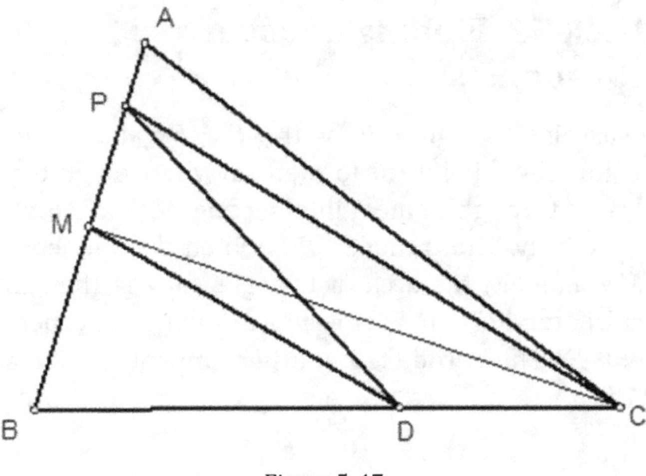

Figure 5-47

Geometrick 50: A Curious Point in the Interior of an Angle

Consider point P in the interior of angle A, as shown in Figure 5-48. Perpendiculars are drawn from point P to the two sides of the angle and extended to the other side of the angle, so that EPC is perpendicular to AC and DPB is perpendicular to AB. The following analogous products evolve: $AD \cdot DC = DP \cdot DB$, and $AE \cdot BE = EP \cdot EC$.

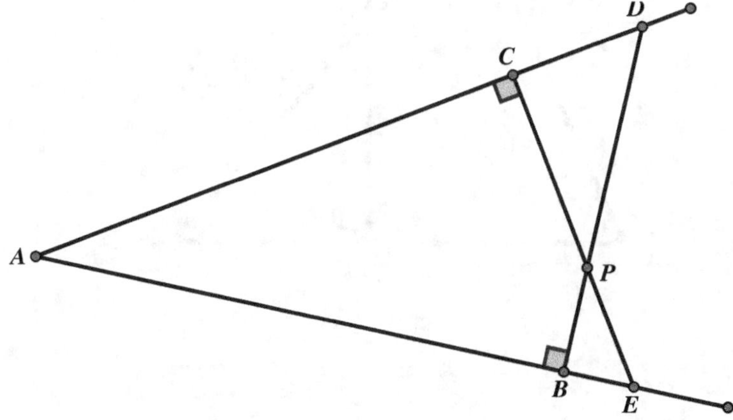

Figure 5-48

Geometrick 51: Finding Unusual Segment Products

Consider isosceles triangle *ABC*, with *AC* = *BC* and altitude *CD*, as shown in Figure 5-49. Point *P* is located anywhere on altitude *CD*, and a line is drawn through point *P* intersecting *AC* and *AB* at points *N* and *M*, respectively. The triangle *ABC* is then partitioned into equal parts by *MN*; namely, the area of triangle *AMN* is the same as the area of quadrilateral *CNMB*. Furthermore, line *NF* is perpendicular to *CD* at point *E*. This produces another unexpected relationship: $PC \cdot PE = PD^2$.

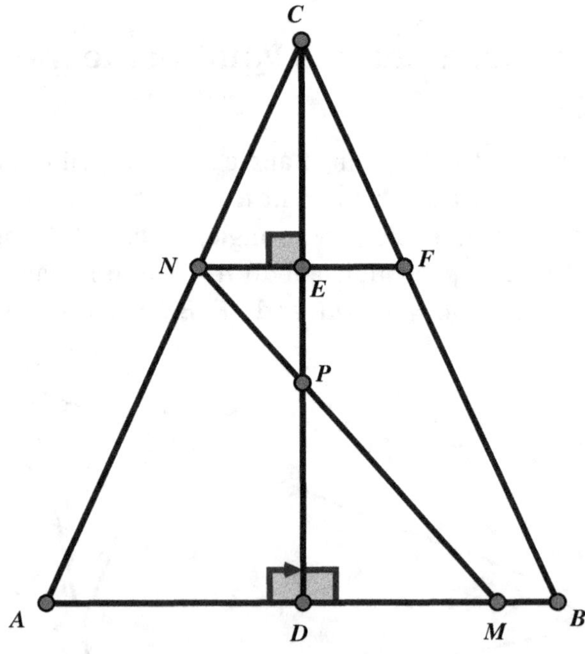

Figure 5-49

Geometrick 52: A Perpendicular Line that Is also a Bisector

A line perpendicular to one side of a cyclic quadrilateral that passes through the intersection of the quadrilateral's perpendicular diagonals bisects the other side. We see this in Figure 5-50, where cyclic quadrilateral $ABCD$ has diagonals $AC \perp BD$, and $EGF \perp AB$. Astoundingly, E is the midpoint of DC.

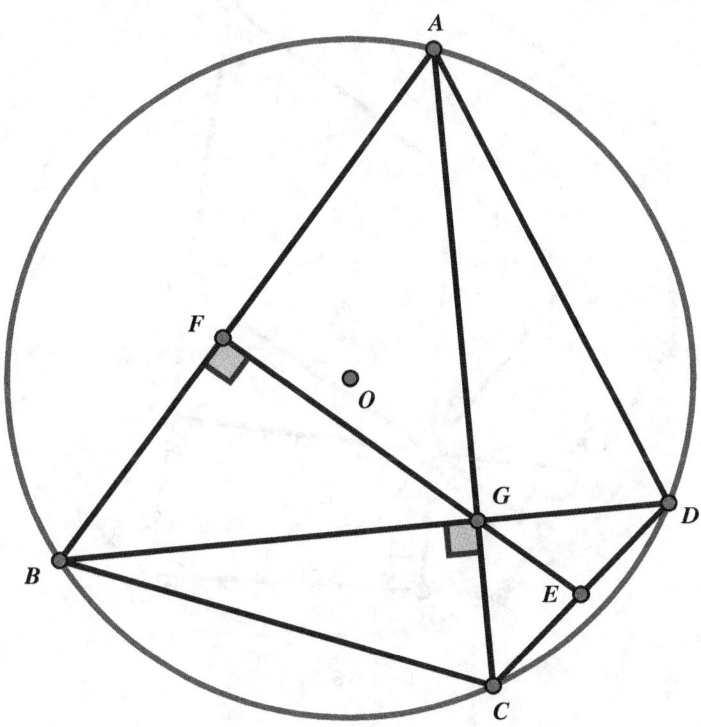

Figure 5-50

Geometrick 53: An Unexpected Concurrency in a Cyclic Quadrilateral

When one locates the midpoint of the four sides of a cyclic quadrilateral and draws a perpendicular from each of those points to the opposite side, these four lines will be concurrent. In Figure 5-51, the lines *EK*, *FL*, *GI*, and *HJ* are surprisingly concurrent at point *N*.

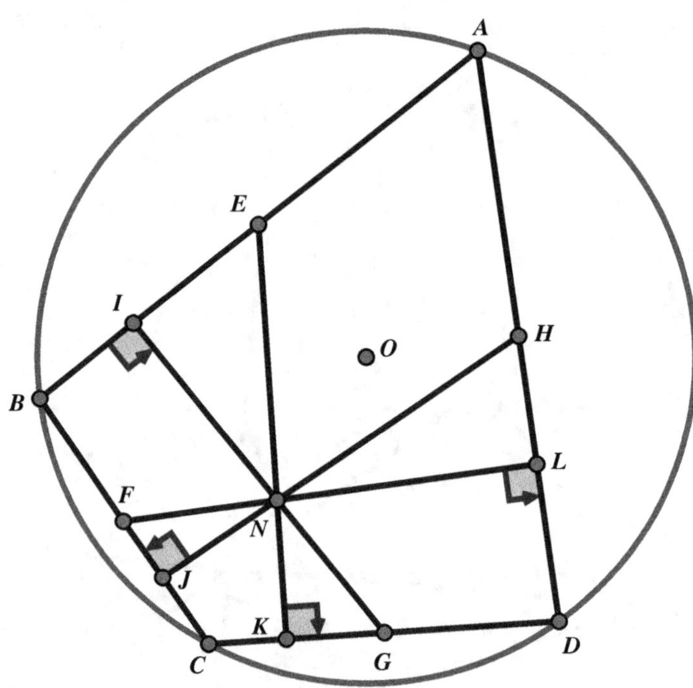

Figure 5-51

Geometrick 54: The Relationship of a Point on the Circumcircle of a Quadrilateral

The product of the distances from a point on the circumcircle of a cyclic quadrilateral to two opposite sides of the quadrilateral is equal the product of the distances to the other pair of opposite sides. In Figure 5-52, we have $ME \cdot MG = MH \cdot MF$.

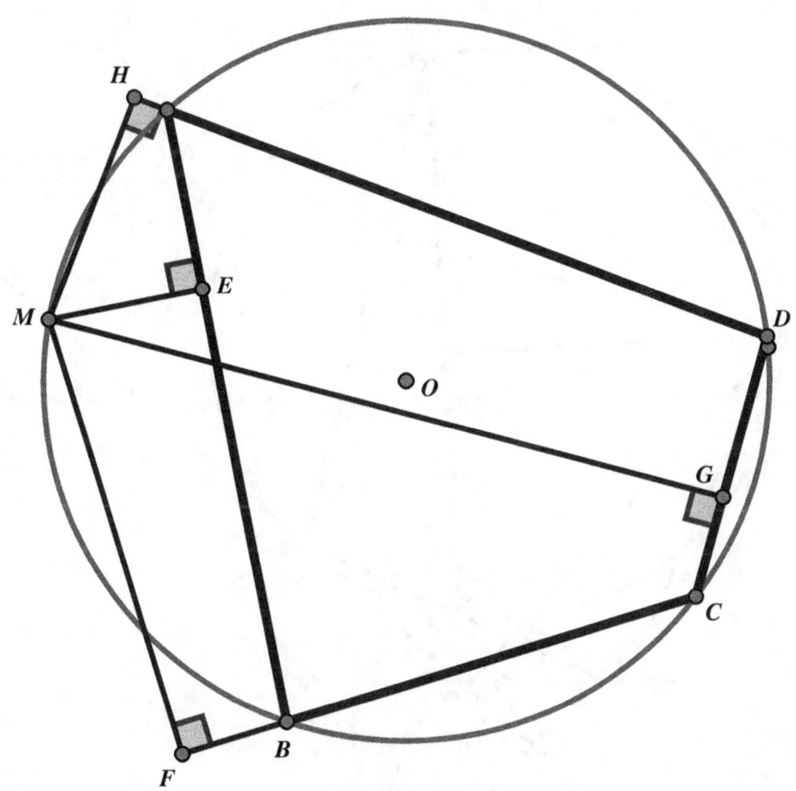

Figure 5-52

Geometrick 55: The Relationship of a Point on the Circumcircle of a Triangle

The product of the distances from a point on the circumcircle of a triangle to two sides of the triangle is equal the product of the distance to the third side and the distance to the tangent line of the remote vertex. In Figure 5-53, we have $ME \cdot MG = MH \cdot MF$.

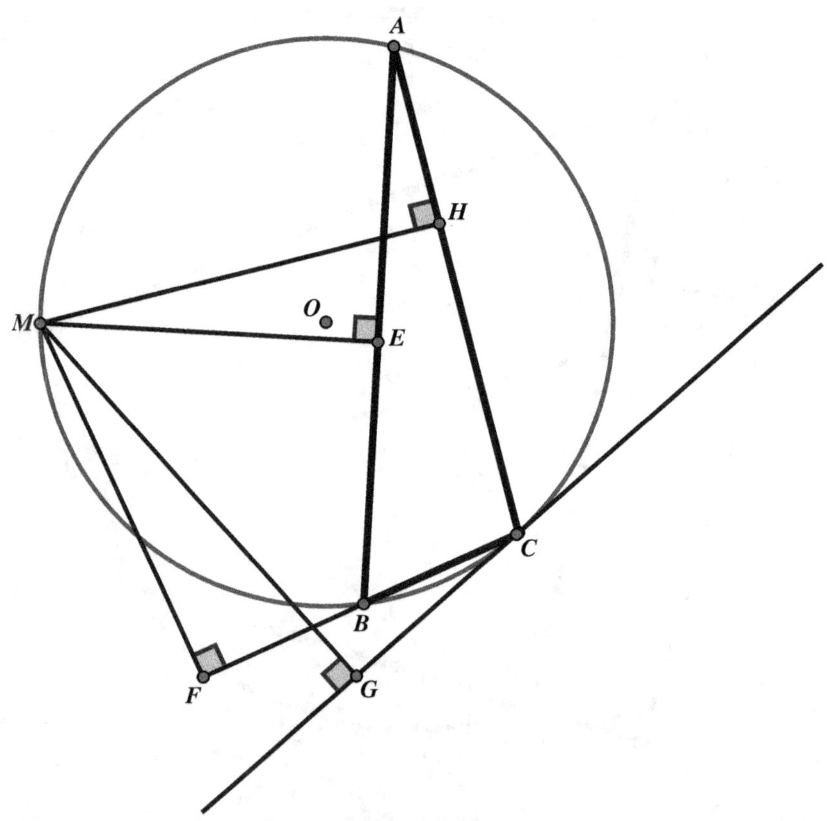

Figure 5-53

Geometrick 56: Determining the Radius of a Circle

In Figure 5-54, AB is tangent to circle O at point A. Point D is in the interior of the circle and BD intersects the circle at point C. If $BC = DC = 3$, $OD = 2$, and $AB = 6$, our challenge is to find the radius of the circle.

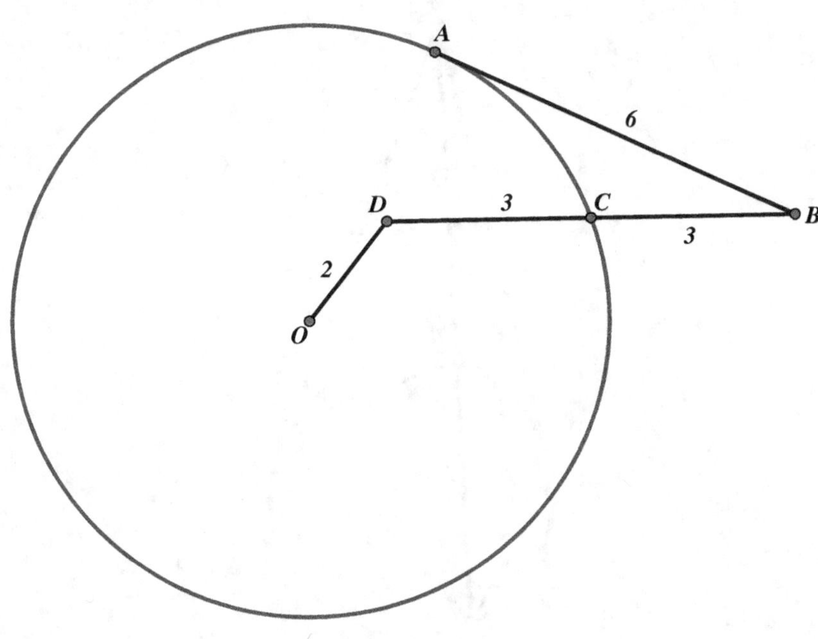

Figure 5-54

Geometrick 57: A Proportion Produced by Tangents and the Secant

Two tangents and a secant to a circle from an external point determine a quadrilateral with adjacent sides that are in an unexpected proportion. We see this in Figure 5-55, where we have drawn tangents MA and MC and secant MBD to circle O. Quadrilateral $ABCD$ has its adjacent sides in the following proportion: $\frac{AB}{AD} = \frac{BC}{DC}$.

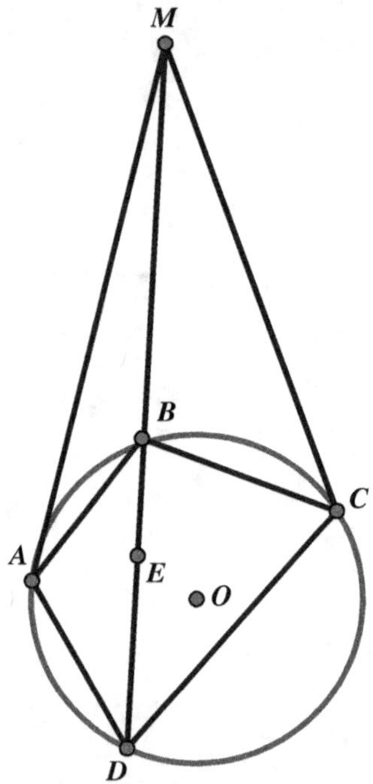

Figure 5-55

Geometrick 58: Parallelogram Comparisons

Here is a problem that has perplexed scores of high school teachers, who may have found it in any 150-year-old geometry book. On the surface, the problem looks very difficult, given the procedures that are typically required to solve geometry problems. However, the solution becomes quite simple when one thinks "out-of-the-box." We are given parallelograms $ABCD$ and $APQR$, with point P on side BC and point *D* on side *RQ*, as shown in Figure 5-56. If the area of parallelogram *ABCD* is 18, what is the area of parallelogram *APQR*?

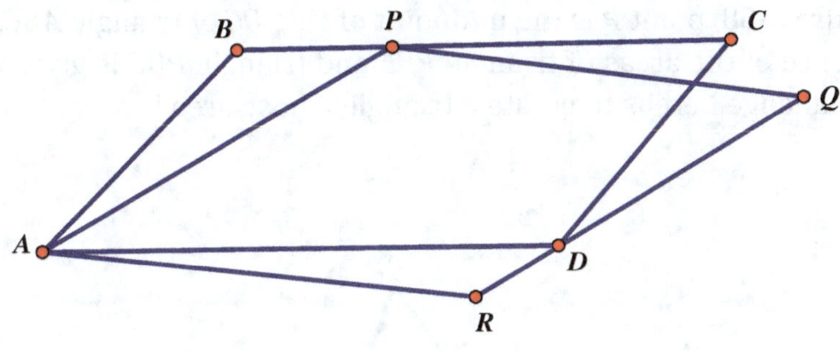

Figure 5-56

Geometrick 59: Constant Area of Two Triangles in a Parallelogram

The sum of the areas of two triangles that have their respective bases on the diagonals of a parallelogram and share a common vertex point on the parallelogram's perimeter is constant. In Figure 5-57, triangle *APC* and triangle *BPD* share a vertex point *P* on side *DC* of parallelogram *ABCD*. Each has a base on one of the diagonals. Regardless of where point *P* is located on side *DC*, area$\triangle APC$ + area$\triangle BPD$ will always be the same.

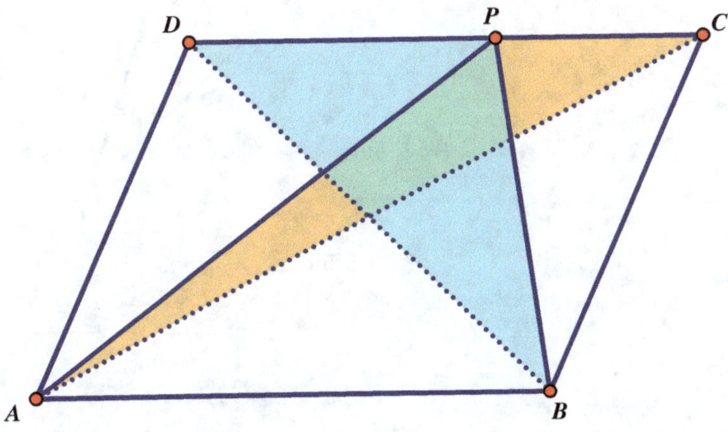

Figure 5-57

Geometrick 60: A Curious Area Difference Between Two Triangles

In Figure 5-58, point *P* is the midpoint of side *BC* of triangle *ABC*. The difference of the areas of triangle *PBE* and triangle *PDC* is given as *K*. The challenge here is to create a triangle whose area is *K*.

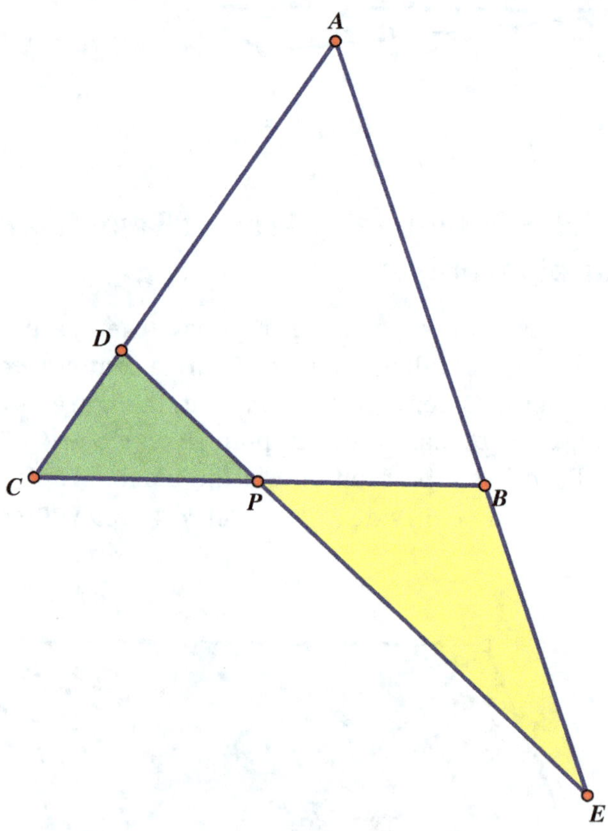

Figure 5-58

Geometrick 61: An Unusual Area Relationship in a Triangle

Triangle *ABC*, shown in Figure 5-59, has $DE \parallel AC$ and $DE = \frac{1}{3} AC$. What part of the area of triangle *ABC* is the area of triangle *DEC*?

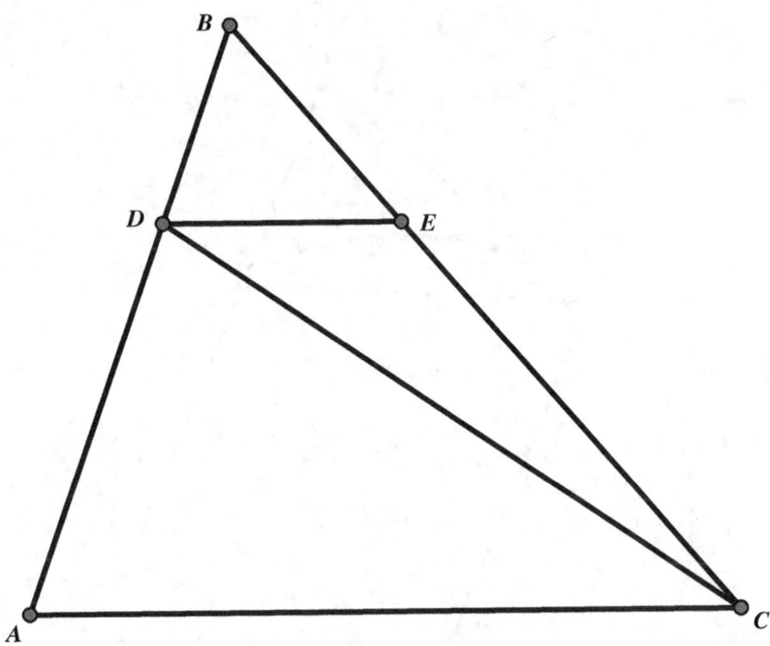

Figure 5-59

Geometrick 62: An Unexpected Parallelogram

A circle with the chord *AB* has a tangent *AC* equal to *AB*. When *CB* is extended, it intersects the circle at point *D*. The midpoint of arc *BDA* is point *M*, as shown in Figure 5-60. Unexpectedly, we find that quadrilateral *ACDM* is a parallelogram.

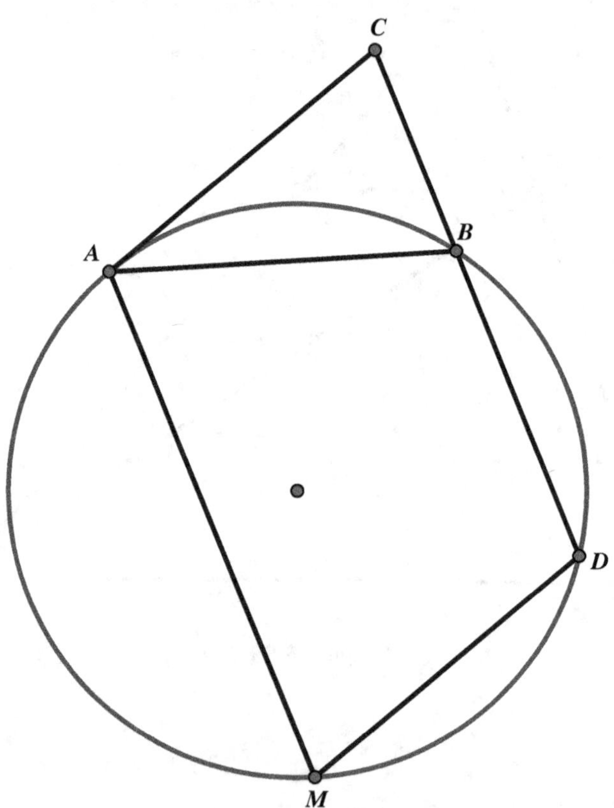

Figure 5-60

Geometrick 63: An Unexpected Rhombus

There are unusual ways to create a rhombus. One such is illustrated in Figure 5-61, where the opposite sides of a cyclic quadrilateral are extended to meet at external points. That is, for cyclic quadrilateral *ABCD*, sides *AB* and *DC* intersect at point *E*, and sides *AD* and *BC* intersect at point *F*. We then draw the bisectors of angles *AED* and *AFB*, which intersect the opposite sides of the cyclic quadrilateral at points *G* and *H* and *I* and *J*, respectively. Surprisingly, the quadrilateral *GIHJ* turns out to be a rhombus.

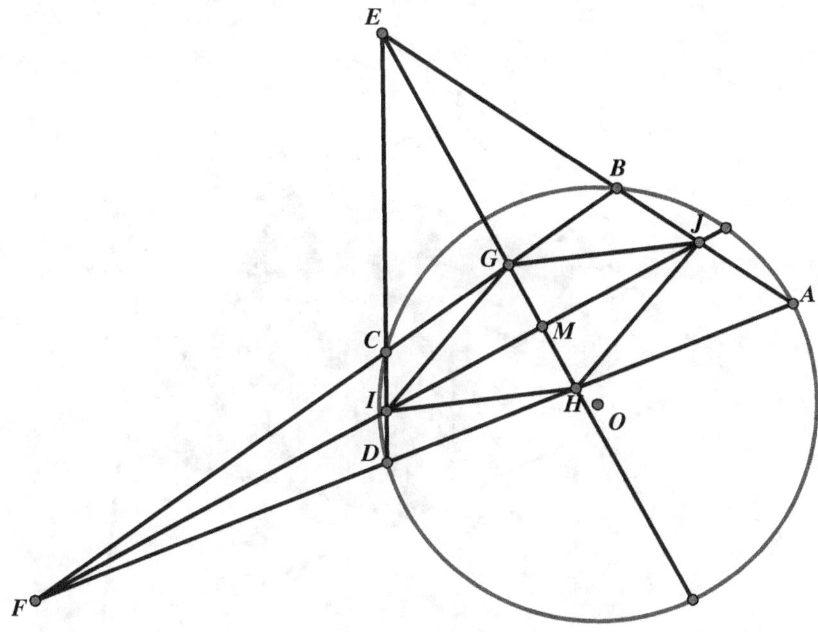

Figure 5-61

Geometrick 64: The Varignon Parallelogram

In the world of quadrilaterals, the parallelogram dominates. The most blatant situation is when we join consecutively the midpoints of any randomly drawn quadrilateral, as we can see in Figure 5-62, where the midpoints *K*, *L*, *M*, and *N* of the sides *AB*, *BC*, *CD*, and *AD* are joined consecutively to form parallelogram *KLMN*. This parallelogram is often referred to as a *Varignon parallelogram*, named after the French mathematician Pierre Varignon (1654–1722). Furthermore, this parallelogram allows us to conclude that the lines joining midpoints of the opposite sides of the original quadrilateral bisect each other because they are the diagonals of the parallelogram.

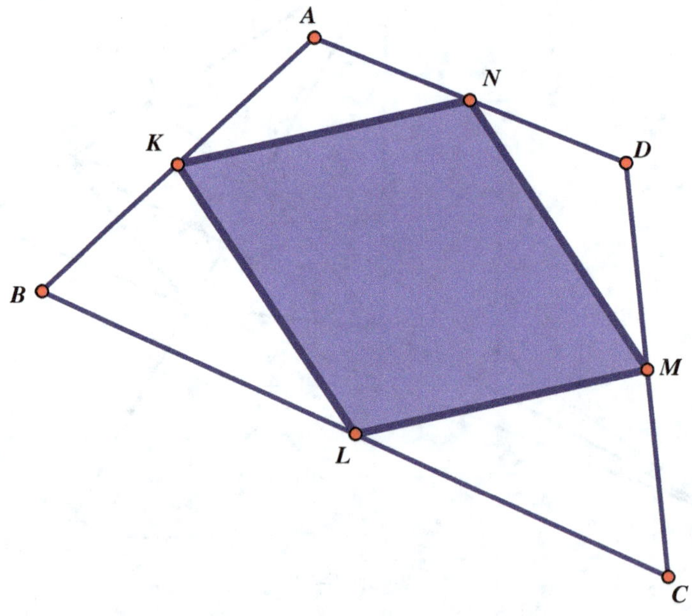

Figure 5-62

However, there is an alternative way of creating a Varignon parallelogram. Let's take a look. Two unrelated triangles *ABC* and *DBC* that share the same base *BC* can determine a parallelogram *PQRS* by joining the midpoints P, Q, R, and S of the sides DC, AC, AB, and BD, respectively, as shown in Figure 5-63.

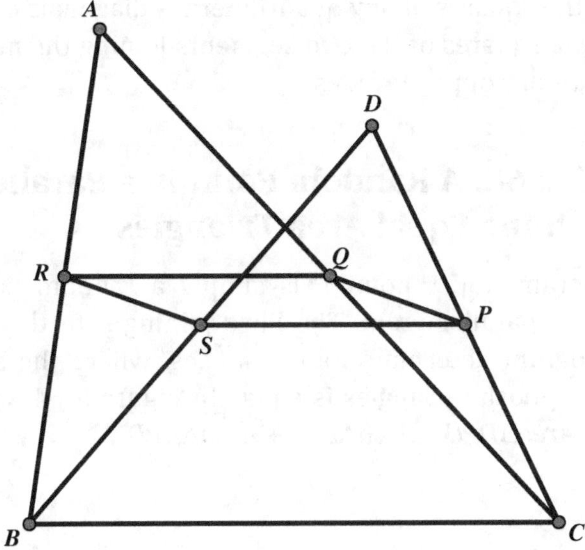

Figure 5-63

Geometrick 65: The Varignon Parallelogram's Special Properties

There are further special properties of the Varignon parallelogram in Figure 5-62. The perimeter of a Varignon parallelogram is equal to the sum of the diagonals of the original quadrilateral. Furthermore, the Varignon parallelogram is half the area of the original quadrilateral. The challenge here is to prove these relationships.

Geometrick 66: An Unexpected Relationship Between the Sides and the Diagonals of a Parallelogram

The sum of the squares of a parallelogram's sides equals the sum of the squares of the diagonals.

Geometrick 67: The Relationship Between the Sides and Diagonals of any Quadrilateral

The sum of the squares of any quadrilateral's diagonals equals twice the sum of the squares of the two segments joining the midpoints of the quadrilateral's opposite sides.

Geometrick 68: A Random Point in a Parallelogram Can Determine Equal Area Triangles

In parallelogram *ABCD*, point *P* is simply a random point in the interior of the parallelogram. The lines joining *P* to the vertices of the parallelogram determine four triangles, where the sum of the areas of the opposite triangles is equal. In Figure 5-64, we have the $Area\triangle APB + Area\triangle DPC = Area\triangle BPC + Area\triangle APD$.

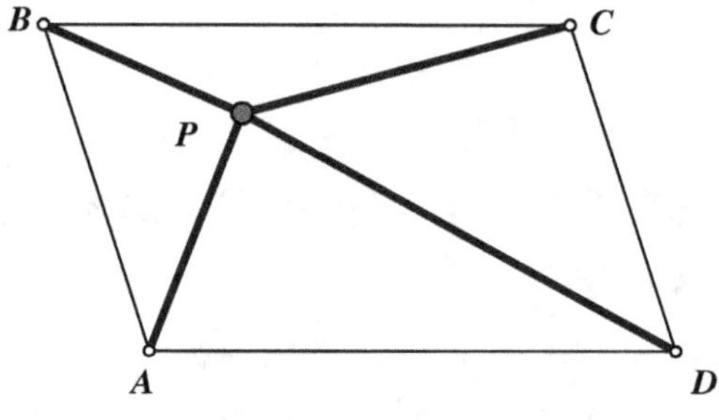

Figure 5-64

Geometrick 69: Equal Area Parallelograms in a Parallelogram

Point *P* is a randomly selected point on diagonal *BD* of parallelogram *ABCD*, as shown in Figure 5-65. Parallel lines are drawn to create two parallelograms, *ADPG* and *ECFP*. Surprisingly, these parallelograms are equal in area regardless of where point *P* is on diagonal *BD*.

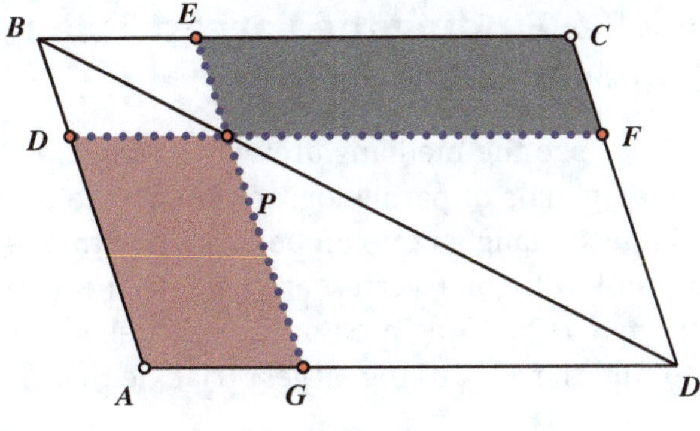

Figure 5-65

Geometrick 70: Detecting a Constant Quadrilateral

A quadrilateral with diagonals of given lengths that intersect at a constant angle will generate a quadrilateral of constant area regardless of where these diagonals intersect.

Geometrick 71: A Surprising Constant Product Created by a Parallelogram

In parallelogram *ABCD*, a line from vertex *A* intersects side *BC* and side *DC* at points *M* and *N*, respectively. The challenge here is to prove that the product of the segments from the adjacent vertices to these points of intersection is constant with the product of the adjacent sides of the parallelogram. That is, in Figure 5-66, $BM \cdot DN = AB \cdot AD$.

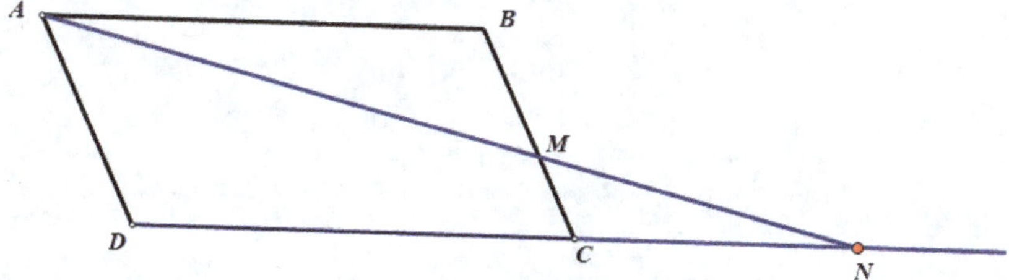

Figure 5-66

Geometrick 72: Finding the Largest Triangle in a Particular Parallelogram Setting

In Figure 5-67, we see the medians drawn in triangle *DOC*, which is defined by the diagonals of parallelogram *ABCD*. The challenge is to generate the largest triangle that can be formed with one side of the parallelogram and with one vertex at point *G*, the intersection of medians of triangle *DOC*. If the area of the parallelogram is 120, our challenge is to find the area of the largest triangle that fits the above definition.

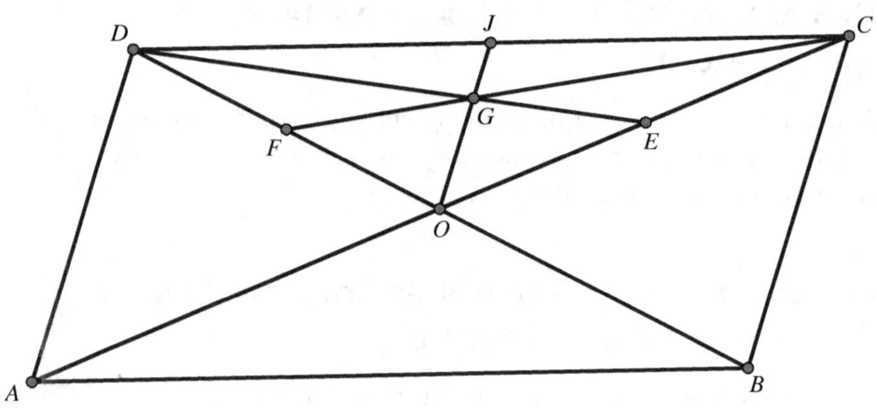

Figure 5-67

Geometrick 73: Relationship of the Perpendiculars from the Vertices of a Parallelogram to a Straight Line Containing One Vertex

Consider a random straight line drawn through one vertex of a parallelogram. The sum of the distances to this random line from the two adjacent vertices is equal to the distance from the remote vertex to the line. In Figure 5-68, line l contains vertex D of parallelogram $ABCD$. Lines AN, BM, and CR are perpendicular to line l. We find that $AN + CR = BM$.

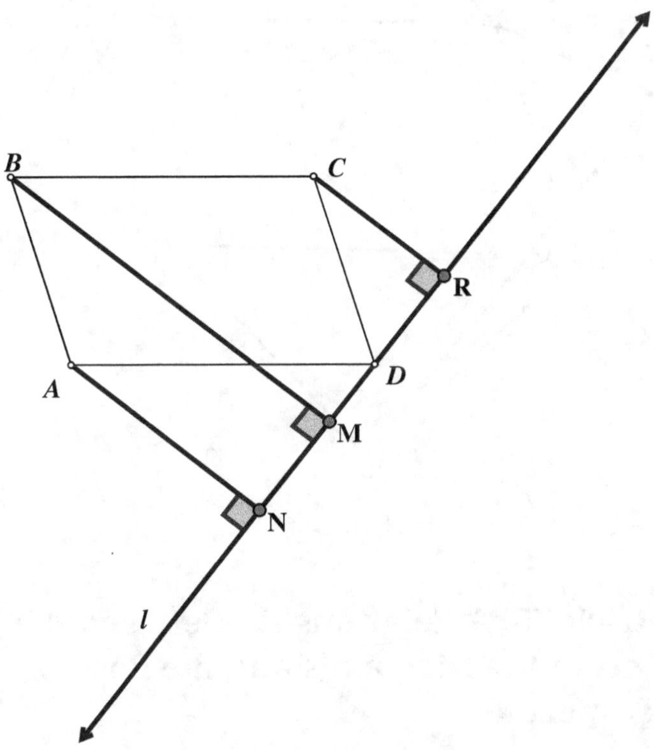

Figure 5-68

Geometrick 74: Isosceles Right Triangles on the Sides of a Parallelogram Create a Square

Isosceles right triangles *ABE*, *BCF*, *DCG*, and *DHA* are drawn on the sides of parallelogram *ABCD*, as shown in Figure 5-69. The result is that quadrilateral *EFGH* is a square.

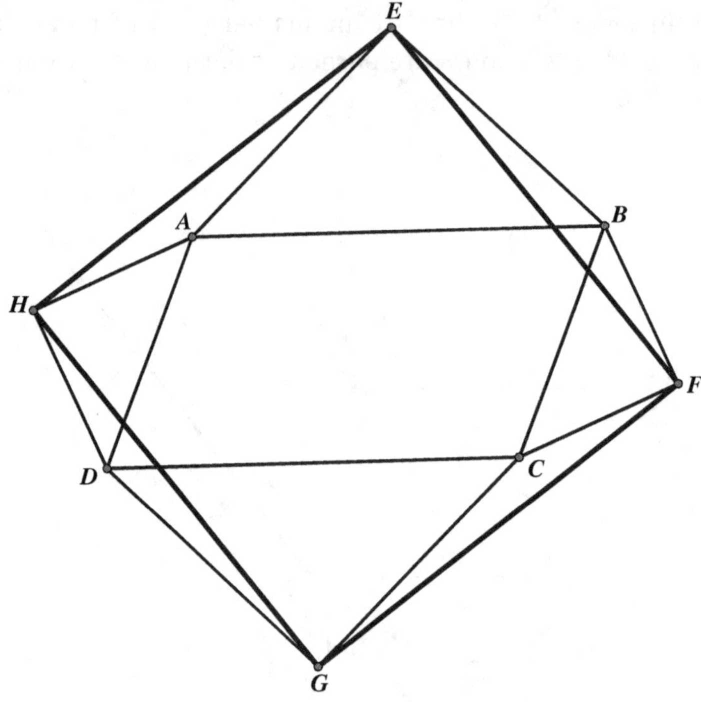

Figure 5-69

Geometrick 75: A Relationship Between the Midpoints of the Diagonals and the Bases of the Trapezoid

Joining the midpoints of the diagonals of the trapezoid produces a line segment, the length of which is half the difference of the length of the bases. In Figure 5-70, points *E* and *F* are midpoints of diagonals *AC* and *BD*, respectively, of trapezoid *ABCD*. We find that $EF = \frac{1}{2}(AD - BC)$.

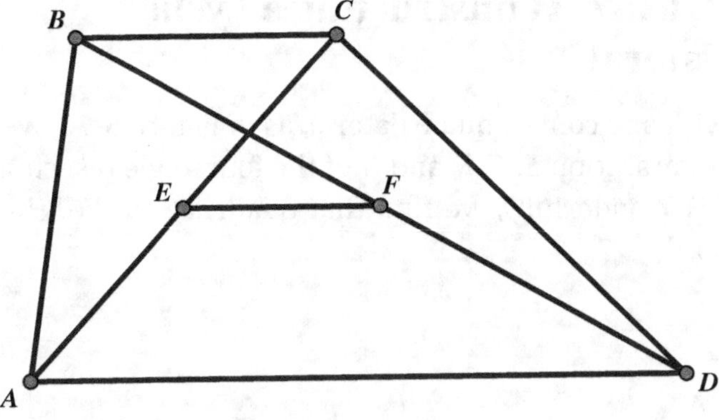

Figure 5-70

Geometrick 76: A Surprise Bisector of a Line in the Quadrilateral

In any convex quadrilateral, the point of intersection of the medians bisects the line segment joining the midpoints of the diagonals. In Figure 5-71, the midpoints of diagonals *AC* and *BD* of quadrilateral *ABCD* are *N* and *M*, respectively. The medians *EF* and *GH*, which are the lines joining the midpoints of opposite sides of the quadrilateral, intersect at point *P*. Amazingly, this point *P* is the midpoint of *MN*.

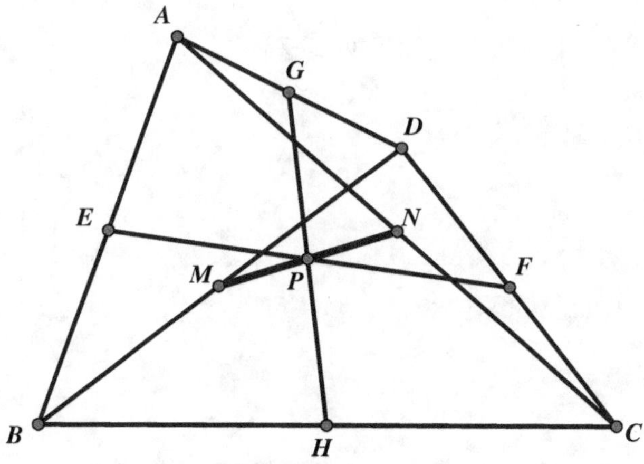

Figure 5-71

Geometrick 77: Constructing a Cyclic Quadrilateral

Starting with any convex quadrilateral, as in Figure 5-72, we note the points of intersection *E*, *G*, *F*, and *H* of the four angle bisectors, *AF*, *BE*, *CE*, and *DF*. Wonderfully, we find that quadrilateral *EGFH* is a cyclic quadrilateral.

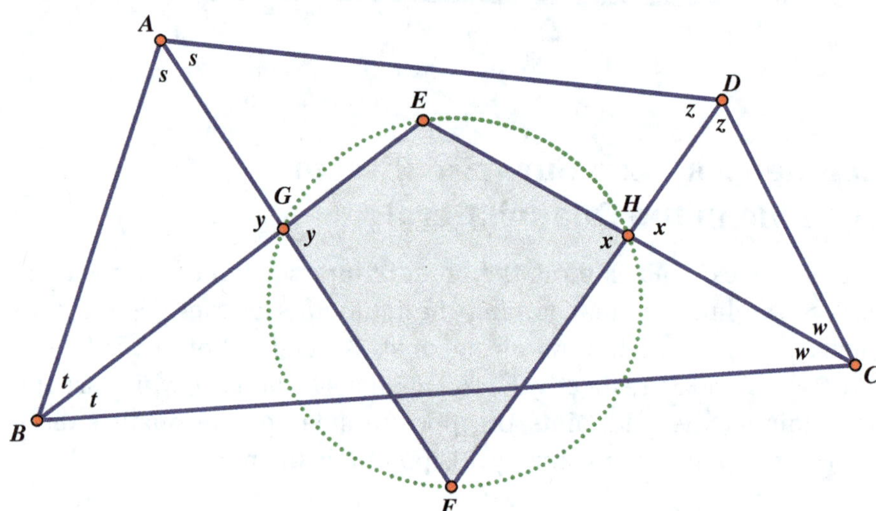

Figure 5-72

Geometrick 78: Relationship Between Orthocenter, Centroid, and Perpendicular Bisectors

In Figure 5-73, three key points are shown for triangle *ABC*. They are the orthocenter *H*, the intersection of the altitudes; the centroid *G*, the intersection of the medians; and the circumcenter *P*, the intersection of the perpendicular bisectors. A little-known relationship between these three points exists, namely, $GH = 2\,PG$.

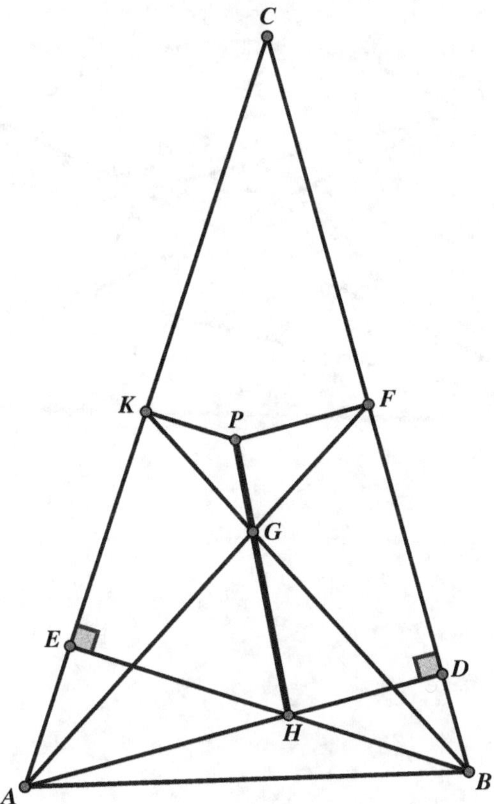

Figure 5-73

Geometrick 79: A Surprise Perpendicularity

On the semicircle with *AB* as diameter and center *O*, a point *C* is randomly selected, as shown in Figure 5-74. Furthermore, points *D* and *E* are selected on the other side on the semicircle so that $\overset{\frown}{DE} = \overset{\frown}{DB}$. Lines *AD* and *BC* intersect at point *F*, and lines *AE* and *CD* intersect point *G*. Curiously, we find that angle *AGF* is a right angle.

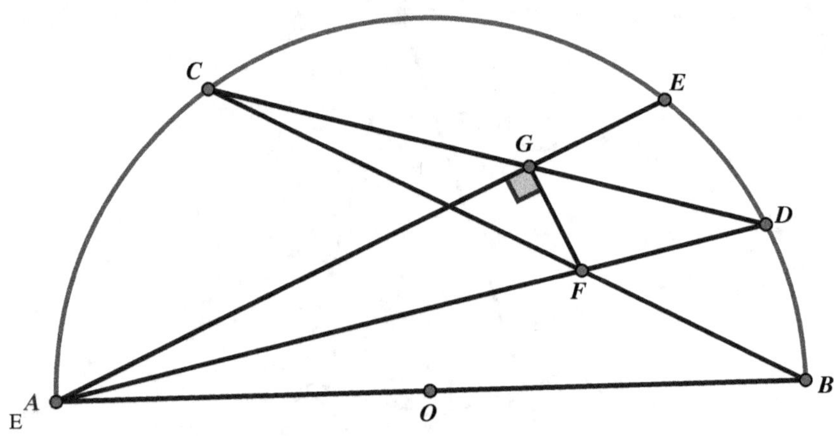

Figure 5-74

Geometrick 80: An Unexpected Cyclic Quadrilateral

Consider a cyclic quadrilateral *ABCD* inscribed in a circle with center *I*, shown in Figure 5-75. On each side of the quadrilateral a circle is drawn that intersects the adjacent circles, creating points *E*, *F*, *G*, and *H*. Unexpectedly, it turns out that these points of intersection create quadrilateral *EFGH*, which is also a cyclic quadrilateral.

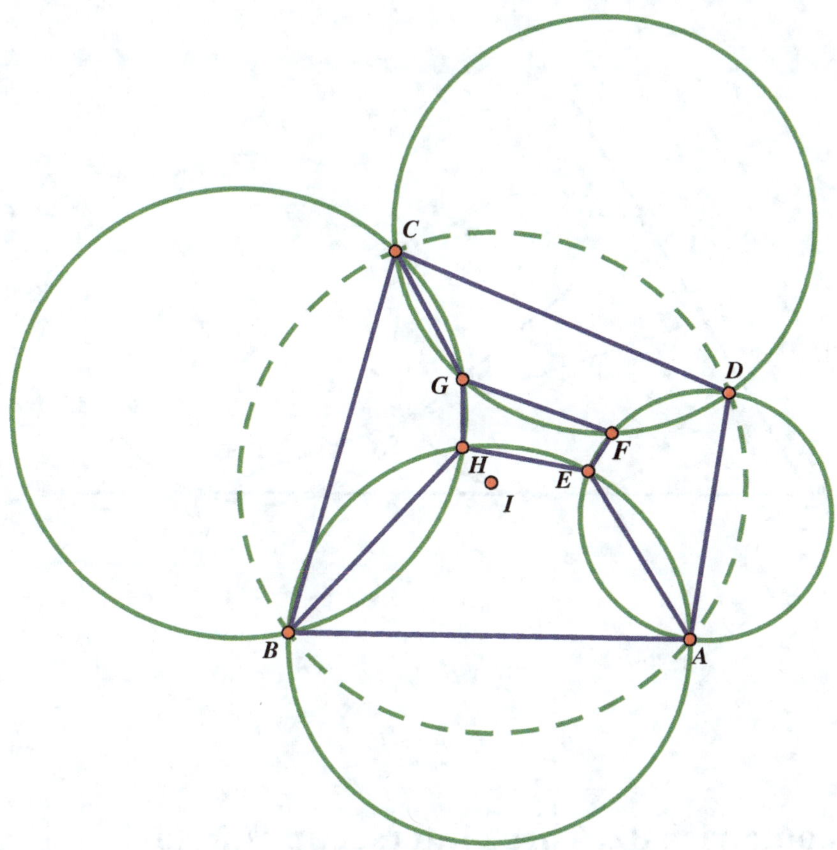

Figure 5-75

Geometrick 81: Another Surprise Cyclic Quadrilateral

Point *B* is randomly placed within angle *XPZ*, as shown in Figure 5-76. Segments *ABC* and *DBE* are drawn so that points *A* and *D* are on *XP*, and points *C* and *E* are on *ZP*. The circumscribed circles of triangle *ABD* and triangle *CBE* intersect at point *M*. Unexpectedly we find that points *A*, *P*, *C*, and *M* are concyclic.

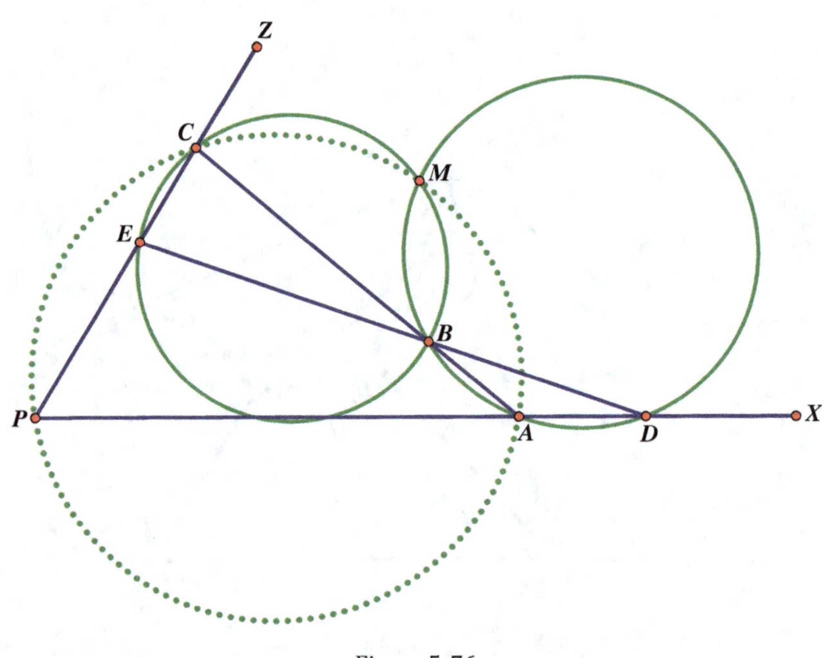

Figure 5-76

Geometrick 82: Three Intersecting Circles

When each of three circles contains a vertex and two predetermined points on the adjacent sides of a triangle, these circles all share a common point. We need to prove that the circles containing three points *FAD*, *DBE*, and *EFC* all share the common point *O*, which is shown in Figure 5-77.

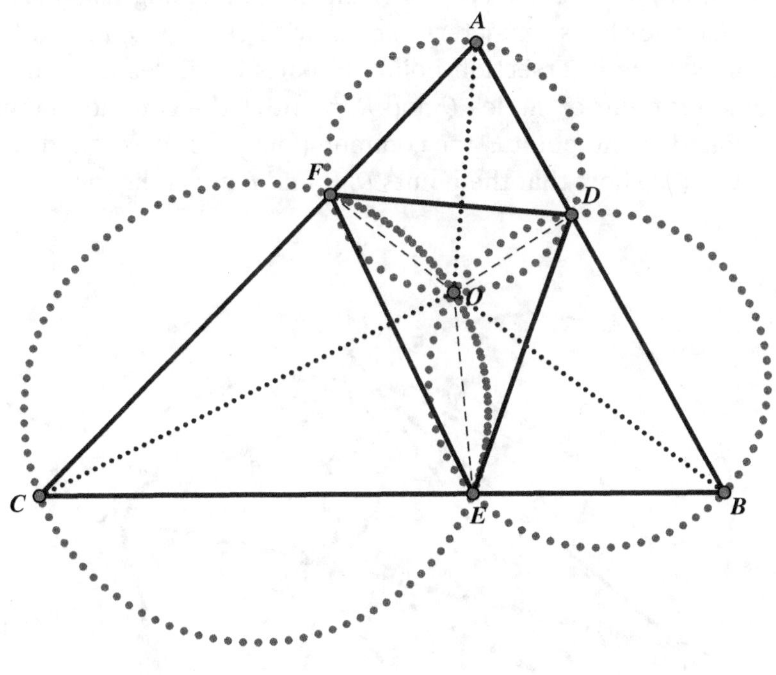

Figure 5-77

Geometrick 83: More with Three Intersecting Circles

Another surprising relationship emerges from Figure 5-77. That is, that $\angle BOC = \angle A + \angle DEF$, $\angle AOC = \angle B + \angle DFE$, and $\angle AOB = \angle C + \angle FDE$. Can you demonstrate that these equalities hold true?

Geometrick 84: Three Intersecting Circles Creating a Straight Line

In Figure 5-78, point *P* is on the circumference of a circle with center *O*, and *PA*, *PB*, and *PC* are chords of the circle forming triangle *ABC*. With these chords as diameters, circles with centers *Q*, *R*, and *S* are constructed. The intersection points of pairs of circles are: point *E*, the common point of circles *Q* and *R*; point *D*, the common point of circles *R* and *S*; and point *F*, the common point of circles *Q* and *S*. The challenge is to show that the points *D*, *F*, and *E* are collinear.

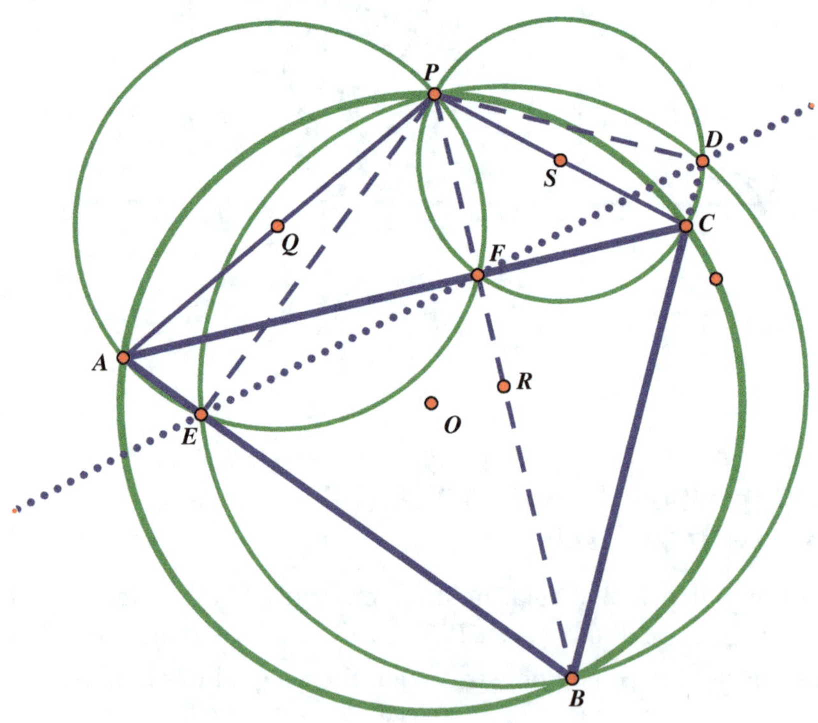

Figure 5-78

Geometrick 85: A Curiosity on a Decagon

A chord, such as *AD*, that subtends an arc three times longer than the arc of a regular decagon side is equal to the sum of the radius of the decagon and a side length. In Figure 5-79, for regular decagon *ABCDEGHJK*, we have $AD = CD + AO$.

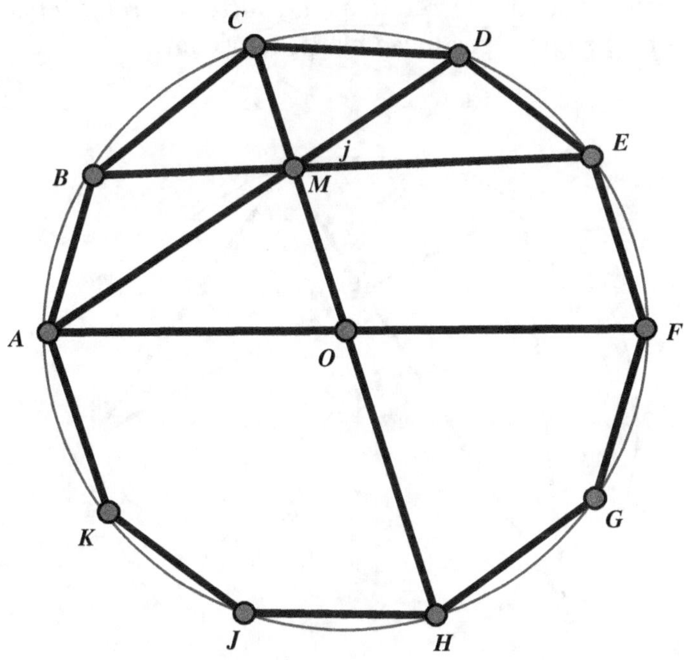

Figure 5-79

Geometrick 86: The Famous Butterfly Relationship

Four well-placed chords can partition a fifth chord of a circle into equal parts. In Figure 5-80, point *M* is the midpoint of chord *AB* in a given circle. Chords *CD* and *EF* are randomly selected and intersect at point *M*. Point *P* is the intersection of *AB* and *CF*, and point *Q* is the intersection of *AB* and *DE*. Quite unexpectedly, we find that *MP = MQ*. This curiosity is often referred to as the *butterfly problem* because of the shape of the figure produced by the four chords.

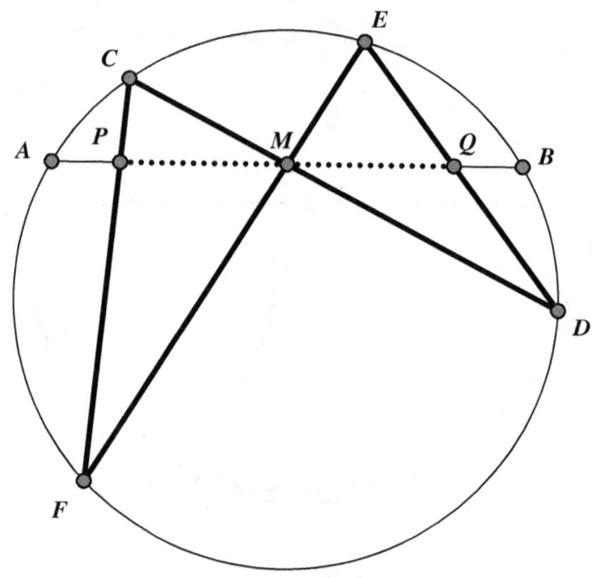

Figure 5-80

Geometrick 87: A Surprise Tangent

In Figure 5-81, on diameter *EOF* of circle *O*, points *M* and *N* are selected so that *MO = NO*. We also select any point *C* on circle *O* and extend *CM* to intersect the circle at point *A*. In addition, *CO* extended intersects the circle at point *T*, and *CN* extended intersects the circle at point *B*. We also extend *AB* to intersect *EF* at point *G*. The challenge is to prove that *GT* is tangent to circle *O*.

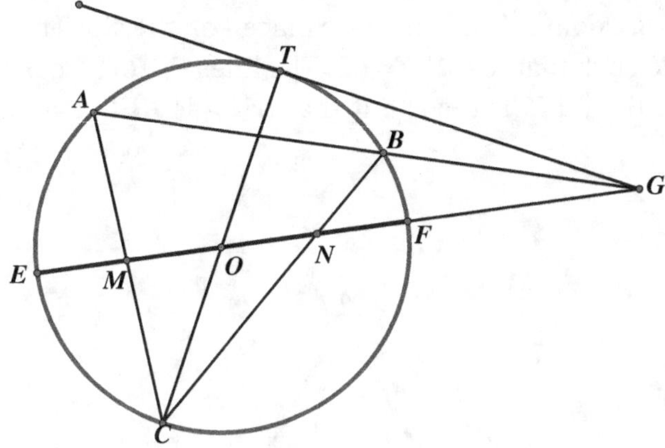

Figure 5-81

Geometrick 88: Equality of Common Tangents

In Figure 5-82, circles *O* and *R* have three common tangents: *AB*, *CD*, and *EF*. We can conclude that $AB = CD = EF$.

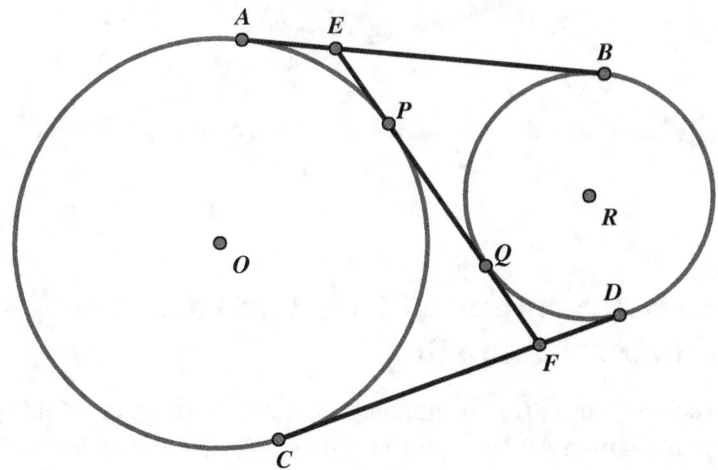

Figure 5-82

Geometrick 89: Appearance of an Equilateral Triangle in an Equilateral Triangle

In Figure 5-83, points *D*, *E,* and *F* are placed on the sides of equilateral triangle *ABC* such that $AD = BF = CE$. The lines *BD*, *CF*, and *AE* intersect at points *G*, *H*, and *K*. It turns out that triangle *GHK* is an equilateral triangle.

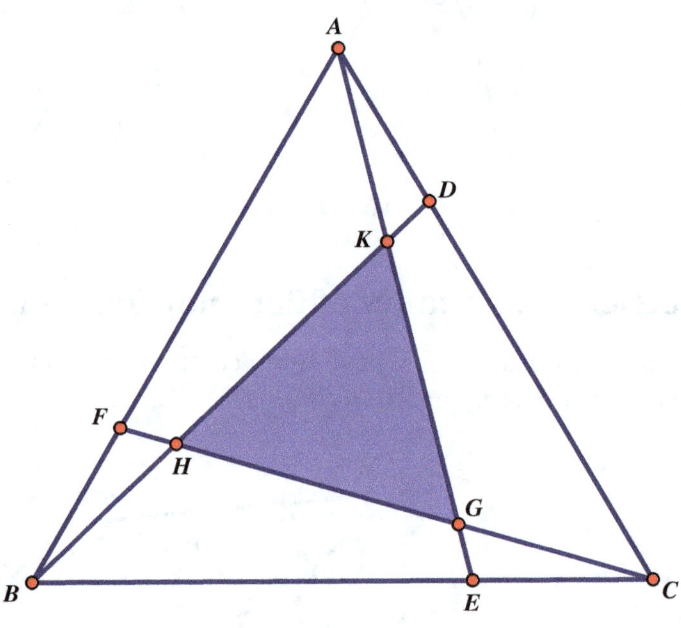

Figure 5-83

Geometrick 90: How a Circle Can Partition an Equilateral Triangle

In Figure 5-84, circle *O* contains vertex *A* and is tangent to side *BC* of equilateral triangle *ABC* at point *M*. The circle also intersects sides *AB* and *AC* at points *D* and *E*, respectively. The challenge is to find out what the ratio is of segments $\frac{BD}{AD}$.

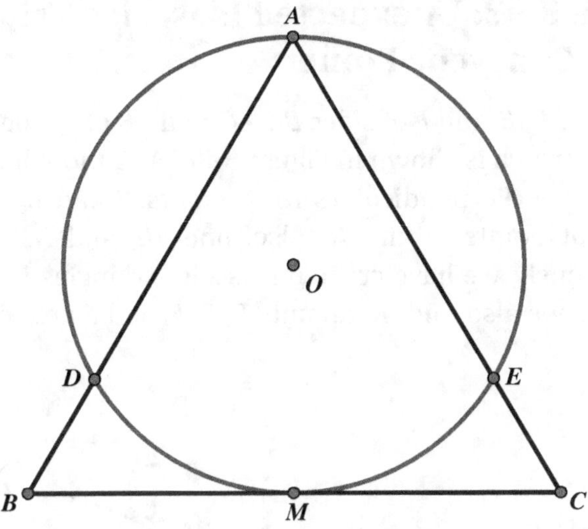

Figure 5-84

Geometrick 91: Finding the Area of an Equilateral Triangle Given the Distances from an Interior Point to the Vertices

In Figure 5-85, point P has a distance to the three vertices of equilateral triangle ABC such that $AP = 5$, $BP = 6$, and $CP = 7$. Our challenge is to find the area of this equilateral triangle.

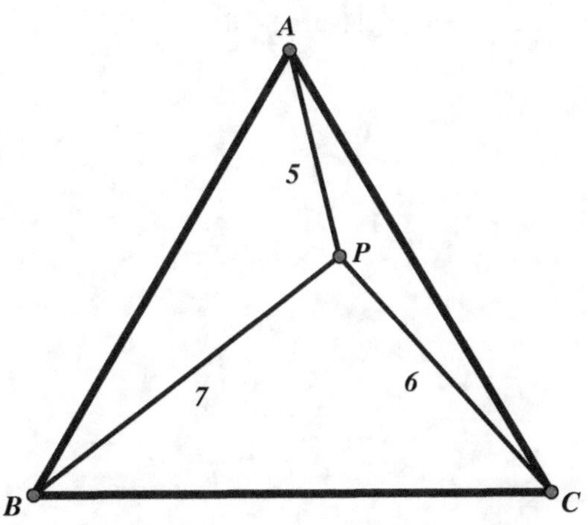

Figure 5-85

Geometrick 92: Unexpected Isosceles Triangles Generate Concyclic Points

The midpoints *D*, *E*, and *F* of sides *BC*, *AC*, and *AB* of triangle *ABC* form triangle *DEF*, which is shown in Figure 5-86. A random line *l* is drawn through point *A*. Perpendiculars from points *B* and *C* to this line *l* intersect it at points *M* and *N*. Also, line *MF* and *NE* intersect at point *P*. Curiously, we have created isosceles triangles *MFA* and *NEA*. Furthermore, we also find that points *D*, *P*, *E*, and *F* are concyclic.

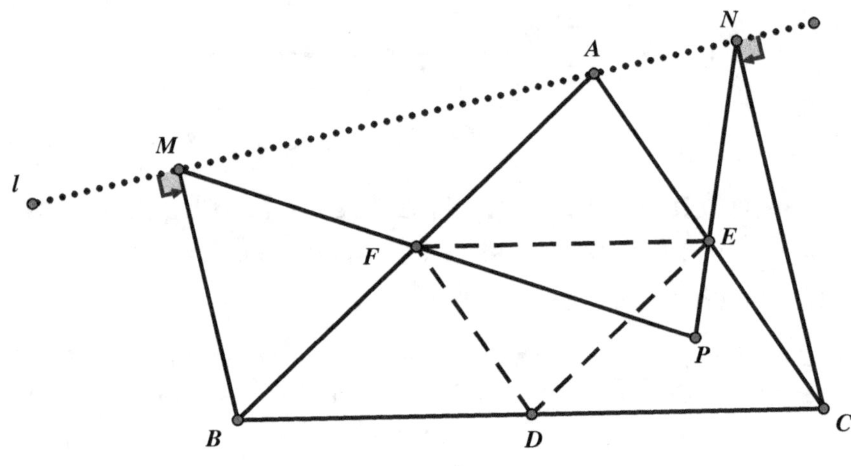

Figure 5-86

Geometrick 93: A Surprise in a Trapezoid

The diagonals of a trapezoid, where the parallel sides are perpendicular to one of the other sides, bisect a perpendicular line drawn from the point of tangency of an inscribed semicircle to side perpendicular to the parallel sides. Simply said, in Figure 5-87, the semicircle with diameter BC is inscribed in trapezoid $ABCD$, where $AB \perp BC$ and $DC \perp BC$, and the semicircle is tangent to side AD at point E. The intersection of the diagonals, G, bisects EF, where $EF \perp BC$.

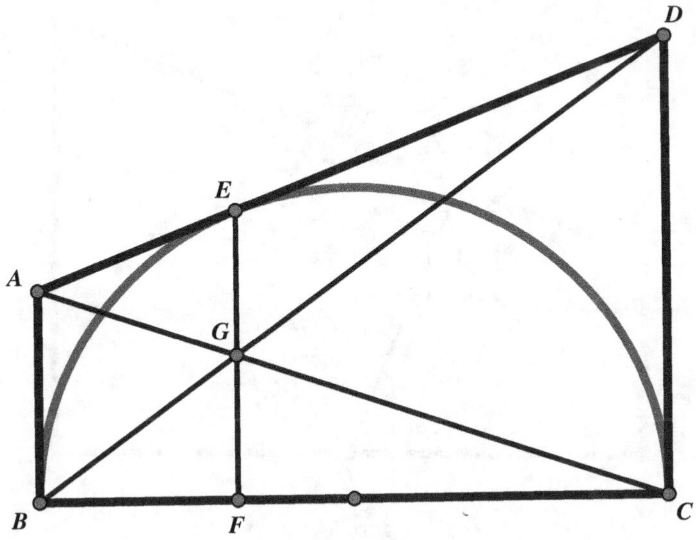

Figure 5-87

Geometrick 94: Another Surprise in a Trapezoid

A semicircle is inscribed in a trapezoid such that the diameter forms one side of the trapezoid and a point of tangency is on the opposite side of the trapezoid. The product of the segments of the tangent is equal to the square of the radius of the semicircle. In Figure 5-88, trapezoid *ABCD*, with parallel lines *AB* and *DC*, circumscribes a semicircle with diameter *BOC* and center *O*. The semicircle is tangent to side *AD* at point *E*. The following is true: $AE \cdot DE = EO^2$.

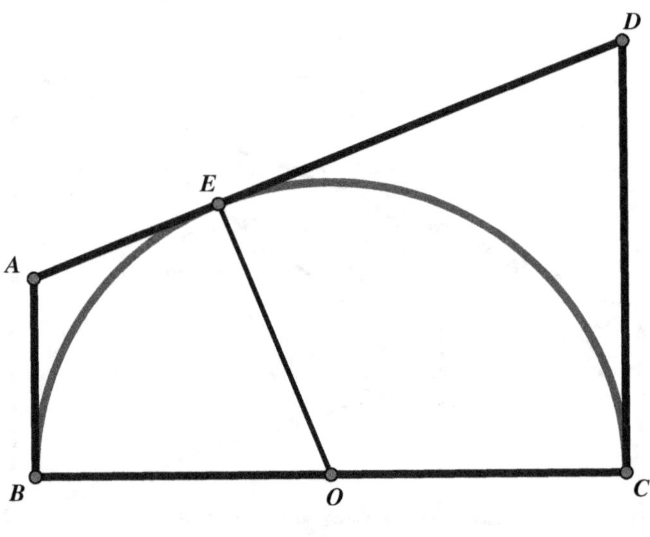

Figure 5-88

Geometrick 95: Partitioning a Quadrilateral into Four Equal Areas

The midpoints *M* and *N* of the diagonals *AC* and *BD* of quadrilateral *ABCD* have lines drawn through them that are parallel to the opposite diagonal and intersect at point *O*, as can be seen in Figure 5-89. The lines joining *O* to the midpoints of the sides of quadrilateral *ABCD* partition the quadrilateral into four regions of equal area.

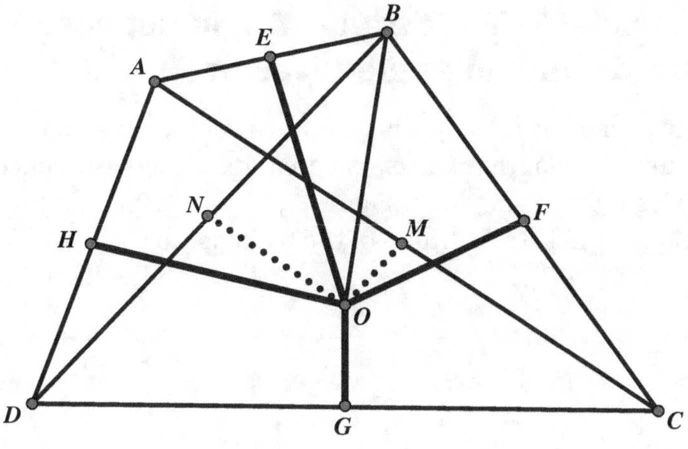

Figure 5-89

Geometrick 96: The Measure of an Angle in the Interior of a Triangle

Line segments *PB* and *PC* bisect two angles of triangle *ABC*, $\angle ABC$ and $\angle ACB$, and intersect at point *P*, as shown in Figure 5-90. Unexpectedly, the measure of $\angle P = 90° + \frac{\angle A}{2}$.

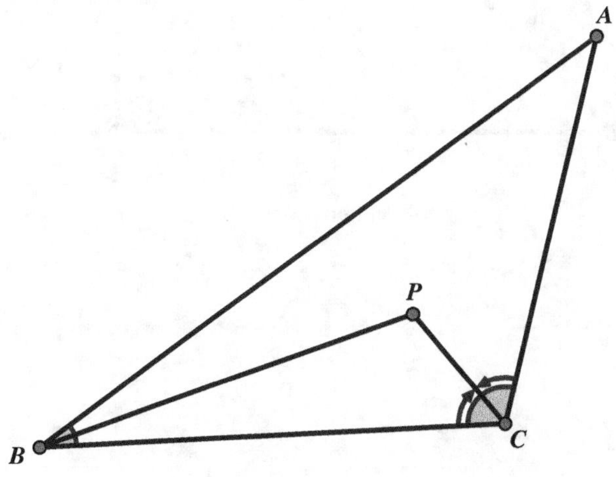

Figure 5-90

Geometrick 97: The Strange Angle Between an Altitude and an Angle Bisector

The angle formed by the angle bisector of a vertex of a triangle and the altitude drawn from that vertex is equal to half the difference of the base angles of the triangle. In Figure 5-91, line *AD* is the bisector of angle *BAC* and *AE* is the altitude of the triangle, so we must show that $\angle DAE = \frac{1}{2}(\angle C - \angle B)$.

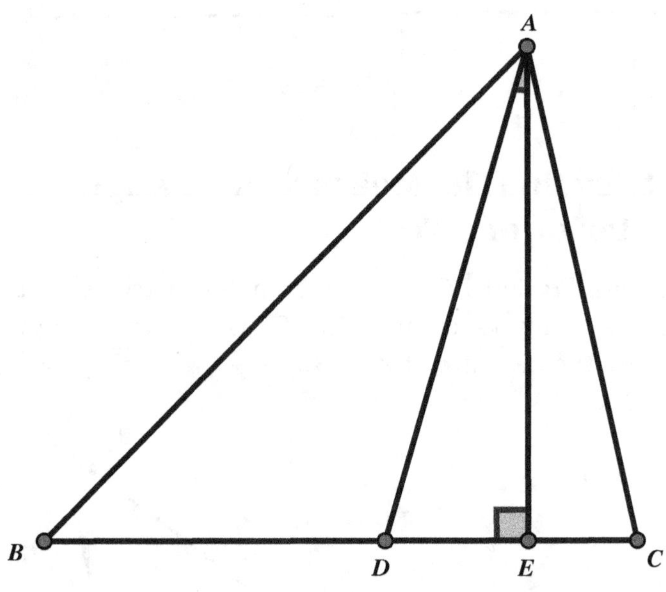

Figure 5-91

Geometrick 98: Another Curiosity Formed by the Altitude and Angle Bisector

The angle between the altitude and angle bisector is related in a curious way to the angles formed by the angle bisector and the base of the triangle. In Figure 5-92, we have $\angle DAE = \frac{m-n}{2}$, where $\angle ADB = m$, and $\angle ADC = n$.

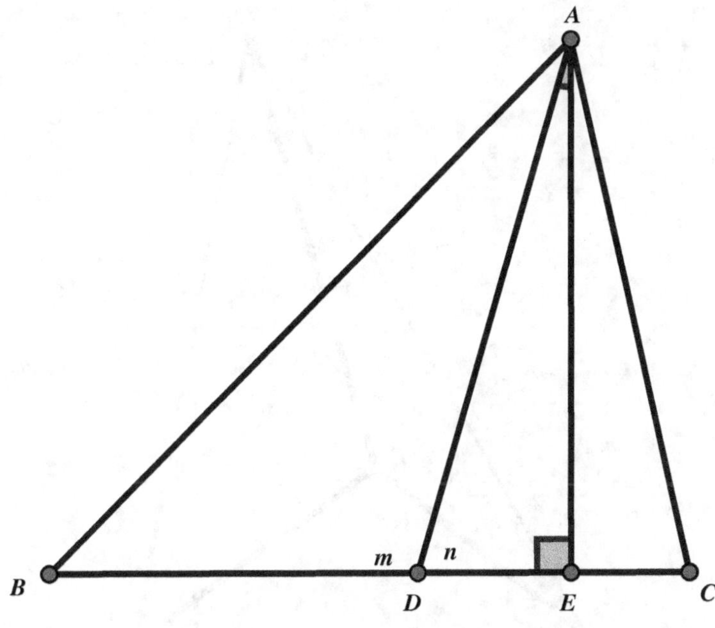

Figure 5-92

Geometrick 99: Comparison Between the Distance from the Vertices to a Point in a Triangle to the Sum of the Two Longest Sides of the Triangle

The sum of the two longest sides of the triangle is greater than the sum of the distances from any point in the triangle to the three vertices. In Figure 5-93, for triangle *ABC*, we have $AC + BC > PA + PB + PC$.

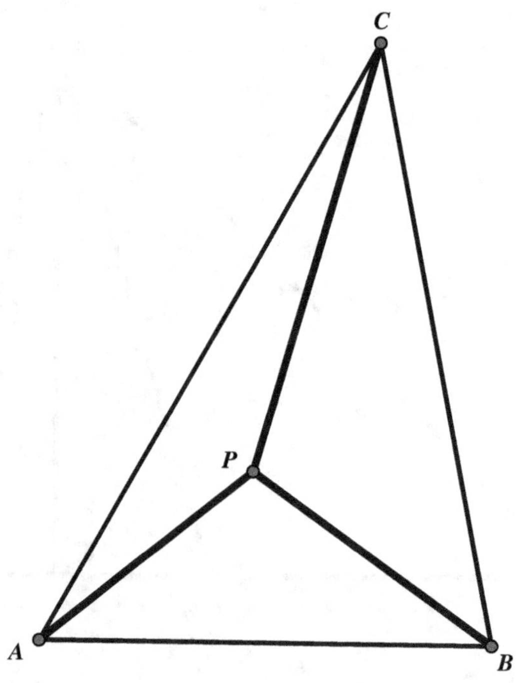

Figure 5-93

Geometrick 100: Results of Perpendiculars from a Random Point in the Triangle

The feet of the perpendiculars from any point in a triangle to the three sides determine six segments along the sides of the triangle. The sum of the squares of the alternate segments are equal. In Figure 5-94, a random point O is selected in triangle ABC, from which perpendiculars OD, OE, and OF are drawn to the three sides. The sum of the squares of the alternate segments are equal, that is, $AF^2 + BD^2 + CE^2 = AE^2 + CD^2 + BF^2$.

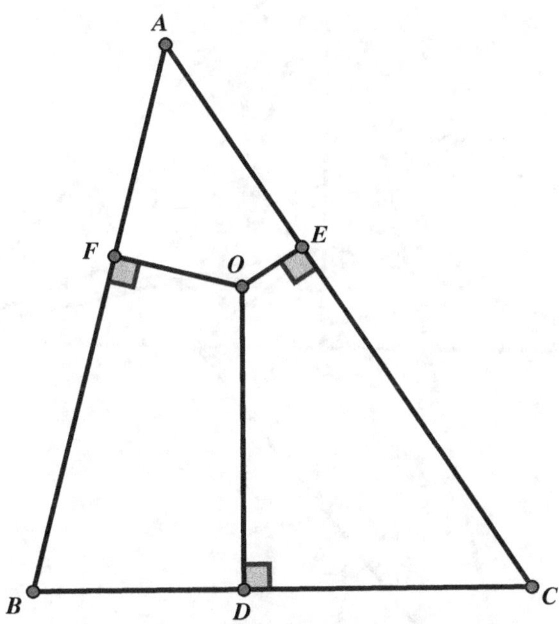

Figure 5-94

Geometrick 101: The Wonders Created by the Altitudes and a Median of a Triangle

In Figure 5-95, altitudes *AD*, *BE*, and *CF* of triangle *ABC* meet at orthocenter *H*. Construct parallelogram *ABKC* such that *AC* is parallel to *BK* and *AB* is parallel to *CK*, and then draw *HP* perpendicular to median *AG*. When *HP* is extended it intersects *BC* at point *Q*. Curiously, point *H* is the orthocenter of triangle *AGQ*, and that points *F*, *E*, and *Q* are collinear.

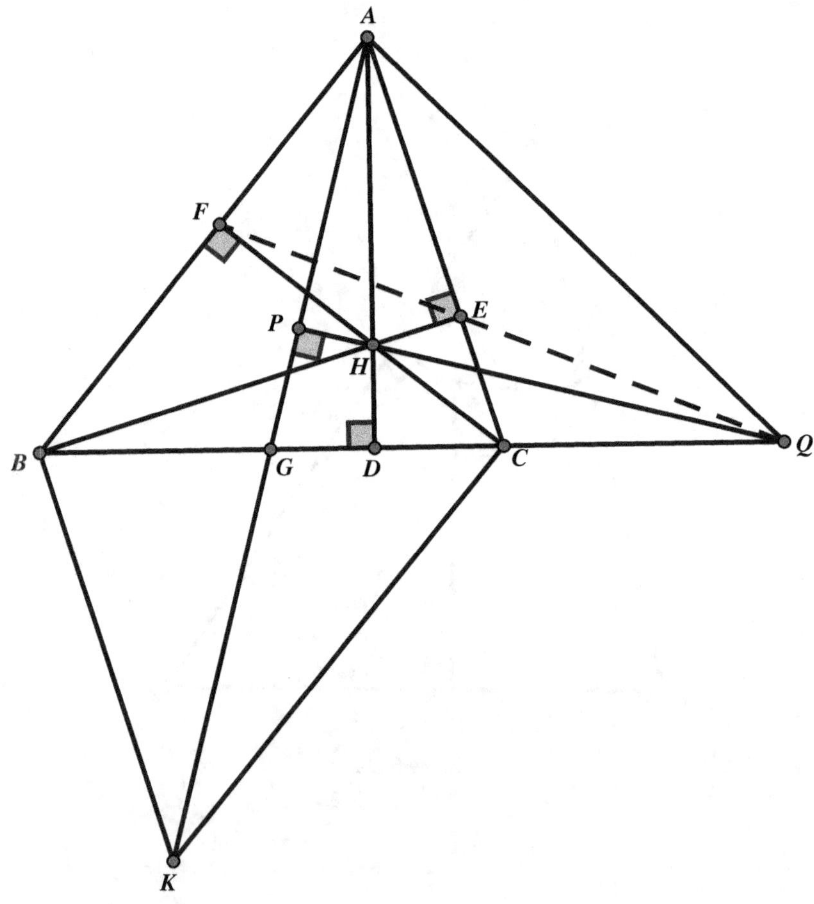

Figure 5-95

Geometrick 102: More Wonders Created by the Altitudes and a Median of a Triangle

In Figure 5-96, we notice a few more relationships that emerge from triangle ABC, such as $\angle CBP = \angle BAP$ and $\angle BCP = \angle CAP$.

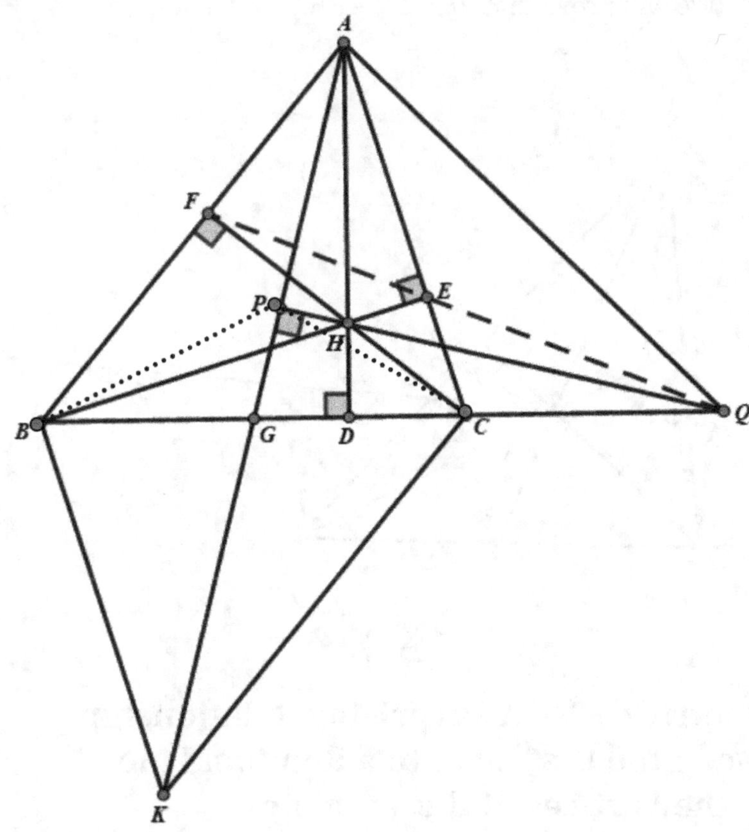

Figure 5-96

Geometrick 103: The Amazing Line Through a Triangle's Centroid

In Figure 5-97, the medians *AM*, *BH*, and *CD* of triangle *ABC* intersect at the centroid *G*. A random line *l* is drawn through *G*. Amazingly, we find that the sum of the distances to line *l* from two vertices on one side, namely, *C* and *B*, is equal to the distance from the third vertex, *A*, to line *l*. We will prove that $AE = CD + BH$.

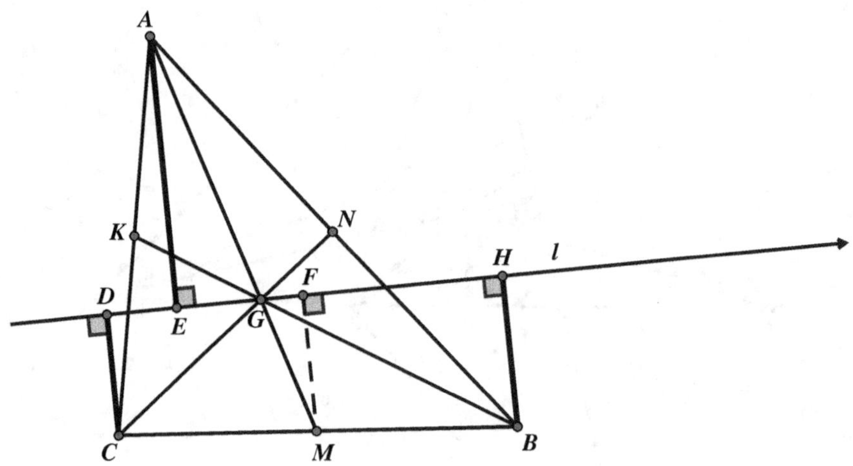

Figure 5-97

Geometrick 104: A Surprising Relationship Between the Distances to a Random Line from the Vertices of the Triangle

The sum of the distances from the three vertices of any triangle to any straight line is three times the distance from the centroid of the triangle to the straight line.

Geometrick 105: An Unusual Relationship Between the Midpoints and the Vertices of the Triangle

The sum of the of the distances from the vertices of a triangle to an external line not parallel to any side of the triangle is equal to the sum of the distances from the midpoints of the sides to the same external line.

Geometrick 106: The Surprising Relationship Between the Circumcenter and Orthocenter of a Triangle

The distance from a vertex of the triangle to the orthocenter is twice that of the distance from the circumcenter to the opposite side. In Figure 5-98, we have triangle ABC with orthocenter H and circumcenter O. We have that $AH = 2OF$, where $OF \perp BC$ at point F.

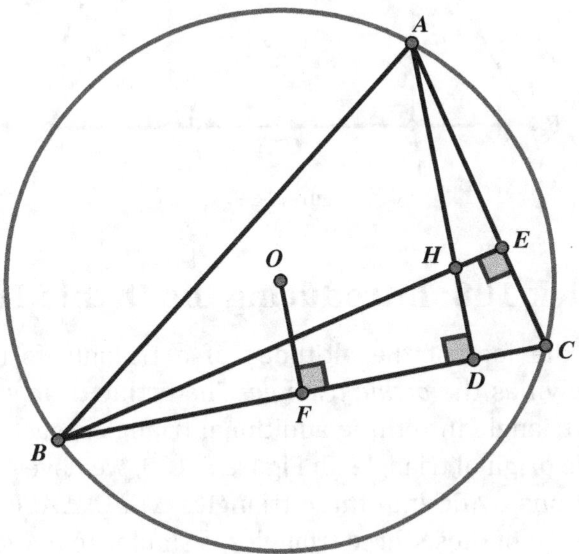

Figure 5-98

Geometrick 107: An Equilateral Triangle and Semicircle Create a Square

A semicircle with diameter *BC* intersects sides *AB* and *AC* at points *D* and *E*, respectively, of equilateral triangle *ABC*. Amazingly, when perpendiculars *DF* and *EG* are drawn to *BC*, as shown in Figure 5-99, the resulting quadrilateral *EDFG* is a rectangle.

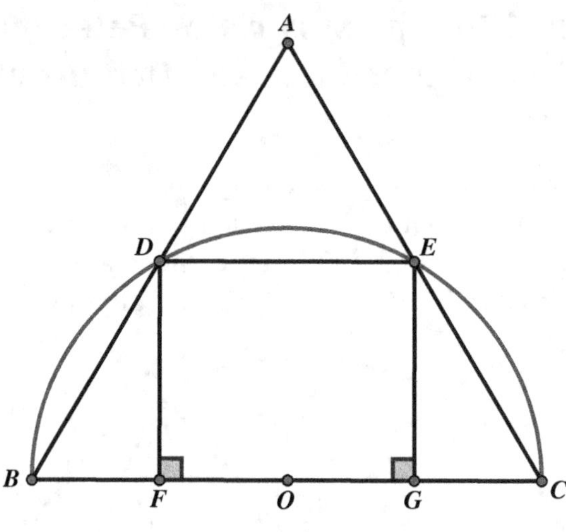

Figure 5-99

Geometrick 108: Introducing the Orthic Triangle

Connecting the feet of the altitudes of a triangle forms another triangle, known as the *orthic triangle*. The orthic triangle partitions the original triangle into three additional triangles, each of which is similar to the original triangle. In Figure 5-100, we have orthic $\triangle DEF$, which partitions $\triangle ABC$ into three triangles $\triangle DEC$, $\triangle AEF$, and $\triangle DBF$. Amazingly, each of these three triangles is similar to $\triangle ABC$.

The property of an orthic triangle established above leads us to even more intriguing relationships. Consider a triangle whose three vertices lie on the sides of a second triangle. This is called the *inscribed triangle* of the second triangle. Now consider the possible inscribed triangles that a given acute triangle can have. Of these, the one with the shortest perimeter is the orthic triangle.

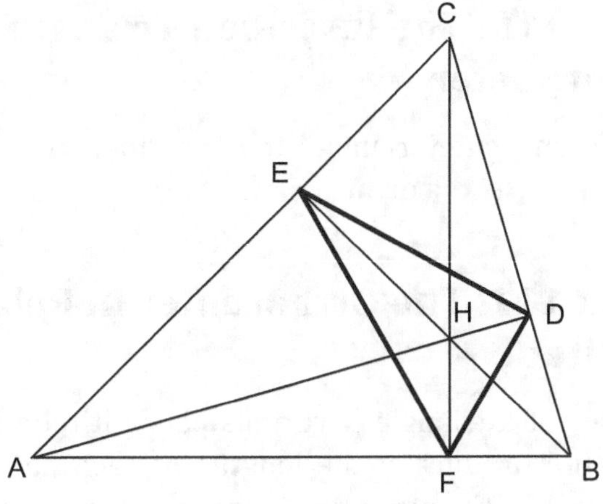

Figure 5-100

Geometrick 109: Triangle Altitudes as Angle Bisectors

The altitudes of a triangle bisect the angles of the orthic triangle. In Figure 5-101, joining the feet D, E, and F of altitudes AD, BE, and CF of triangle ABC form an orthic triangle. The surprising result is that altitudes AD, BE, and CF bisect $\angle EDF$, $\angle DEF$, and $\angle DFE$.

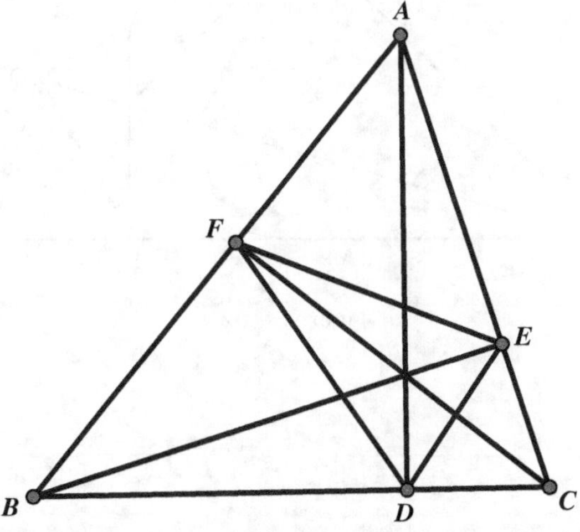

Figure 5-101

Geometrick 110: The Inscribed Triangle with the Shortest Perimeter

Show that, for an acute triangle, the inscribed triangle with the minimum perimeter is the orthic triangle.

Geometrick 111: The Orthocenter in Relation to Each Altitude

The orthocenter of a triangle partitions each altitude into two segments, where the product of the lengths of each pair of segments equals that of each of the other two pairs. In Figure 5-102, one example is that $(CH)(HF) = (AH)(HD)$.

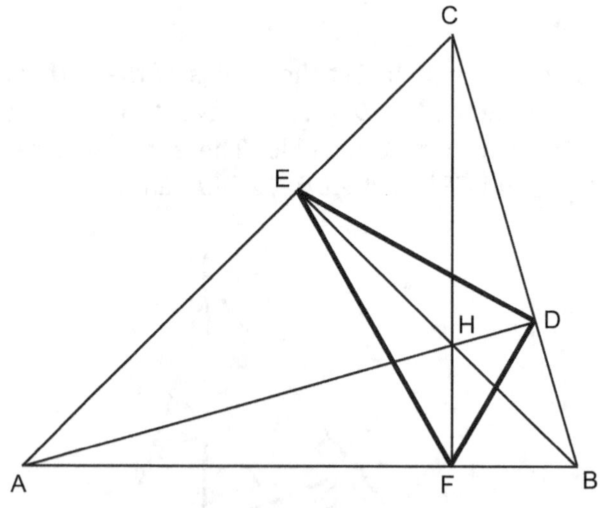

Figure 5-102

Geometrick 112: An Unexpected Bisection

The segment from a triangle's orthocenter to the intersection of the altitude (extended through the foot of the altitude) with the circumcircle is bisected by a side of the triangle. As an example, in Figure 5-103, for $\triangle ABC$, altitude CF intersects the circumcircle O at S. We notice that AB bisects HS.

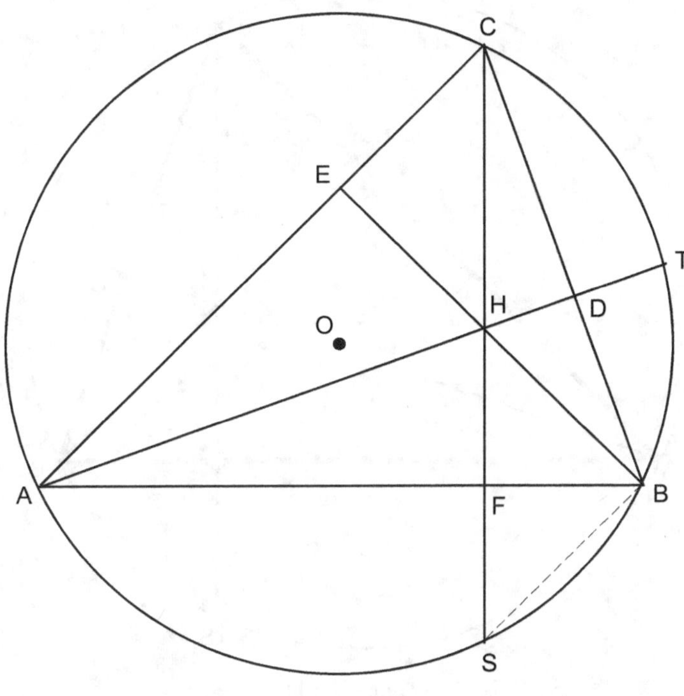

Figure 5-103

Geometrick 113: Triangle Vertices Bisect Arcs on the Triangle's Circumcircle

A vertex of a triangle is the midpoint of the arc of the circumcircle determined by the intersections of two altitudes (extended through their feet) with the circumcircle. Figure 5-104 shows that vertex B is the midpoint of arc TS, where points T and S are the intersections of the circumcircle with altitudes DA and CF, respectively.

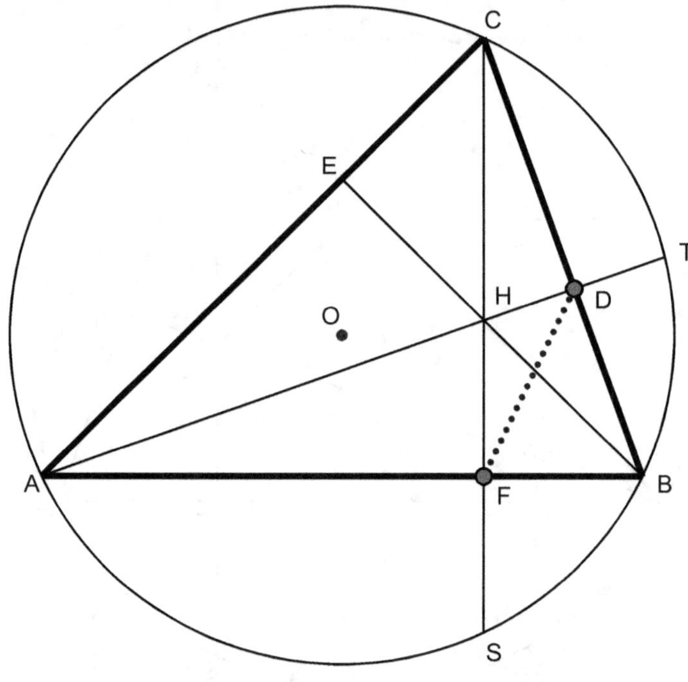

Figure 5-104

Geometrick 114: Proportional Areas

In Figure 5-105, the altitudes of triangle ABC meet at point H, the orthocenter. A right triangle AGB is constructed with AB as the hypotenuse and the right-angle vertex, G, on the extension of CD. It turns out that the area of triangle *AGB* is the mean proportional between the areas of triangle *ABC* and triangle *ABH*.

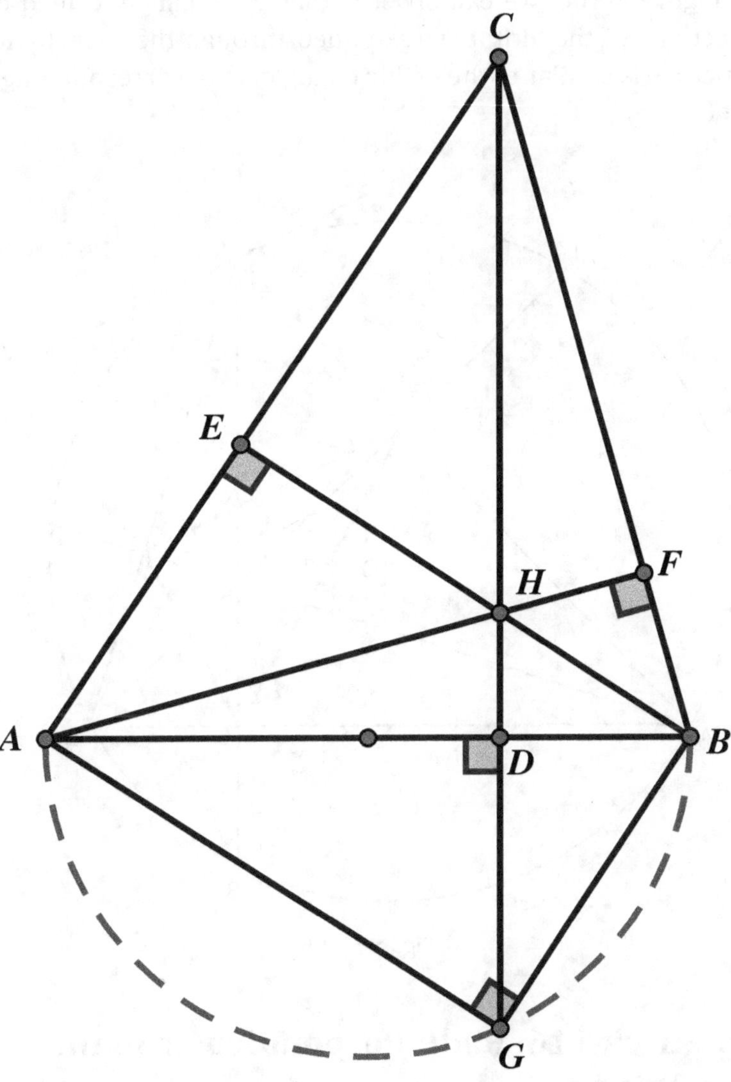

Figure 5-105

Geometrick 115: An Unexpected Similarity to the Orthic Triangle

Using Figure 5-106, we can observe that the triangle formed by the intersections of the altitudes (extended through their feet) with the circumcircle is similar to the orthic triangle, with corresponding sides parallel.

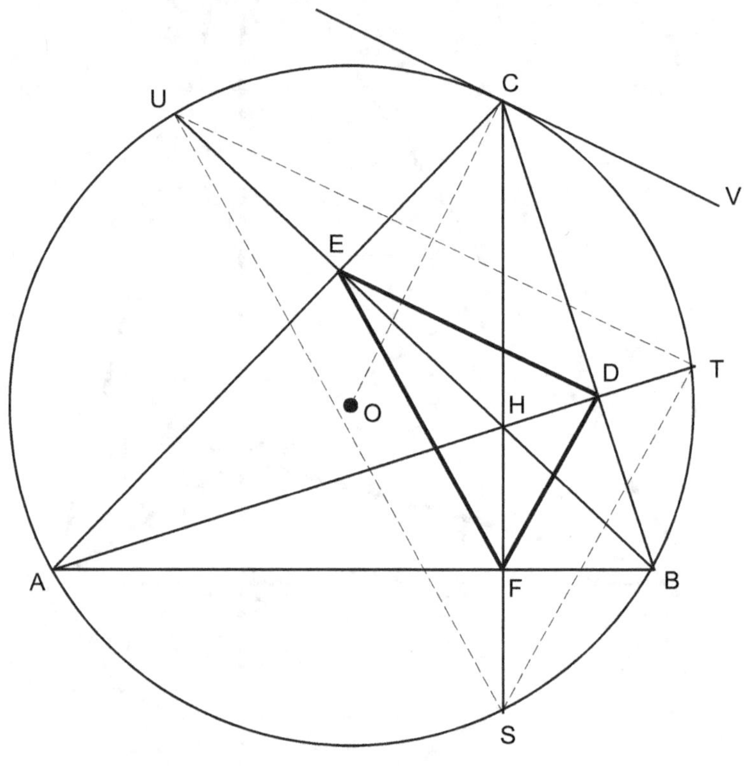

Figure 5-106

Geometrick 116: Radii Perpendicular to the Orthic Triangle

The circumradii, which contain the vertices of the triangle, are perpendicular to the corresponding sides of the orthic triangle. (See Figure 5-106.)

Geometrick 117: Lines Parallel to the Orthic Triangle

The tangents to the circumcircle of a triangle at the vertices of the triangle are parallel to the corresponding sides of its orthic triangle. (See Figure 5-106.)

Geometrick 118: Relationships of the Incenter, Circumcenter, and Orthocenter of a Triangle

In Figure 5-107, triangle *ABC* has incenter point *I*. Lines *AI*, *BI*, and *CI* intersect the circumcircle of *ABC* at points *P*, *Q*, and *R*, respectively, and intersect the sides *QR*, *RP*, and *PQ* of triangle *PQR* at points *X*, *Y*,

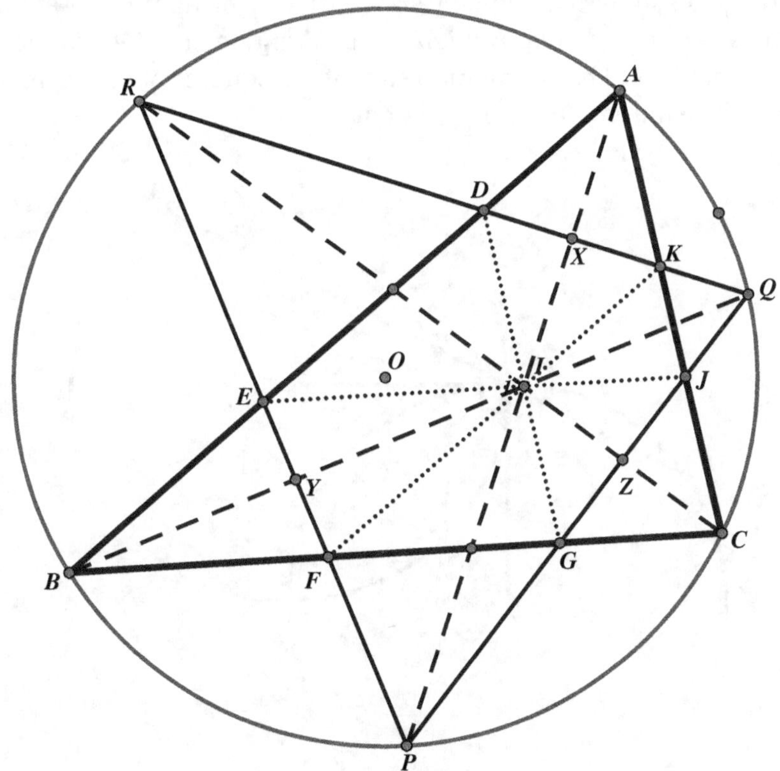

Figure 5-107

and *Z*, respectively. The sides of triangle *PQR* intersect the sides of triangle *ABC* at points *F*, *G*, *J*, *D*, and *E*. Several things evolve from this configuration, all of which involve fascinating proofs. For example, *EJ*, *DG*, and *FK* are all concurrent at point *I*. Point *I* is also the orthocenter of triangle *PQR*. There are several rhombuses to be seen, such as quadrilateral *ADIK*. Also, points *P*, *Q*, and *R* are the circumcenters of triangles *BIC*, *CIA*, and *AIB*, respectively. We should stumble on other unanticipated relationships when doing the proofs, such as that points *Q*, *R*, *Y*, and *Z* are concyclic.

Geometrick 119: A Surprise Constant from a Common Chord of Two Intersecting Circles

Figure 5-108 shows two equally sized circles which intersect at point *A* and have perpendicular radii $AB \perp AC$. A random line *NAM* is a chord for each circle and has a common sum of the squares of the segments. That is, the sum $AN^2 + AM^2$ is constant.

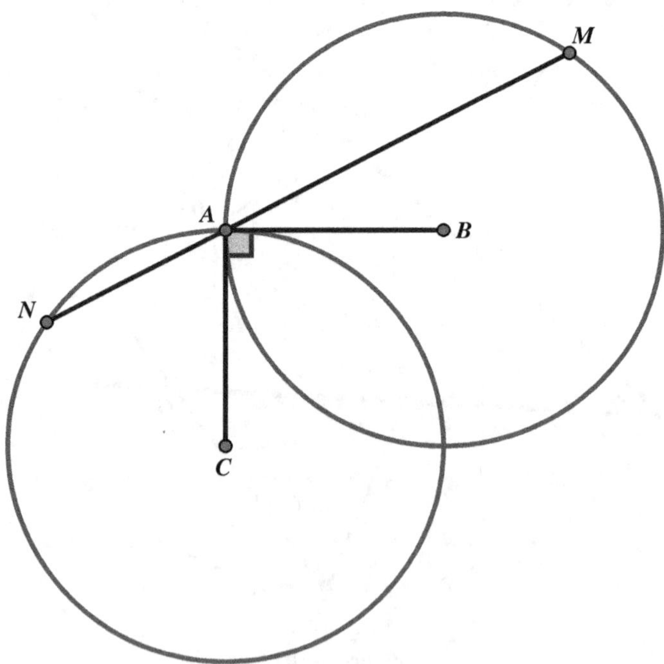

Figure 5-108

Geometrick 120: A Special Chord Partitioned by the Diameter

In Figure 5-109, chord AC in circle O intersects at point B the diameter MN at a 45° angle. Regardless of the place where intersection occurs along with diameter, the relationship $AB^2 + BC^2 = 2r^2$ holds, where r represents the radius of the circle.

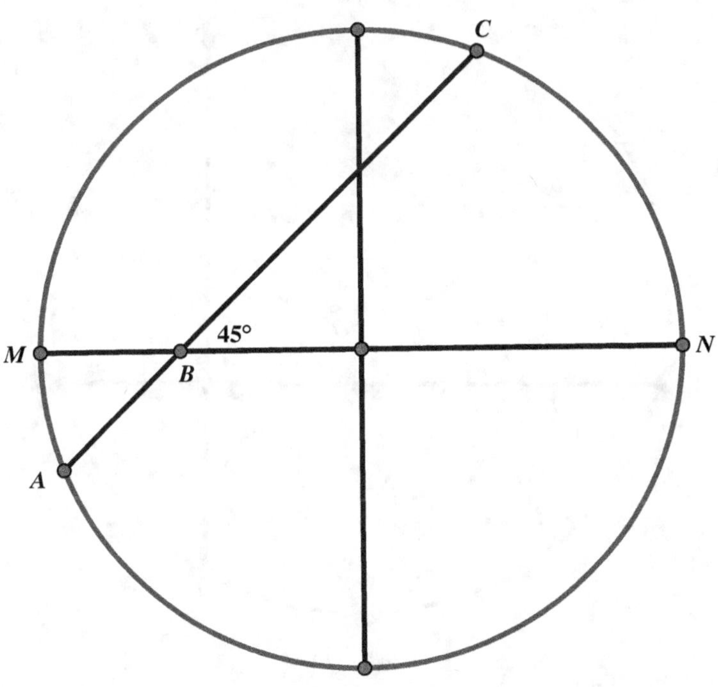

Figure 5-109

Geometrick 121: Perpendicular Chords Intersect with the Constant Sum

The sum of the squares of the four segments formed by two perpendicular chords is four times the square of the radius. In Figure 5-110, where chords $AB \perp CD$, and where $AM = a$, $BM = b$, $CM = c$, and $DM = d$, we have $a^2 + b^2 + c^2 = d^2 = 4r^2$.

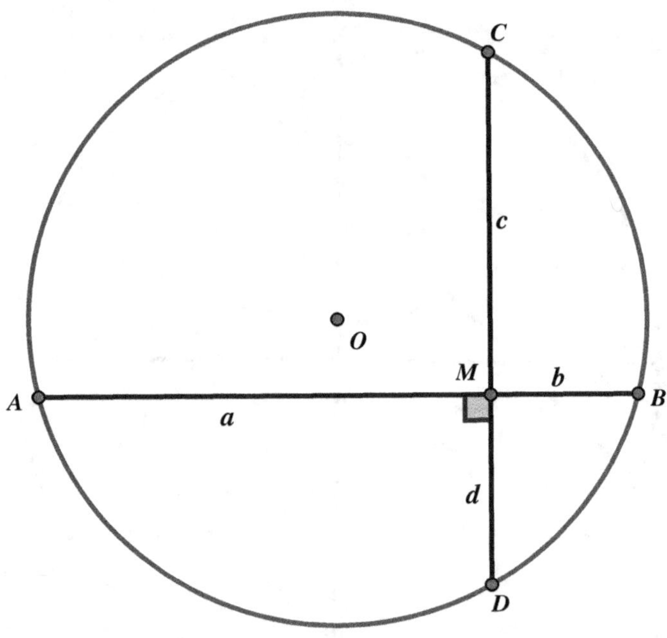

Figure 5-110

Geometrick 122: A Surprising Semicircular Path

We begin with right triangle ABC, where on hypotenuse BC a perpendicular is erected that intersects side AB at point E and side AC extended at point F, as shown in Figure 5-111. A point P is selected on FD such that $DP^2 = DE \cdot DF$. Regardless of where the perpendicular is placed on hypotenuse BC, the point P will always be located on the semicircle of triangle ABC.

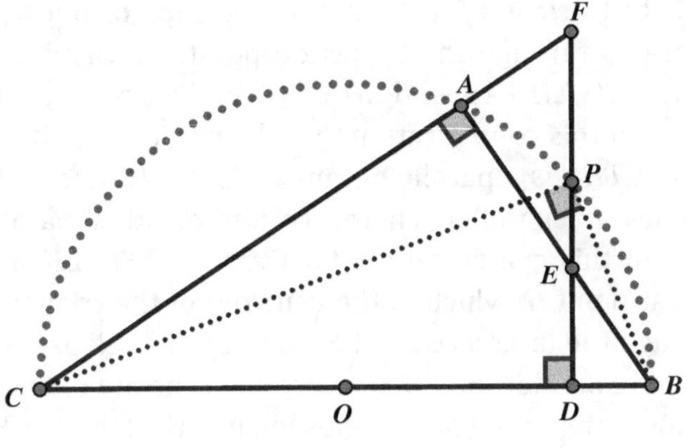

Figure 5-111

Geometrick 123: The Unexpected Property of the Lines Joining the Midpoints of the Bases of the Trapezoid

It is not obvious that the line joining the midpoints E and F of the bases AD and BC of non-isosceles trapezoid $ABCD$ is concurrent with the diagonals of the trapezoid. That is, EF, AC, and BD are concurrent at point G, as seen in Figure 5-112.

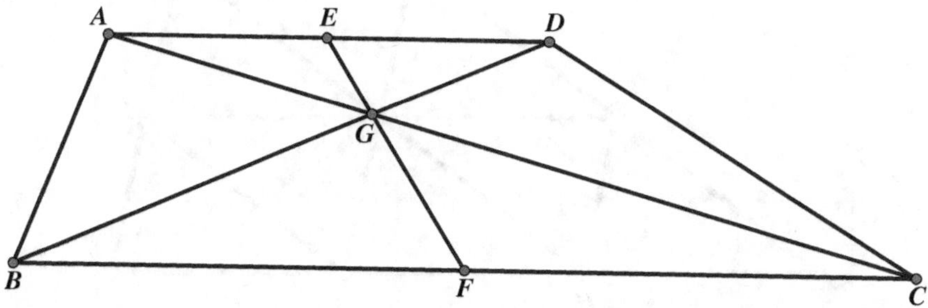

Figure 5-112

Geometrick 124: A Multitude of Properties in the Triangle

In Figure 5-113, triangle *ABC* has three groups of lines parallel to the sides drawn through the trisection points on the sides. That is, $FL \parallel ED \parallel BC$, $DK \parallel LH \parallel AB$, and $EH \parallel EF \parallel AC$. There are a host of noteworthy items in this configuration that the reader is invited to prove. For example, *ALHE* is a parallelogram, as is *AFKD*, since each pair of opposite sides are parallel. There are lots of other parallelograms to be found in this configuration. Furthermore, *FK*, *LH*, and *ED* are concurrent at point *G*, which is the centroid of the triangle. Also, *AG* passes through the intersection of *FD* and *EL*, as well as the intersection of *HD* and *EK*. Additionally, *G* is also the centroid of triangles *FHD* and *ELK*. More interesting discoveries include the fact that *HD* and *EK* are concurrent with the centroid of triangle *HGK*. To conclude, we notice that point *G* is a centroid of a triangle whose vertices are the centroids of triangles *HGK*, *DGL*, and *EGF*.

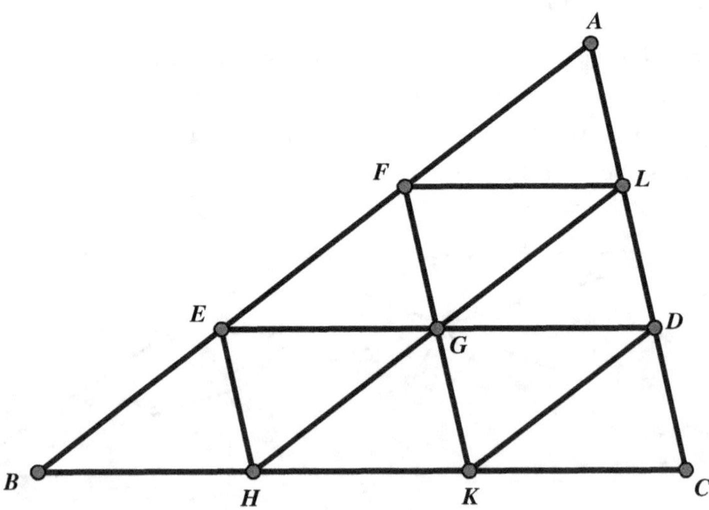

Figure 5-113

Geometrick 125: A Relationship Between the Circumradius and the Inradius

In Figure 5-114, the incenter and circumcenter of triangle ABC are I and O, respectively. Also, radii $IX = r$ and $ON = R$. The distance between these centers is $IO = d$. We find that $d^2 = R^2 - 2rR$.

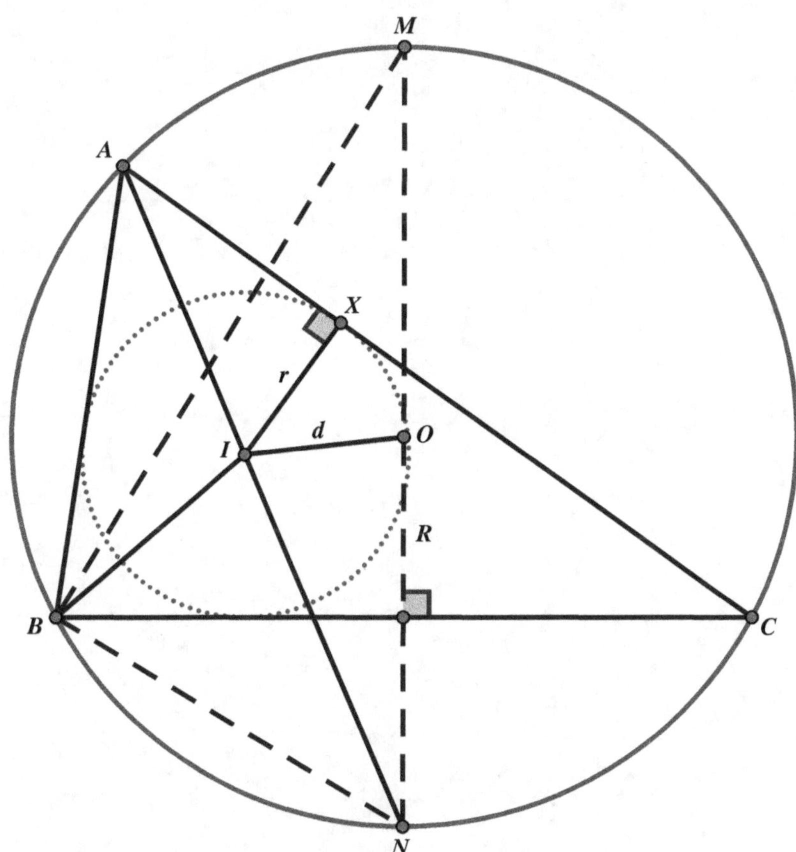

Figure 5-114

Chapter 6

Proofs of the Geometricks

Geometrick 1: Comparing Lengths in a Clever Way

We start off with a rather simple problem and a clever proof. In Figure 6-1, when we draw radius OB, we form triangle ABO. Clearly, $AD + DO < AB + BO$. However, since radii $DO = BO$, the desired result is obvious, namely, that $AD < AB$.

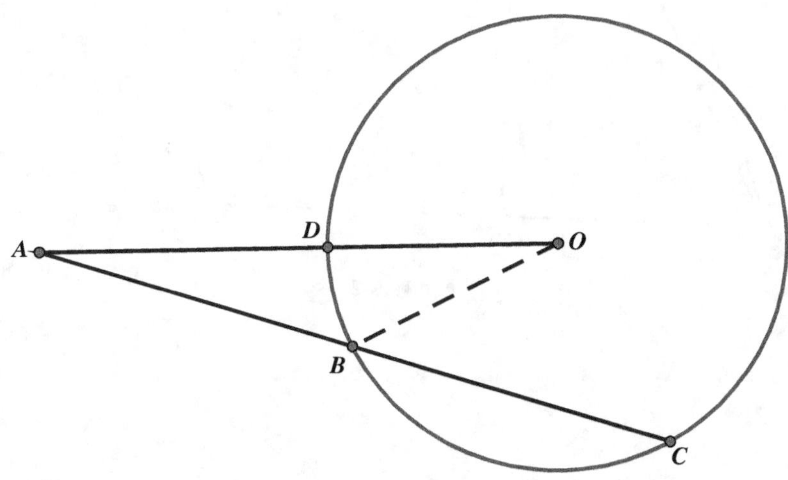

Figure 6-1

Geometrick 2: An Easily Overlooked Situation

Do not be distracted by the various triangles that appear in the figure. This challenge is quite simple. Begin by drawing the perpendicular *PHK* to the two bases, as shown in Figure 6-2. The $area\triangle APB = \frac{1}{2}DC \cdot PH$, and $area\triangle DPC = \frac{1}{2}AB \cdot PK$.

Because $AB = DC$, the difference of these areas is simply $\frac{1}{2}AB(PK - PH) = \frac{1}{2}AB \cdot HK$, which is one-half the area of the parallelogram.

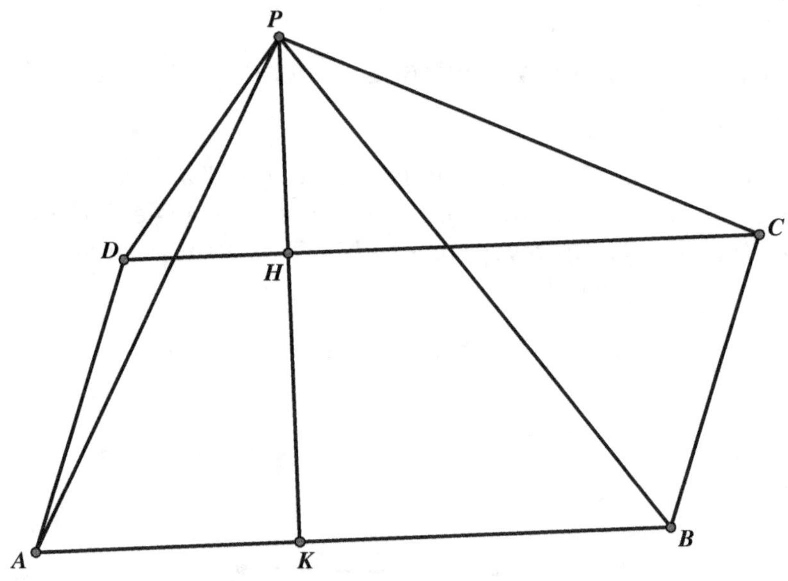

Figure 6-2

Geometrick 3: How a Circle's Diameter Relates to the Altitude of an Inscribed Triangle

When we draw CE, we find that triangle ACE is a right triangle because it is inscribed in a semicircle. Furthermore, as shown in Figure 6-3, we have $\angle ABC = \frac{1}{2}\overset{\frown}{AC} = \angle AEC$. Therefore, $\triangle ADB \sim \triangle ACE$, and so $\frac{AB}{AE} = \frac{AD}{AC}$, or $AB \cdot AC = AD \cdot AE$.

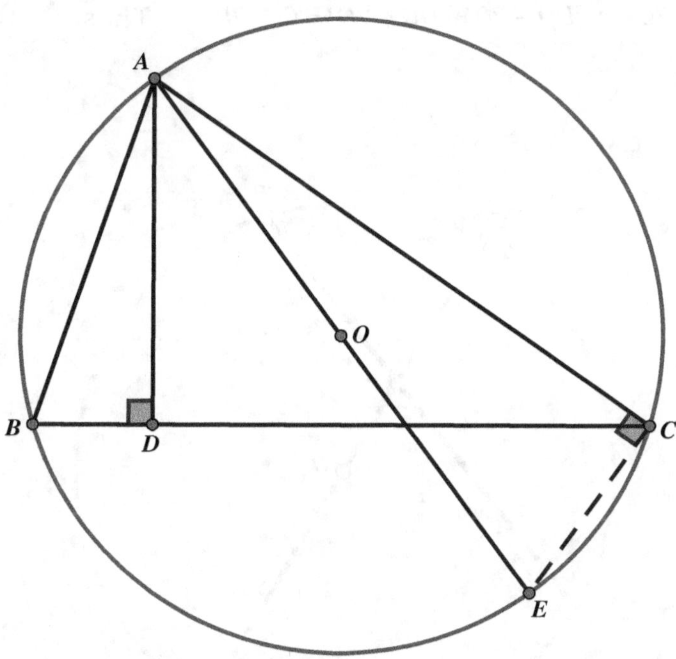

Figure 6-3

Geometrick 4: A Surprise at the Intersection of Two Chords of the Circle

When two chords in a circle intersect, the product of the segments of one chord is equal to the product of the other chord's segments. Therefore, in Figure 6-4, we have $AP \cdot BP = CP \cdot DP$. If we consider the second half of our required equation and use the algebraic of difference of two squares, namely, $(DO)^2 - (PO)^2 = (DO - PO) \cdot (DO + PO)$, we find that $DO - PO = CO - PO = CP$ and $DO + PO = DP$. Therefore, $(DO)^2 - (PO)^2 = (DO - PO) \cdot (DO + PO) = CP \cdot DP$. Thus, $AP \cdot BP = (DO)^2 - (PO)^2$.

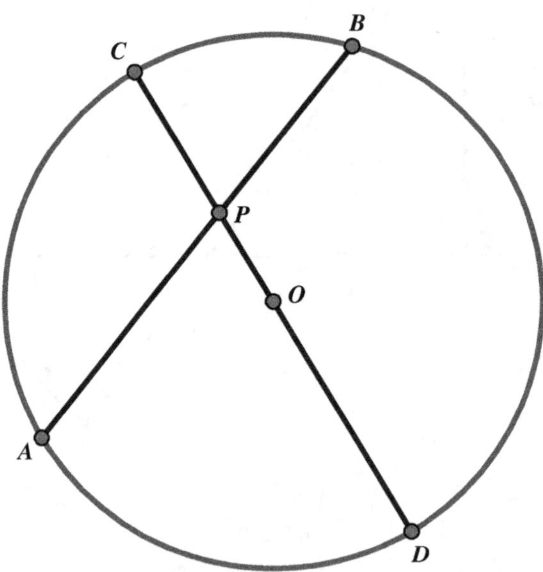

Figure 6-4

Geometrick 5: Two Tangents to a Circle Creating Parallel Lines

In Figure 6-5, the two tangents from a common external point generate equal angles: $\angle DOE = \angle BOE$. Therefore, $\angle GOB = \overset{\frown}{GB} = \frac{1}{2}\overset{\frown}{DB}$. We also know that the inscribed $\angle DAB = \frac{1}{2}\overset{\frown}{DB}$. Thus, we have $\angle GOB = \angle DAB$, and so $OE \parallel AC$.

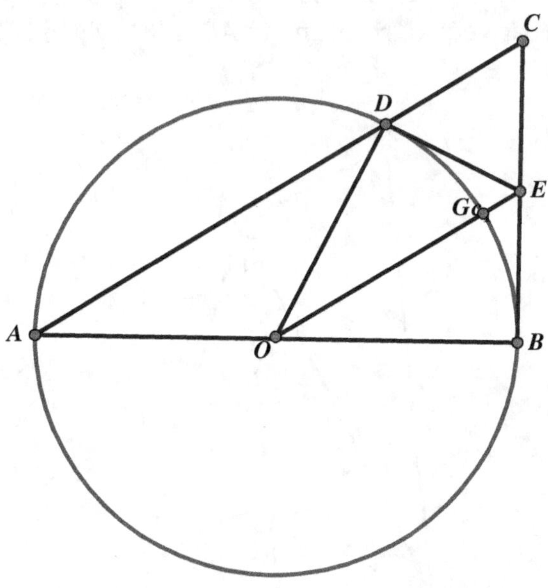

Figure 6-5

Geometrick 6: Finding the Radius of a Circle in an Unusual Fashion

The radius BO, drawn to the point of tangency, is perpendicular to tangent, and therefore, $BO \perp AB$. In addition, we have $AD \perp AB$, as seen in Figure 6-6, and then $BO \parallel AD$. We can show that $\triangle ABO \cong \triangle ACO$ since the two tangents $AB = AC$, and the two radii $BO = CO$. Therefore, $AO \perp BC$, and it follows that since BD and AO are both perpendicular to BC, they must be parallel. We then have quadrilateral $ADBO$ with opposite sides parallel. Therefore, it is a parallelogram and the opposite sides are equal, resulting in $AD = BO$, which is the radius of the circle.

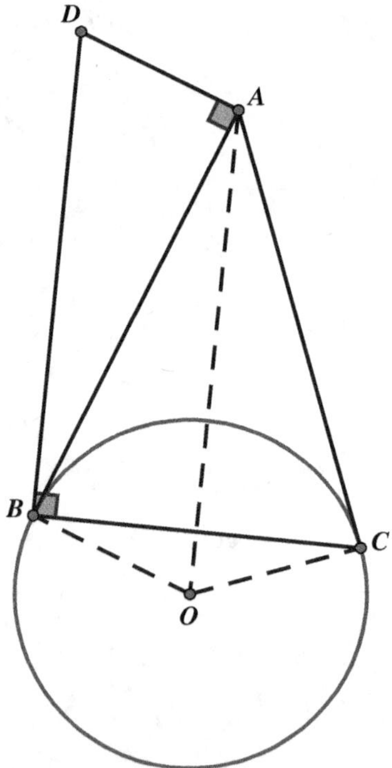

Figure 6-6

Geometrick 7: Overlapping Triangles

We begin by noting in Figure 6-7 that $BD = AB - AD = AC - AE = CE$, and since the supplements of equal angles are equal, $\angle BDG = \angle CEF$. Also, $\angle ABC = \angle ACB$, $\triangle BDG \cong \triangle CEF$ (by ASA). We now have $FC = GB$, which leads us to our desired conclusion, in that $FB = FC - BC = GB - BC = GC$.

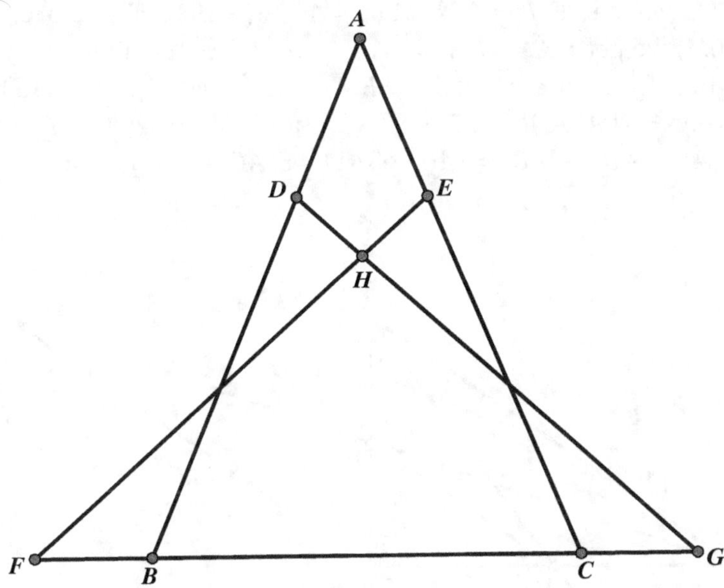

Figure 6-7

Geometrick 8: A Curious Extension of the Pythagorean Theorem

We begin by drawing MN, shown in Figure 6-8, which is parallel to BC since it joins the midpoints of two sides of the triangle ABC. Since $MN \parallel BC$ and $ED \parallel BC$, we then have $MN \parallel ED$. In triangle EAC, since MN bisects AC and is parallel to ED, we have M as the midpoint of EC, or $CM = EM$. The same argument can be made for triangle ABD so that $BN = DN$. By applying the Pythagorean theorem to triangle ACM, we get $CM^2 = AC^2 + AM^2 = AC^2 + (\frac{1}{2}AB)^2$. Therefore, since $CE = 2CM$, we get $CE^2 = 4CM^2 = 4AC^2 + AB^2$. In triangle NAB, the Pythagorean theorem yields $BN^2 = BA^2 + AN^2 = BA^2 + (\frac{1}{2}AC)^2$, or $BD^2 = 4BN^2 = 4BA^2 + AC^2$. Thus, we ultimately get $CE^2 + BD^2 = 4AC^2 + AB^2 + 4BA^2 + AC^2 = 5AC^2 + 5AB^2 = 5(AC^2 + AB^2) = 5BC^2$.

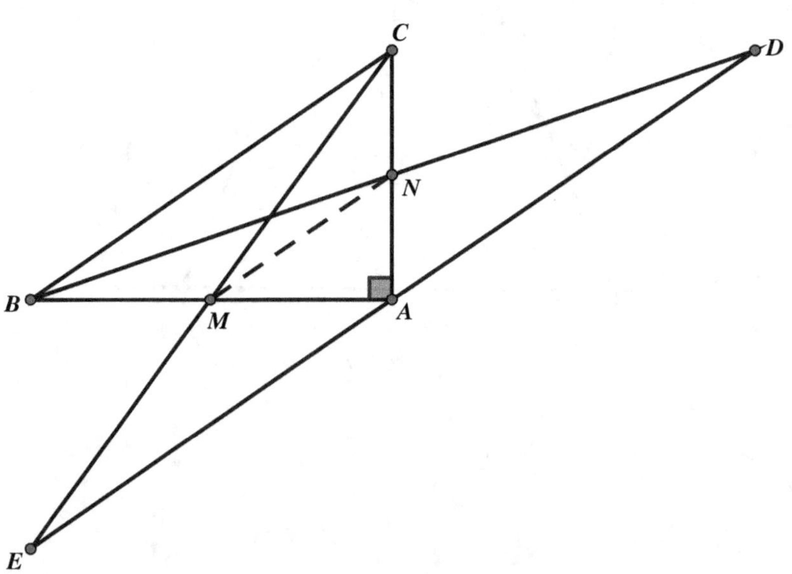

Figure 6-8

Geometrick 9: A Tangent That Bisects a Secant

In Figure 6-9, we have an angle formed by secant and the tangent from an external point is equal to half the difference of the two intercepted arcs, which in this case is $\angle C = \frac{1}{2}(\overset{\frown}{AEB} - \overset{\frown}{EB}) = \frac{1}{2}\overset{\frown}{AE}$. Also, we find that $\angle AEF = \frac{1}{2}\overset{\frown}{AE}$. But since $\angle DEC = \angle AEF$, we have $\angle DEC = \angle C$, so that triangle CDE is isosceles and $CD = ED$. However, the two tangents from point D to circle O are also equal, so that $BD = ED$, and therefore, $BD = CD$, which is what we set out to prove.

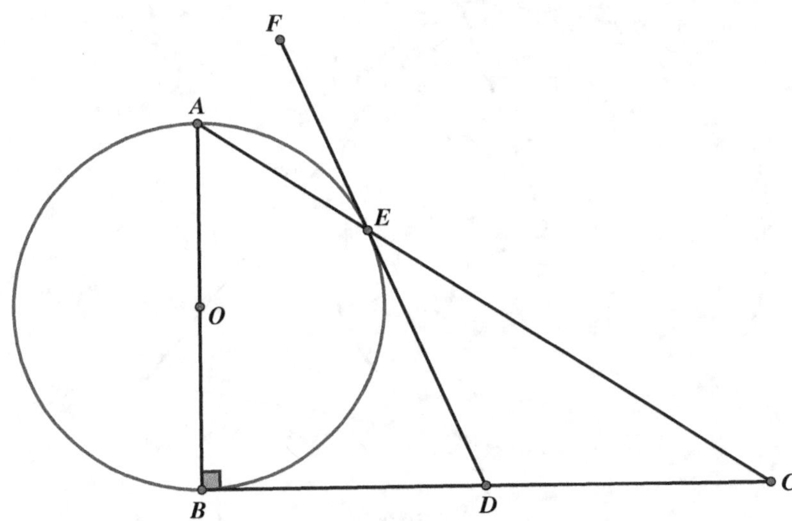

Figure 6-9

Geometrick 10: An Uexpected Placement of the Circle

Since $AB \parallel CD$, we have $\angle ABC + \angle BCD = 180°$. As OB and OC bisect angles ABC and BCD, respectively, we have $\angle OBC + \angle OCB = 90°$, and so $\angle BOC = 90°$, as shown in Figure 6-10. This tells us that the circumscribed circle of triangle BOC with diameter BQC contains point O.

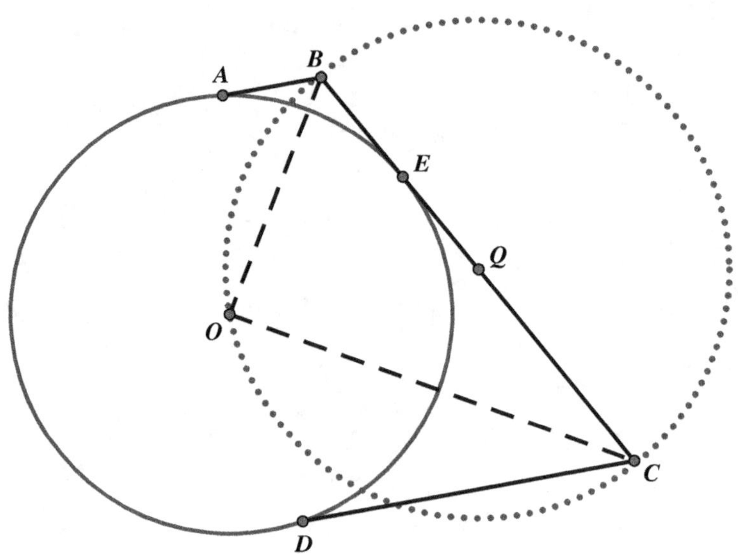

Figure 6-10

Geometrick 11: A Hidden Isosceles Triangle

In Figure 6-11, inscribed angles $\angle ABD = \angle ACD = \frac{1}{2}\overset{\frown}{AED} = 72°$ and $\angle AFB = \angle DFC = \frac{1}{2}(\overset{\frown}{AB} + \overset{\frown}{DC}) = 72°$. Therefore, $\angle ABF = \angle AFB$ and $\angle DCF = \angle DFC$. Thus, the two triangles FDC and FAB are isosceles.

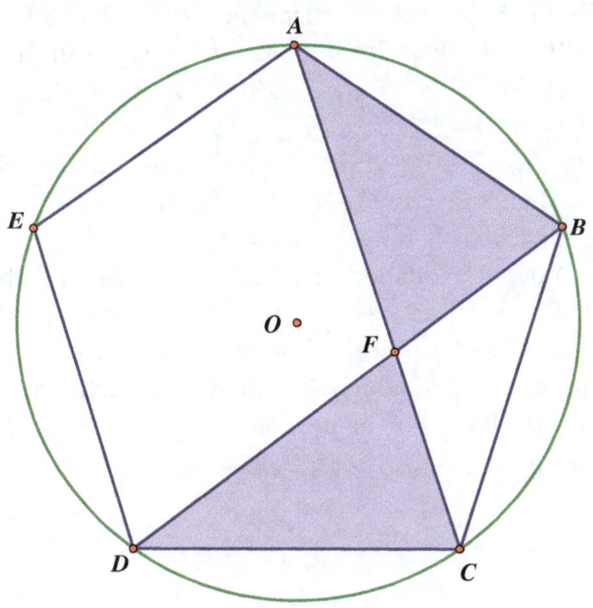

Figure 6-11

Geometrick 12: Sides of a Triangle Are Related to the Angle Bisector

In Figure 6-12, we draw DF so that $BF = BE$, extend AC to point E such that $CD = CE$, and then draw DE. To simplify the proof, we will refer to the angles by the numbers assigned to them in Figure 6-12. Since $\angle 8 = \angle 7$, angle 4 is an exterior angle for triangle BDF, and so $\angle 4 = \angle 7 + \angle B = \angle 8 + \angle B$. In triangle ADF, $\angle 3 = 180° - \angle 2 - \angle 4 = 180° - \angle 2 - \angle 8 - \angle B$. Now let's focus on angle 9 in triangle DBF. The exterior angle yields $\angle 6 + \angle 3 = \angle 8 + \angle B$, or $\angle 6 = \angle 8 + \angle B - \angle 3 = 2(\angle 8) + 2(\angle B) - 180° + \angle 2$.

Furthermore, $180° - \angle 5 = \angle 6 + \angle 1 = \angle 6 + \angle 2 = 2(\angle 8) + 2(\angle B) - 180° + 2(\angle 2)$. Then, $\angle 5 = 360° - 2(\angle 8) - 2(\angle B) - 2(\angle 2)$. Finally, we have $\angle 9 = \frac{1}{2}\angle 5 = 180° - \angle 8 - \angle B - \angle 2 = \angle 3$.

We have now established that $\angle AED = \angle ADF$, so that $\triangle DAE \sim \triangle FAD$. It follows that $\frac{AE}{AD} = \frac{AD}{AF}$ or $\frac{p+b}{AD} = \frac{AD}{c-q}$ and then $AD^2 = (p+b) \cdot (c-q) = pc - pq + bc - bq$. Recall that the angle bisector of a triangle partitions a side to which is drawn proportional to the two adjacent sides, so that $\frac{p}{q} = \frac{b}{c}$, or $pc = bq$. Therefore, $AD^2 = bc - pq$, and so $AD = \sqrt{bc - pq}$.

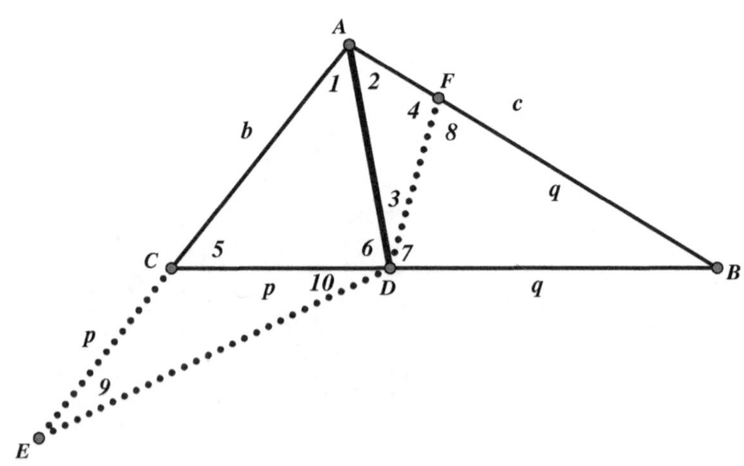

Figure 6-12

Geometrick 13: An Unexpected Angle Bisector

From Figure 6-13, since $AP = AN$, the given information $AB \cdot AM = AN^2$ can be written as $AB \cdot AM = AP^2$, or $\frac{AB}{AP} = \frac{AP}{AM}$, which enables us to establish the similarity $\triangle APM \sim \triangle APB$. Therefore, $\angle AMP = \angle APB$ and $\angle APM = \angle ABP$. The exterior angle of triangle PBM is $\angle AMP = \angle MPB + \angle ABP = \angle MPN + \angle NPB + \angle APM$. Since $AP = AN$, we have isosceles triangle PAN, so the equal base angles yield $\angle APM + \angle MPN = \angle MNP$. Now considering triangle MPN, the exterior angle $\angle AMP = \angle MNP + \angle MPN$. Therefore, $\angle MPN + \angle NPB + \angle APM = \angle MNP + \angle MPN$ (base angles of isosceles triangle PAN), and when we subtract $\angle MPN$ from both sides of the previous equation, we get $\angle NPB + \angle APM = \angle MNP$. However, $\angle MNP = \angle APM + \angle MPN$. Thus, $\angle NPB = \angle MPN$, which indicates that PN bisects $\angle MPB$.

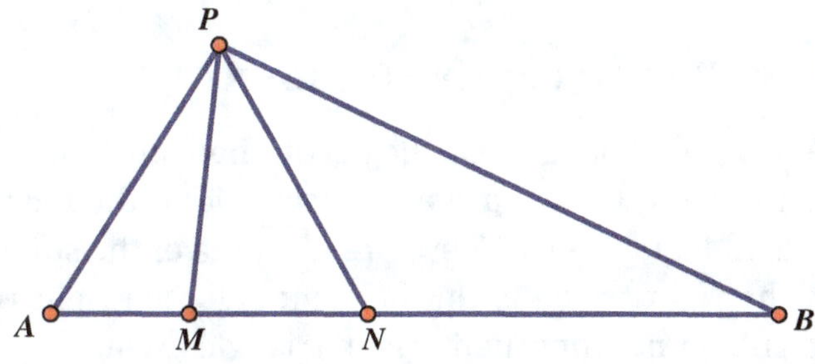

Figure 6-13

Geometrick 14: A Geometric Area Challenge

In Figure 6-14, by joining the midpoints E, J, and H, we find that the area of triangle $ECJ = \frac{1}{4}$ area triangle ABC.

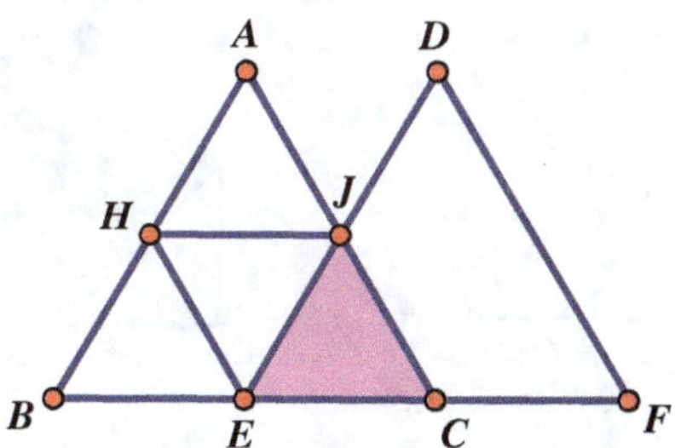

Figure 6-14

There are 4 equilateral triangles congruent to triangle *ECJ*, shown in Figure 6-14, which is the equivalent of one of the large equilateral triangles (see Figure 6-15). The total area covered is equivalent of the area of five equilateral triangles minus one equilateral triangle, that is: $5\left(\frac{2^2\sqrt{3}}{4}\right) - \left(\frac{2^2\sqrt{3}}{4}\right) = 4\left(\frac{2^2\sqrt{3}}{4}\right) = 4\sqrt{3}$.

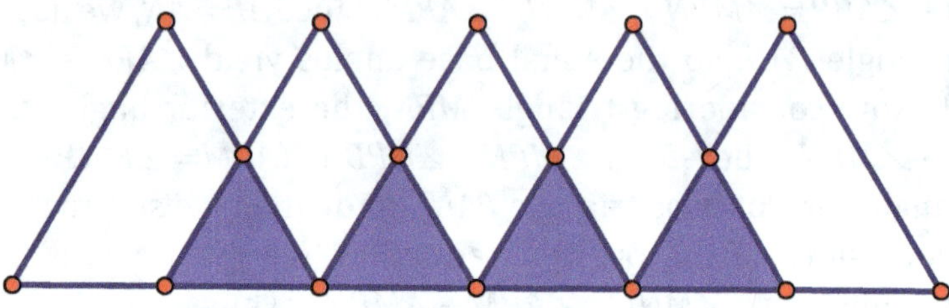

Figure 6-15

Geometrick 15: Deceptive Geometry

Each of the two right triangles can be partitioned into congruent right triangles. On the left, the square comprises two of the four right triangles shown in Figure 6-16. Therefore, the area of the square is $\frac{2}{4} = \frac{1}{2}$ the area of the right triangle. On the right side, the right triangle is partitioned into nine congruent right triangles, where the square comprises four of the nine right triangles. Therefore, the area of the square is $\frac{4}{9}$ of the area of the large right triangle, which is less than $\frac{1}{2}$. Thus, the larger square is the one on the right.

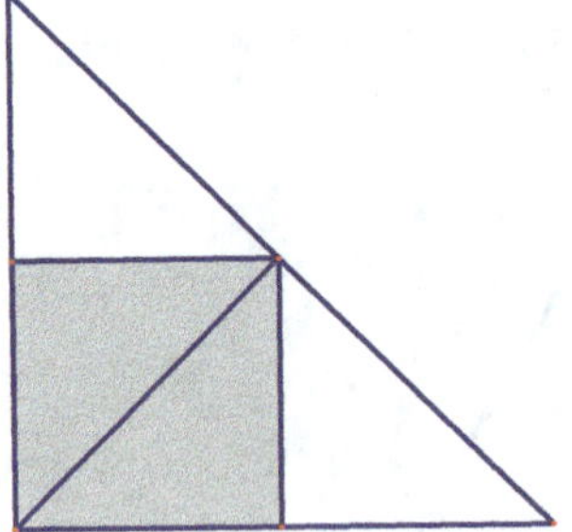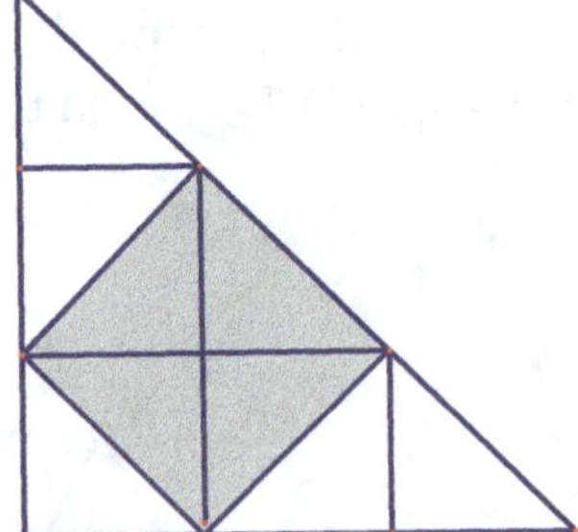

Figure 6-16

Geometrick 16: Determining the Area Between Two Concentric Circles

We will first consider the traditional method of solving this problem and then compare it to a more elegant method. We know that a radius is perpendicular to a tangent at the point of contact, T. Furthermore, a radius perpendicular to a chord divides the chord into two equal segments, so that $AT = BT = 4$. We also know that the area of the region between the two circles (the "doughnut" shape) can be found by obtaining the difference between the areas of the two circles. Thus, the area of this region between the two circles equals $\pi R^2 - \pi r^2 = \pi(R^2 - r^2)$. Now, since $OC = R$ and $OT = r$, $CT = (R - r)$ and $DT = (R + r)$, as can be seen in Figure 6-17.

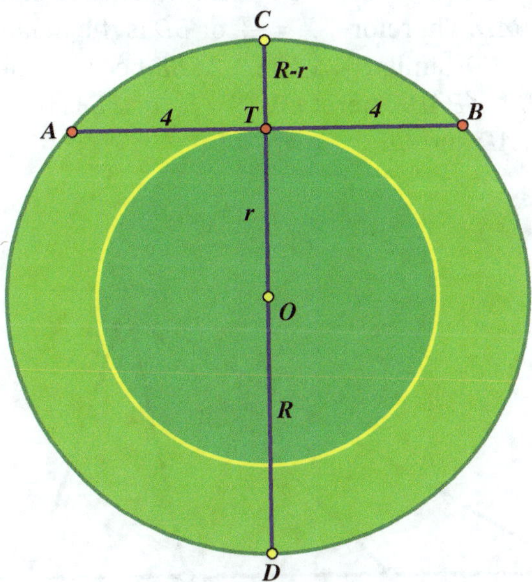

Figure 6-17

Since the product of the segments of two intersecting cords are equal, we have $(R - r)(R + r) = 4 \cdot 4$, so that $R^2 - r^2 = 16$. Therefore, the area of the region between the two circles equals 16π square units.

We can also solve this problem by drawing line segment OA. This creates a right triangle ATO, in which $(OA)^2 = (OT)^2 + 4^2$, and once again $R^2 - r^2 = 16$.

However, we can also look at this problem by considering an extreme case. Let's assume that the smaller circle gets smaller and smaller, until it becomes a point and coincides with point O. Then AB becomes a diameter of the larger circle, and the area of the region between the two circles is merely the area of the larger circle, $\pi R^2 = 16\pi$, which is a considerably more elegant solution to the problem.

Geometrick 17: Unexpected Mean Proportional

We find in Figure 6-18 that $\triangle ABC \sim \triangle CBD$, as they share angle ADB and $\angle CBD = \angle BAD$. Therefore, $\frac{AD}{BD} = \frac{BD}{CD}$, or BD is the mean proportional between AD and CD. Similarly, $\triangle ABE \sim \triangle CBE$ because they share angle AEB, and $\angle CBE = \angle BAD$. Therefore, $\frac{AD}{BE} = \frac{BE}{CD}$, or BE is the mean proportional between AD and CD.

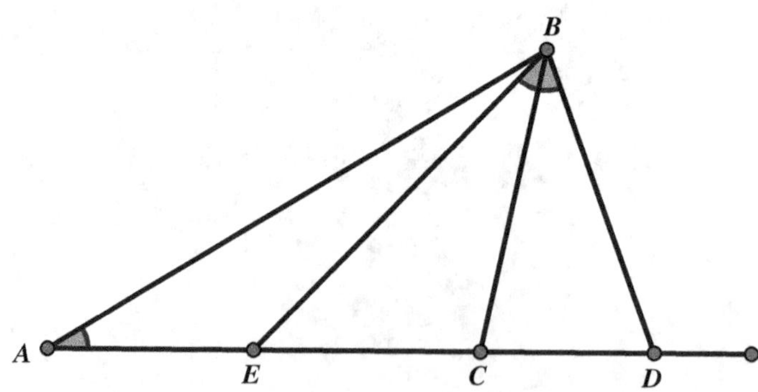

Figure 6-18

Geometrick 18: The Surprise Angle Between an Altitude and Angle Bisector

The angle bisector AE of triangle ABC yields $\angle BAE = \angle CAE$, as can be seen in Figure 6-19. Also, $\angle DAE = 90° - \angle AED = \angle ADE - \angle AED$. The exterior angle $\angle ADE$ of triangle ABD yields $\angle ADE = \angle B + \angle BAD$. The exterior angle $\angle AED$ of triangle AEC yields $\angle AED = \angle C + \angle CAE$. Therefore, $\angle DAE = 90° - \angle AED = \angle ADE - \angle AED = \angle B + \angle BAD - \angle C - \angle CAE$. Since $\angle BAD = \angle BAE - \angle DAE$ and AE is an angle bisector so that $\angle CAE = \angle BAE$, we then have $\angle DAE = \angle B + \angle BAE - \angle DAE - \angle C - \angle BAE = \angle B - \angle DAE - \angle C$. Adding $\angle DAE$ to both sides of the previous equation gives us $2\angle DAE = \angle B - \angle C$, or $\angle DAE = \frac{1}{2}(\angle B - \angle C)$.

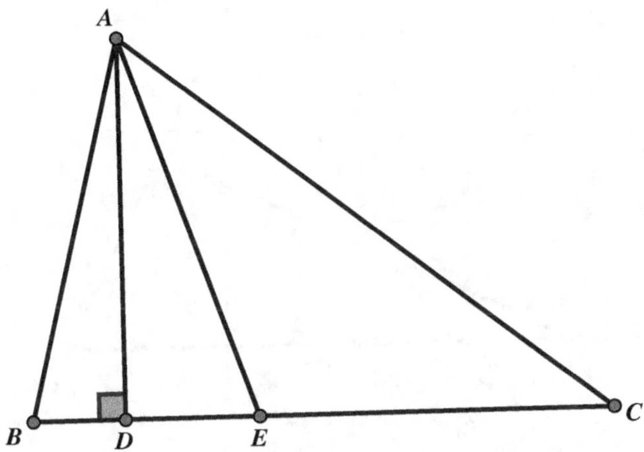

Figure 6-19

Geometrick 19: The Perpendicular to the Hypotenuse of a Right Triangle

In Figure 6-20, since angles M and C are both complementary to angle B, we have $\angle M = \angle C$. Therefore, right triangles $\triangle BPM \sim \triangle PCN$. Thus, $\frac{PM}{PB} = \frac{PC}{PN}$, which gives us the sought-after product $PN \cdot PM = PB \cdot PC$.

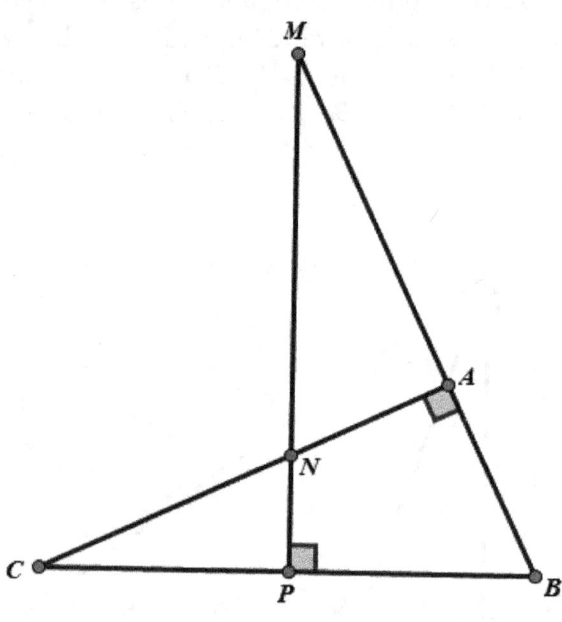

Figure 6-20

Geometrick 20: An Unexpected Chord Product

Begin by drawing chord *BE* and noticing that triangle *CBE* is a right triangle because it is inscribed in a semicircle, as can be seen in Figure 6-21. Furthermore, angle *A* and angle *E* are both measured by one-half arc *CB*. Therefore, $\triangle CBE \sim \triangle CDA$ and $\frac{AC}{CD} = \frac{CE}{BC}$, or $AC \cdot BC = CD \cdot CE$.

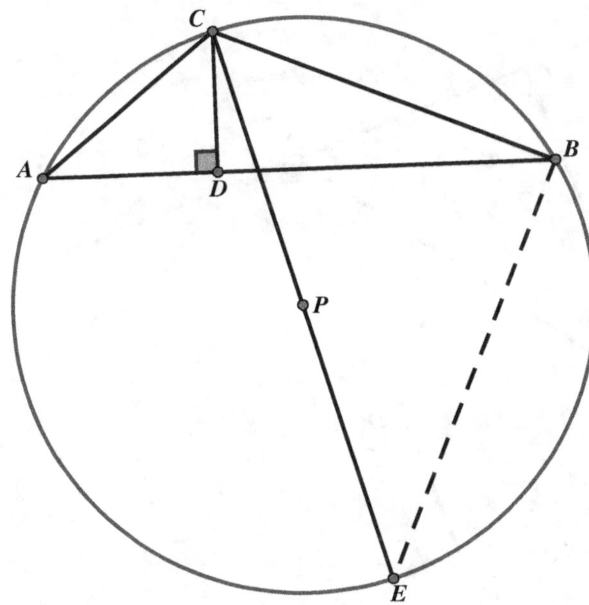

Figure 6-21

Geometrick 21: The Power of a Point on a Circle

In Figure 6-22, we begin by drawing BD and noticing that $\angle CAD$ and $\angle ABD$ both measure one-half of $\overset{\frown}{AD}$, and so they are equal. This allows us to establish that right triangles $\triangle ACD \sim \triangle ABD$. Therefore, we have $\frac{CD}{AD} = \frac{AD}{AB}$.

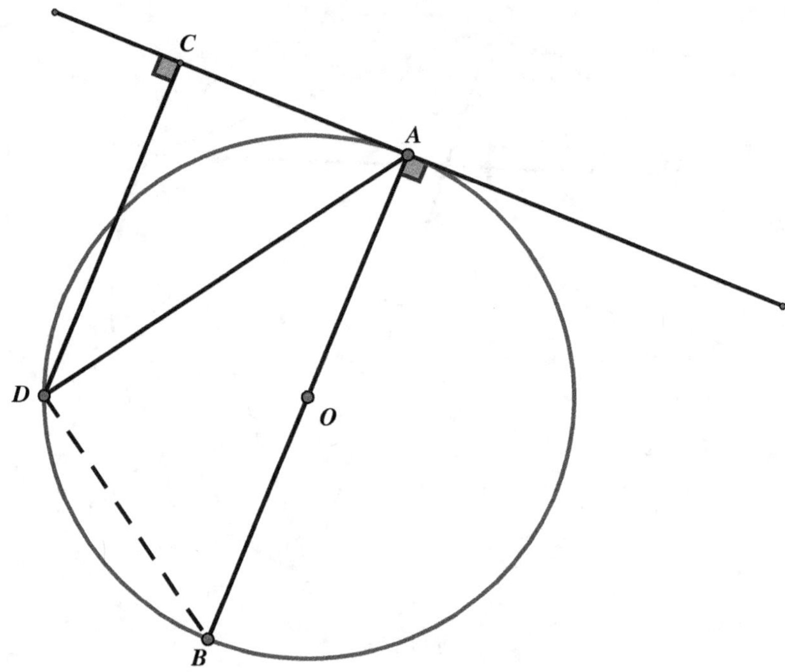

Figure 6-22

Geometrick 22: A Strange Relationship of Circle Lines

We begin by extending OC to point E on circle O, and then drawing AE, as shown in Figure 6-23. When we draw AO and BO, we form a right angle AOB since it is inscribed in a semicircle. Similarly, angle CAE is also a right angle because it is also inscribed in a semicircle. We notice that AO and OC are radii of circle O. Therefore, $AO = OC$, which makes an isosceles triangle AOC and $\angle CAO = \angle ACO$. In right triangle AOB, angle ABO is complementary to angle OAC. Similarly in right triangle EAC, angle AEC is complementary to angle OCA, and therefore, $\angle AEC = \angle ABO$. Thus, $\triangle EAC \sim \triangle OAB$, and then $\frac{AB}{EC} = \frac{OA}{AC}$, which is $\frac{AB}{2OC} = \frac{OC}{AC}$. Therefore, $AB \cdot AC = 2OC^2$.

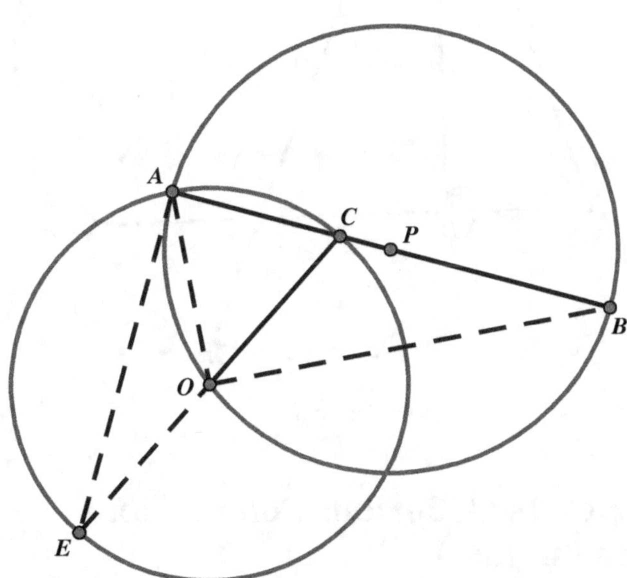

Figure 6-23

Geometrick 23: An Unusual Product of Triangle Sides

In Figure 6-24, triangle ABC is inscribed in a circle with diameter AE. Also, AD is the altitude to BC. We notice that $\angle ABD = \angle AEC$, since both angles are measured by $\overset{\frown}{AC}$. Therefore, the two right triangles ABD and AEC are similar, which gives us $\frac{AB}{AE} = \frac{AD}{AC}$, or $AB \cdot AC = AD \cdot AE$. This is what we originally needed to prove.

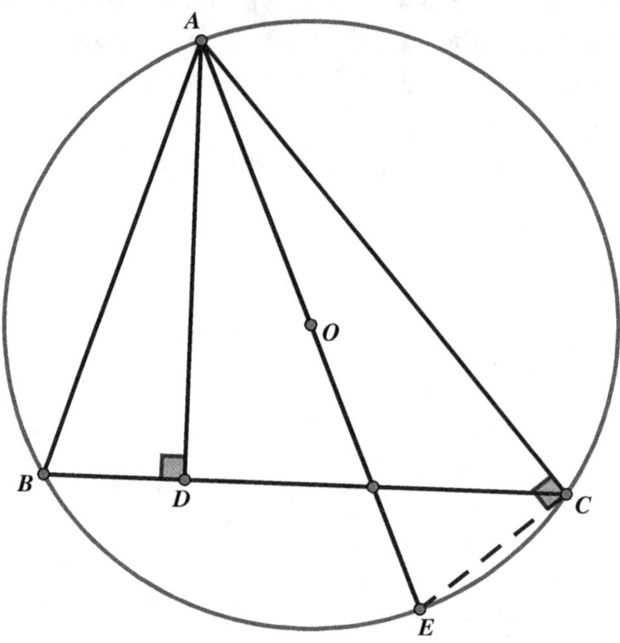

Figure 6-24

Geometrick 24: A Curious Point on an Inscribed Circle

In Figure 6-25, triangle ABC is inscribed in circle O with radius r, and the circle is inscribed in triangle EFG. Using the relationship established in Geometrick 23, (that is, the product of two sides of the triangle is equal to the product of the altitude to the third side and the diameter of the circumscribed circle) we get the following relationships with the lines measures as labeled in figure: $2ra = MA \cdot MB$, $2rb = MA \cdot MC$, and $2rc = MB \cdot MC$. When we multiply the three equations together, we get $8r^3abc = MA^2 \cdot MB^2 \cdot MC^2$.

Now using Geometrick 21, which states that from any point on a circle the distance to the point of contact of a tangent line is the mean proportional between the diameter and the distance from that point to the tangent line, we get the following: $2rd = MA^2$, $2re = MB^2$, and $2rf = MC^2$. When we multiply these together, we get $8r^3def = MA^2 \cdot MB^2 \cdot MC^2$. Therefore, $8r^3abc = 8r^3def$, so that $abc = def$, which is what we had to prove.

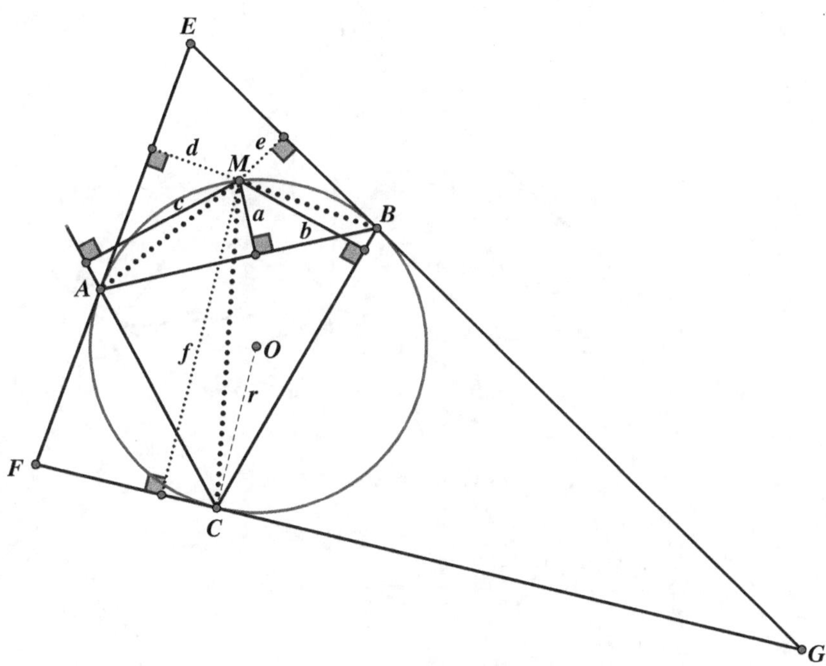

Figure 6-25

Geometrick 25: A Common Chord Is a Mean Proportional

The angle formed by a tangent and a chord from the point of tangency is one-half the measure of the intercepted arc. Therefore, in Figure 6-26, $\angle CAB = \frac{1}{2}\overparen{AGB}$ and inscribed angle $\angle ADB = \frac{1}{2}\overparen{AGB}$. Therefore, $\angle CAB = \angle ADB$, and in like fashion we get $\angle BAD = \angle ACB$. Thus, $\triangle ABC \sim \triangle DAB$ and so $\frac{BC}{AB} = \frac{AB}{BD}$, which establishes that AB is a mean proportional between BC and BD.

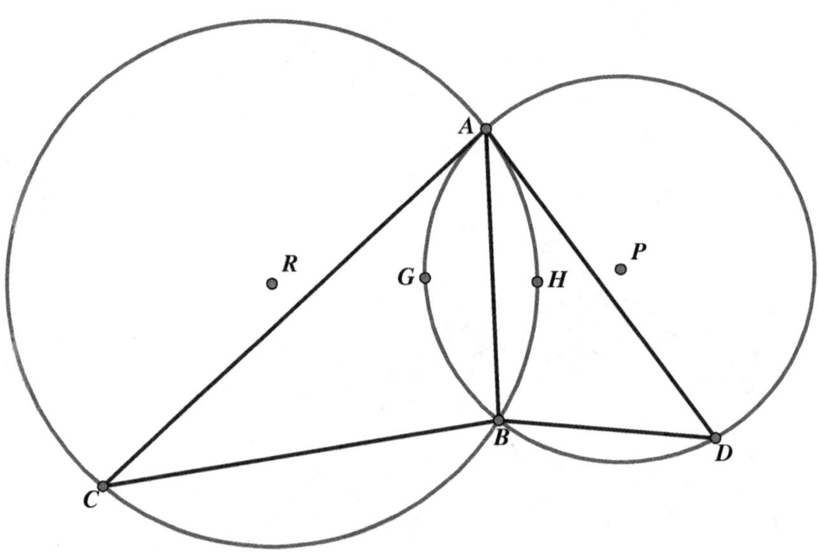

Figure 6-26

Geometrick 26: Unusual Bisection of a Right Angle

Since the diagonals of the square are perpendicular, $\angle AGB = 90°$. Therefore, quadrilateral $ACBG$ is cyclic because the opposite angles are right angles, as shown in Figure 6-27. Since the diagonals of the square bisect each other, $AG = BG$ and, therefore, $\overset{\frown}{AG} = \overset{\frown}{BG}$. We then have $\angle ACG = \frac{1}{2}\overset{\frown}{AG}$, and $\angle BCG = \frac{1}{2}\overset{\frown}{BG}$, and so $\angle ACG = \angle BCG$.

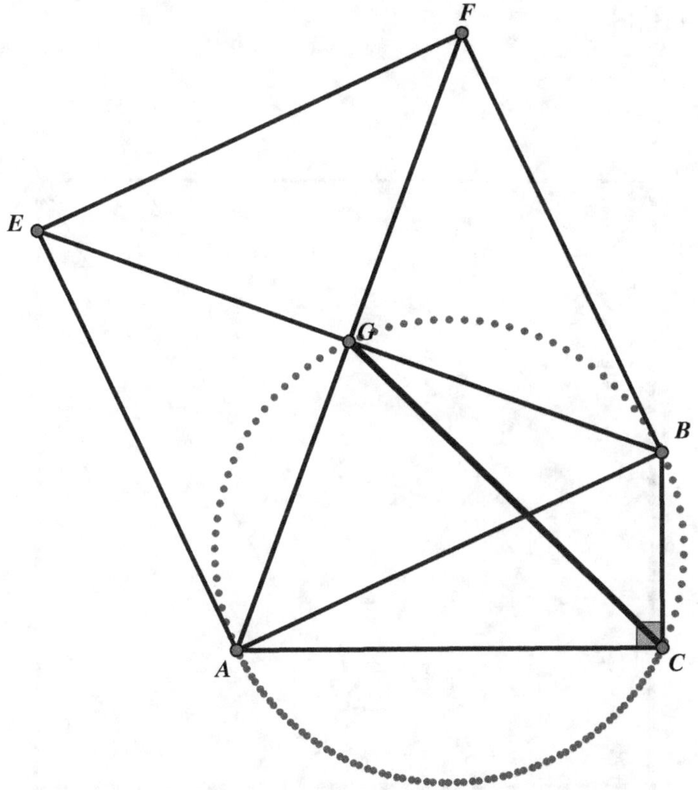

Figure 6-27

Geometrick 27: The Relationship Between an Equilateral Triangle and a Square

Using Figure 6-28, we notice that the angle measures are $\angle DCE = 60°$, $\angle FCB = 30°$, and $\angle BFC = 60°$. If we let $BF = 1$, then for the 30-60-90 triangle BCF, we have $BC = \sqrt{3}$, and $CF = 2$. Therefore, the area of triangle $BCF = \frac{1}{2}BC \cdot BF = \frac{\sqrt{3}}{2}$. Since $DC = BC = \sqrt{3}$, the area of equilateral triangle $CDE = \frac{\left(\sqrt{3}\right)^2 \sqrt{3}}{4} = \frac{3\sqrt{3}}{4}$. Therefore, we arrive at the required ratio $\frac{\Delta CDE}{\Delta BCF} = \frac{\frac{3\sqrt{3}}{4}}{\frac{\sqrt{3}}{2}} = \frac{3}{2}$.

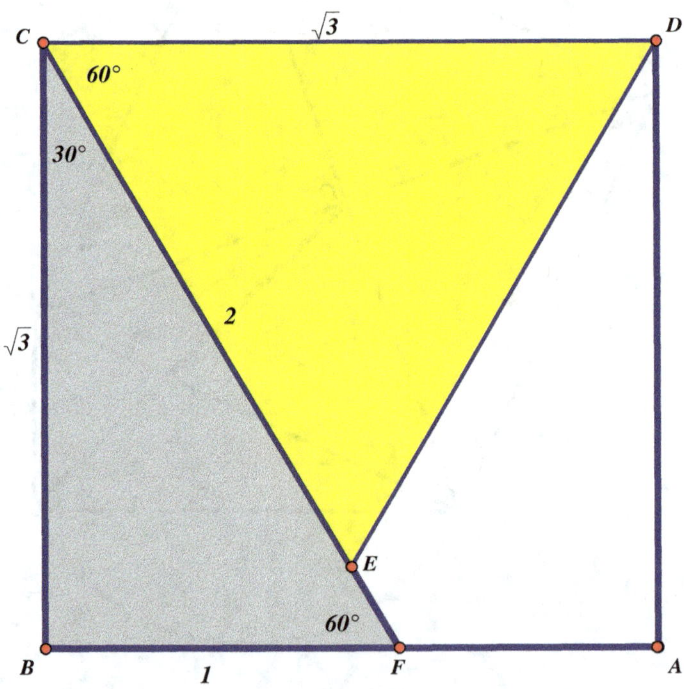

Figure 6-28

Geometrick 28: Discovering an Equilateral Triangle in a Square

This proof is different from most other proofs in the book. We are going to use symmetry to show that the equilateral triangle can coexist with the given triangle *AFB*. We begin by drawing two quarter circles centered at points *C* and *D*, as shown in Figure 6-29. Clearly, triangle *BCF* and triangle *ADF* are isosceles, since they have two equal sides of the same length as the sides of the given square. Then, $\angle CFB = \angle CBF = 90° - 15° = 75°$. Therefore, $\angle BCF = 180° - 150° = 30°$, and it follows that $\angle DCF = 90° - 30° = 60°$. Since triangle *CDF* is isosceles with the base angle of 60°, the vertex angle is also 60°. Therefore, triangle *CDF* is equilateral.

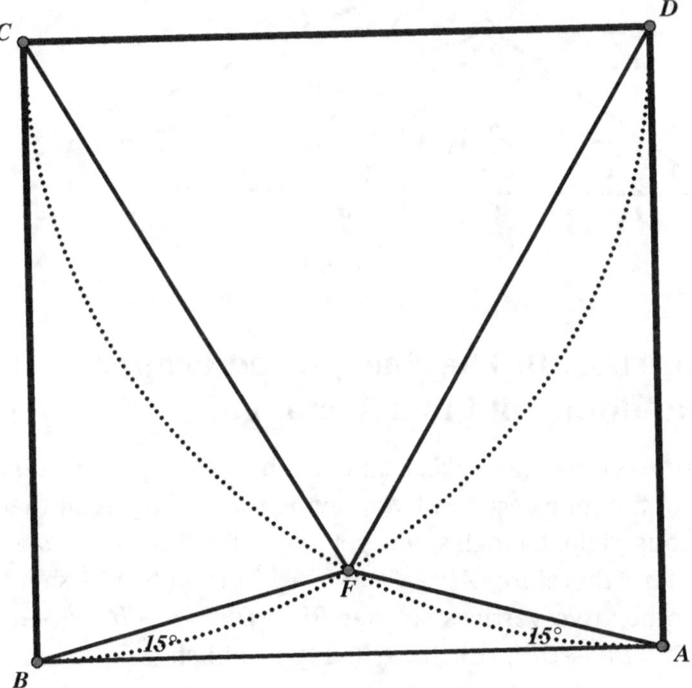

Figure 6-29

Geometrick 29: A Significant Point on the Median of a Triangle

Since the areas of the triangles we seek to compare share the same base *BC*, we need to only compare their altitude lengths. In Figure 6-30, we construct *DPE* parallel to *BC*, which then tends to bisect *AB*, *AF*, *AM*, and *AC*. Therefore, $PG = \frac{1}{2}AF$. Thus, when two triangles share the same base and one altitude is half the length of the other, the smaller triangle has half the area of the larger triangle.

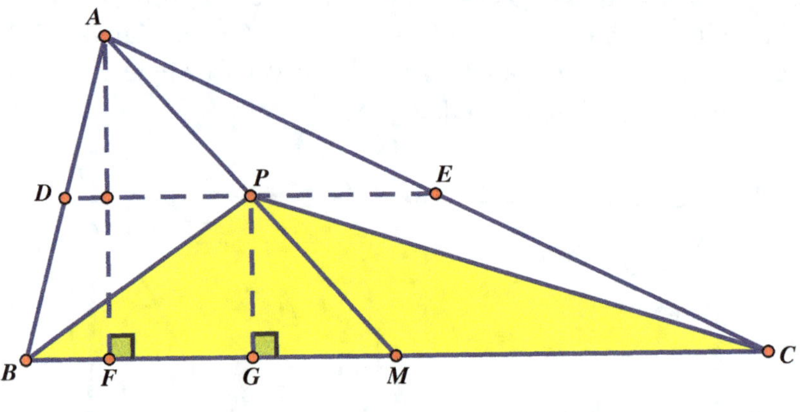

Figure 6-30

Geometrick 30: The Unexpected Property of a Random Point in a Rectangle

In Figure 6-31, we have *a*, *b*, *c*, and *d* representing the distance from point *P* to the four vertices. By applying the Pythagorean theorem to the various right triangles, we get the following: $AP^2 = a^2 + b^2$, and $CP^2 = c^2 + d^2$; therefore, $AP^2 + CP^2 = a^2 + b^2 + c^2 + d^2$. The same is true for the other two vertices so that $BP^2 + DP^2 = a^2 + b^2 + c^2 + d^2$. Thus, $AP^2 + CP^2 = BP^2 + DP^2$, which we had to establish.

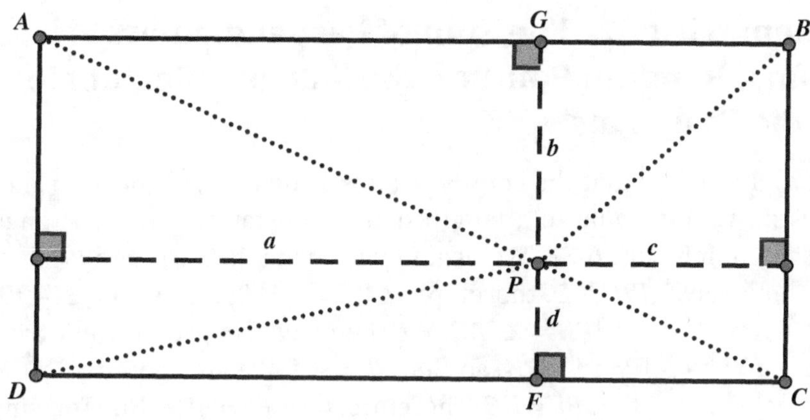

Figure 6-31

Geometrick 31: Another Peculiar Property of a Random Point in a Rectangle

The line MP in Figure 6-32 is a median of triangle APC and of triangle DPB. This helps us get the following: $AP^2 + CP^2 = 2AM^2 + 2MP^2$, and $BP^2 + DP^2 = 2BM^2 + 2MP^2$. However, $AM = BM$, and $4AM^2 = AC^2$. Therefore, $AP^2 + CP^2 + BP^2 + DP^2 = AC^2 + 4MP^2$.

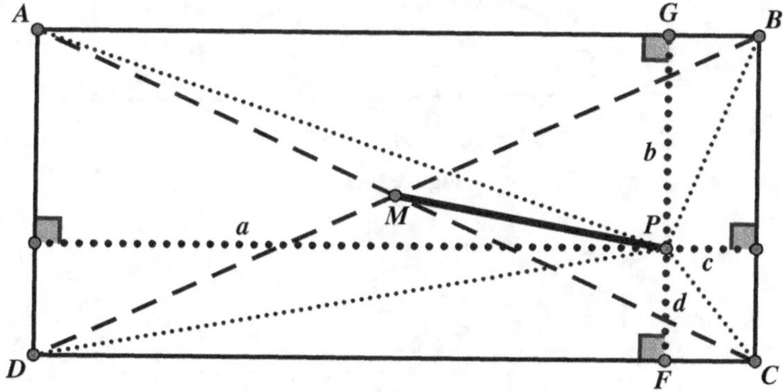

Figure 6-32

Geometrick 32: The Sum of Perpendiculars from a Random Point on the Side of a Rectangle to the Diagonals

In order to show that the sum $PE + FE$ is a constant, we need to show that the sum is equal to a length that is constant for this rectangle, as shown in Figure 6-33. We begin by drawing $AG \perp BD$ and $PH \perp AG$. We now have $PH \parallel BD$, therefore, $\angle APH = \angle ABD$. However, $\angle ABD = \angle BAC$, and thus, $\angle APH = \angle PAF$. We then have congruent right triangles, namely, $\triangle APF \cong \triangle APH$, so that $PF = AH$. We also note that $EGHP$ is a rectangle, so that $PE = HG$. Therefore, $PE + PF = AH + HG$. And since $AH + HG$ is a constant, $PE + PF$ is a constant regardless of the location of point P on AB.

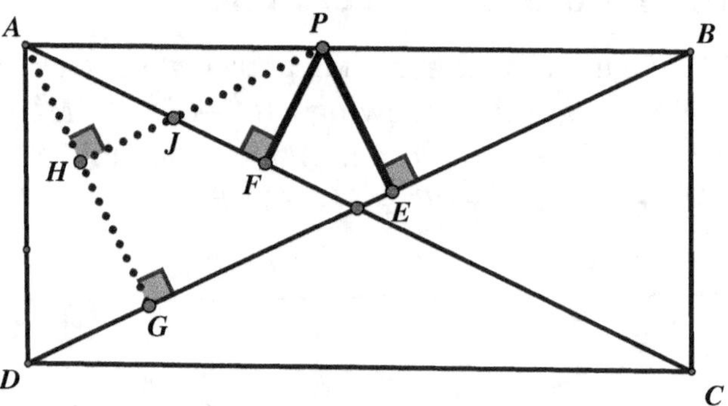

Figure 6-33

Geometrick 33: Trisection Points of a Semicircle Arc Can also Trisect the Diameter

In Figure 6-34, we construct altitudes EB and AK of triangles DEF and BDA, respectively, as well as line segments AD and AB. We first need to establish that triangle BDA is an equilateral triangle. Since central angle $\angle DBA = \overset{\frown}{AD} = 60°$, and $\angle ADF = \frac{1}{2}\overset{\frown}{ACF} = 60°$, triangle BDA is equilateral. Triangle DEF is four times as large as triangle BDA, and $\triangle AGK \sim \triangle EBG$, and therefore, $\frac{GK}{BG} = \frac{1}{2}$. It follows that $BG = \frac{2}{3}BK$. Since $GH = 2BG$, we double the previous relationship to get $GH = \frac{2}{3}BD$, which is equivalent to $GH = \frac{1}{3}DF$. By symmetry, we know that $DG = FH$, and therefore, $DG = GH = HF$.

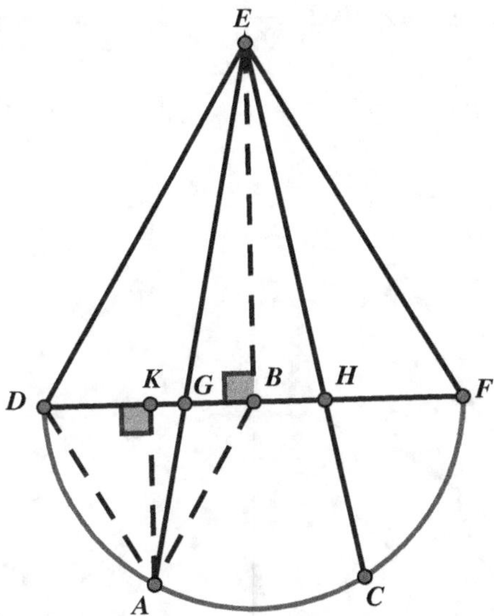

Figure 6-34

Geometrick 34: Combinations of Circular Arcs

We notice immediately in Figure 6-35 that triangles ADC and BDC are both isosceles right triangles, so that $\angle DBC = 45°$. By the Pythagorean theorem, since $BC = DC$, we have $BD = \sqrt{2}$. Furthermore, since $\angle ADB = 90°$, also $\angle EDF = 90°$. Since $BE = 2$, we have $DE = BE - BD = 2 - \sqrt{2}$. Also, sectors ABE and BAF are congruent. The area bounded by the bold curves is equal to the

$$area(\text{sector } EDF) + 2(area \text{ sector } ABE) -$$

$$area\triangle ABD - area(\text{semicircle } ABE)$$

$$= \frac{90}{360}\pi\left(2-\sqrt{2}\right)^2 + 2\left(\frac{45}{360}\pi\cdot 2^2\right) - \frac{1}{2}\cdot 2\cdot 1 - \frac{1}{2}\pi\cdot 1^2$$

$$= \left(\frac{3}{2}-\sqrt{2}\right)\pi + 2\left(\frac{\pi}{2}\right) - 1 - \frac{\pi}{2} = 2\pi - \pi\sqrt{2} - 1.$$

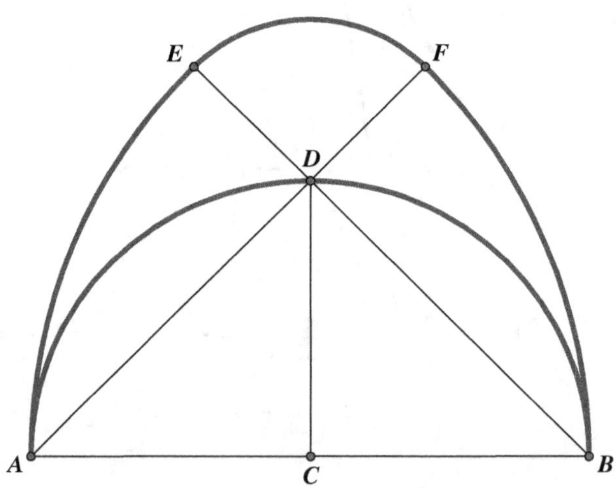

Figure 6-35

Geometrick 35: Comparing Unusual Combinations of Circular Regions

In Figure 6-36, we can represent the sought after area indicated by the bold semicircle lines as follows: *Area*Semicircle $AB + Area$ Semicircle $CD - 2Area$Semicircle AC. This is equivalent to

$$\frac{\pi r^2}{2} + \frac{\pi s^2}{2} - \frac{\pi (r-s)^2}{4} = \frac{\pi (r^2 + s^2)}{2} - \frac{\pi (r^2 + s^2 - 2rs)^2}{4}$$

$$= \frac{2\pi (r^2 + s^2)}{4} - \frac{\pi (r^2 + s^2) + 2\pi rs}{4} = \frac{\pi (r^2 + s^2) + 2\pi rs}{4} = \frac{\pi (r+s)^2}{4}.$$

The area of the circle with radius $FE = r + s$ is $\frac{\pi(r+s)^2}{4}$, which shows that the area of the bold line region is equal to the area of the circle.

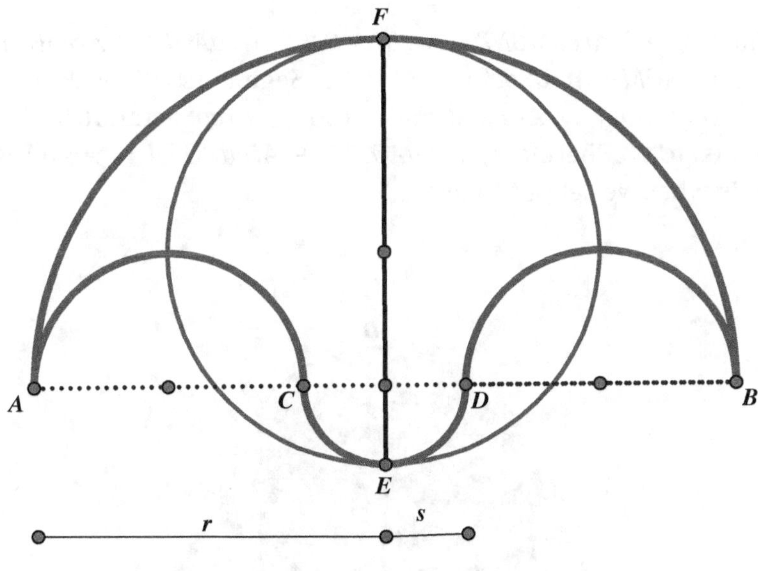

Figure 6-36

Geometrick 36: A Most Unusual Relationship

In Figure 6-37, we recognize immediately that the $Area\Delta MON = \frac{1}{2}MO^2 = \frac{1}{2}AO^2$. Furthermore, we can show that the area of the circle is equal to the area of the semicircle. In right triangle MON, we have $MN = AO\sqrt{2}$. Then $Area\,CircleMOND = \pi\left(\frac{AO\sqrt{2}}{2}\right)^2 = \frac{\pi AO^2}{2}$, and $Area\text{Semicircle}ACB = \frac{\pi AO^2}{2}$.

Therefore, the semicircle MDM is half the area of semicircle ACB.

Then, $\overparen{MEO} = \frac{1}{4}Circle MDNO$, and $\overparen{MCN}\frac{1}{2}Semicircle ACB$. That is, that $\overparen{MEO} = \frac{1}{2}\overparen{MCN}$.

$$Area MCND = Area\text{Semicircle}MND - Area\overparen{NCM}$$

$$= \frac{1}{2}Area\text{Semicircle}MON - Area\text{Segments}OEM - OFN.$$

Therefore, $Area MCND = Area\Delta MON$. $Area AOEM + Area BOFN = Area\text{Sector}AOM + Area\text{Sector}BOM - Area\text{Segments}(OEM + OFN)$.

Consequently, $Area(AOEM + BOFN) = Area\text{Sector}MONC - Area\text{Segments}MCN$. Therefore, $Area AOEM + Area BOFN = Area\Delta MON$, which is what we set out to prove.

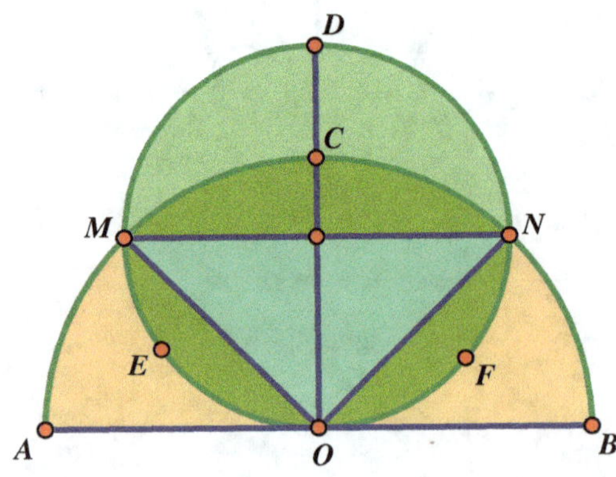

Figure 6-37

Geometrick 37: Finding the Area of a Quarter of a Quadrilateral

Begin by inserting the midpoints P and R of sides AB and CD, respectively, as shown in Figure 6-38. We then draw PY, PX, RY, RX, and RW. We then connect the midpoints of two sides of the triangle DCB, and find that $RY \parallel BC$. The extended line RY bisects the other diagonal DY of quadrangle $RWYD$. Since the diagonal bisects the other diagonal, it bisects the area of the quadrangle. Therefore, $area\triangle WYR = area\triangle DRY = \frac{1}{4}area\triangle DCB$. Analogously, $area\triangle XWR = \frac{1}{4}area\triangle ADC$. A quadrilateral formed by joining successively the midpoints of the sides of a quadrilateral creates a parallelogram whose area is half that of the original quadrilateral. Applying that to quadrangle $ABDC$, we get $area\triangle YXR = \frac{1}{2}areaXRYP = \frac{1}{4}areaABDC = \frac{1}{4}area\triangle CAB + \frac{1}{4}area\triangle BDC = \frac{1}{4}area\triangle ABC - \frac{1}{4}area\triangle DBC$. By adding the last three inequalities, we get $area\triangle WXY = area\triangle YXR + area\triangle WYR + area\triangle XWR = \frac{1}{4}area\triangle ABC - \frac{1}{4}area\triangle DBC + \frac{1}{4}area\triangle DBC + \frac{1}{4}area\triangle ADC = \frac{1}{4}area\triangle ABC + \frac{1}{4}area\triangle ADC$.

Therefore, $area\triangle WXY = \frac{1}{4}areaABCD$.

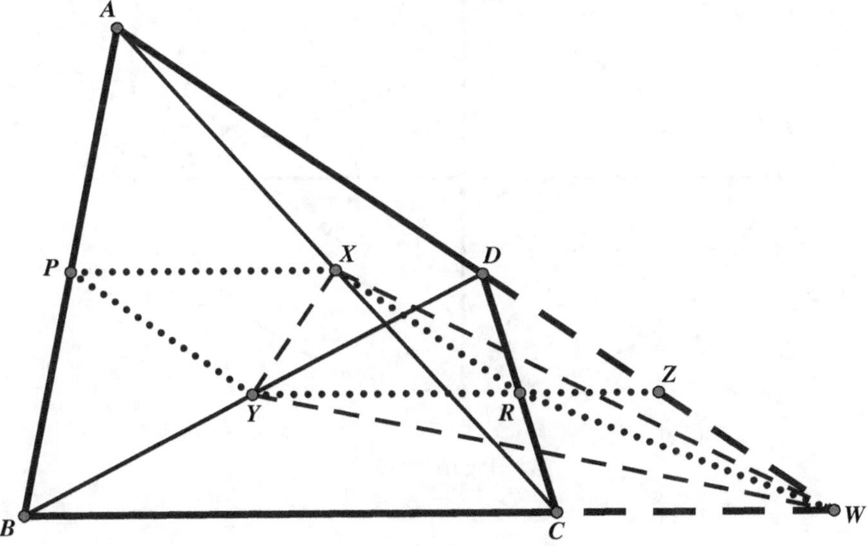

Figure 6-38

Geometrick 38: An Isosceles Trapezoid Generates Concyclic Points

We notice in Figure 6-39 that the exterior angle of isosceles triangle $\angle BGC = 2\angle BDC$. Furthermore, as inscribed $\angle BDC = \frac{1}{2}\overset{\frown}{BC}$, while central $\angle BPC = \overset{\frown}{BC}$; therefore, $\angle BPC = 2\angle BDC$. Consequently, $\angle BGC = \angle BPC$. These two equal angles could be measured by arc BC if they were points on a circle. In other words, we now have points B, P, G, and C concyclic.

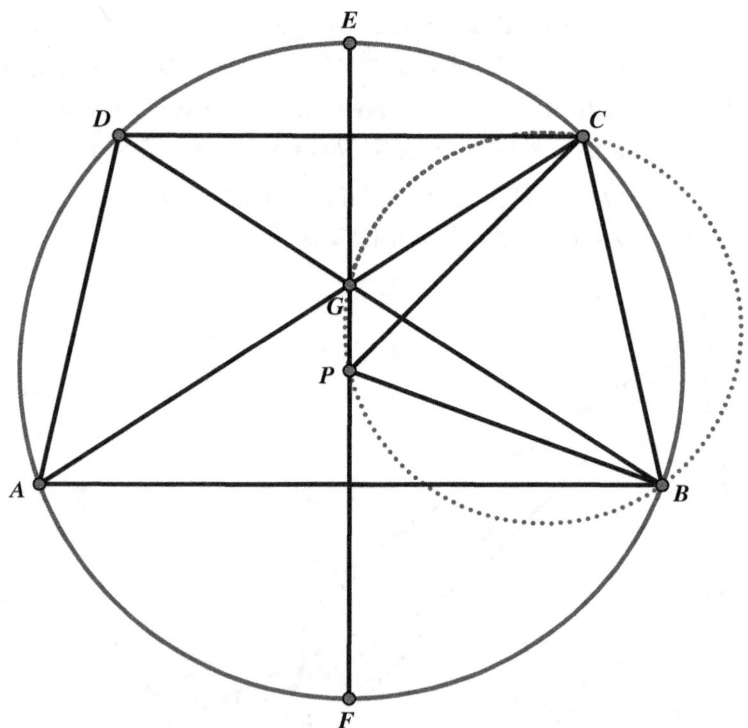

Figure 6-39

Geometrick 39: An Unanticipated Angle Bisector

We begin by constructing a tangent to the two circles at point A, and then drawing PD, as seen in Figure 6-40. Then $\angle EAD = \frac{1}{2}\overset{\frown}{AD}$ and $\angle EAB = \frac{1}{2}\overset{\frown}{AB}$. However, since $\angle EAD = \angle EAB$, then $\overset{\frown}{AD} = \overset{\frown}{AB}$. Furthermore, $\angle ACB = \angle APD$ because they are measured by equal arcs. Next, we have $\angle ADP = \frac{1}{2}\overset{\frown}{AP} = \angle APC$. Therefore, $\triangle APD \sim \triangle APC$, and so $\angle DAP = \angle CAP$, which shows that AP bisects $\angle BAC$.

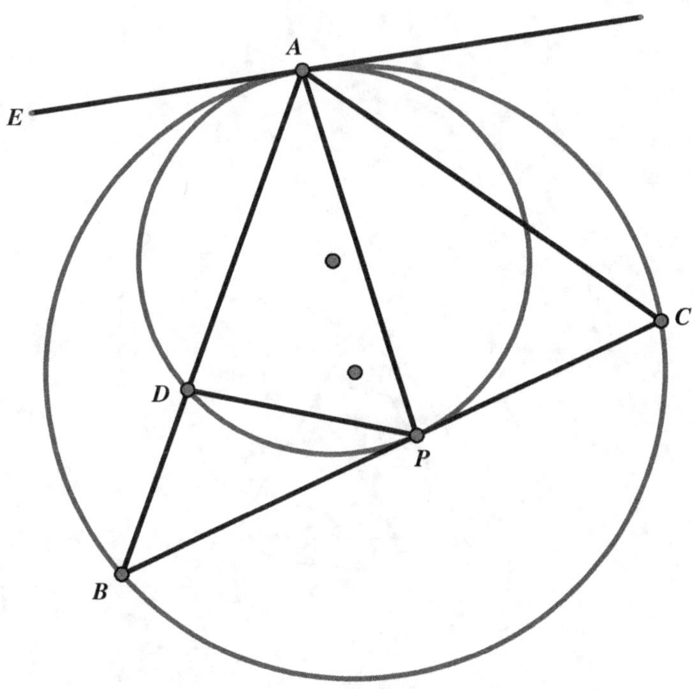

Figure 6-40

Geometrick 40: To Find the Common Area Determined by Three Intersecting Circles

As can be seen in Figure 6-41, the points of intersection of the equal circles create an equilateral triangle *ABC*. To get the common area of the three intersecting circles (shaded region), simply calculate the sum of the three sectors and subtract twice the area of the equilateral triangle. The area of each sector is one-sixth of the area of one of the circles with radius *r*, or $Area Sector = \frac{1}{6}\pi r^2$. Therefore, the area of the shaded region is $3\left(\frac{1}{6}\pi r^2\right) - 2\left(r^2\frac{\sqrt{3}}{4}\right) = \frac{\pi r^2}{2} - \frac{r^2\sqrt{3}}{2} = \frac{r^2}{2}\left(\pi - \sqrt{3}\right)$.

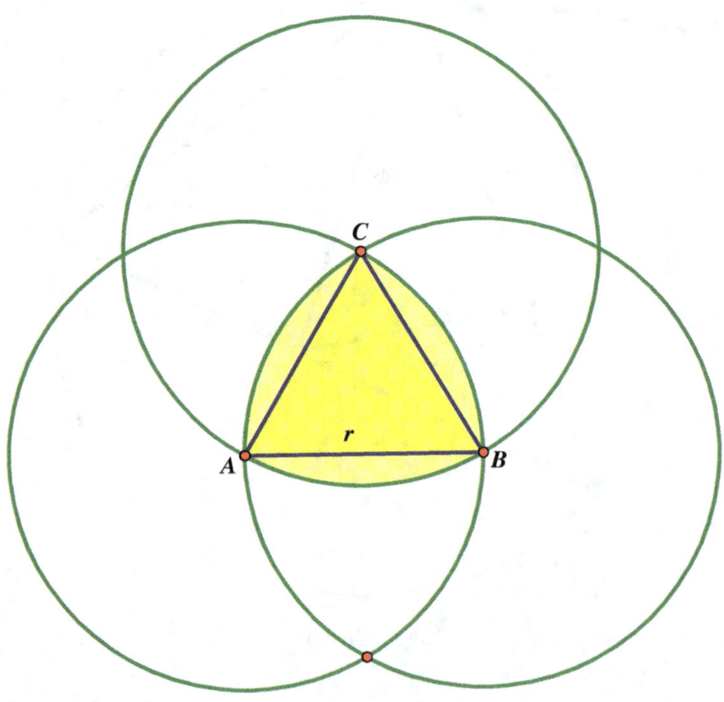

Figure 6-41

Geometrick 41: Surprise Angle Sum

In Figure 6-42, we have $\angle AHF = 180° - \angle G$, and $\angle CHF = 180° - \angle E$, then $\angle AHC = 360° - (\angle AHF + \angle CHF) = 360° - (180° - \angle G + 180° - \angle E) = \angle G + \angle E = 180° - \angle B$. Therefore, point H lies on the circumscribed circle of triangle ABC, as well as on the circumcircles of the other two triangles.

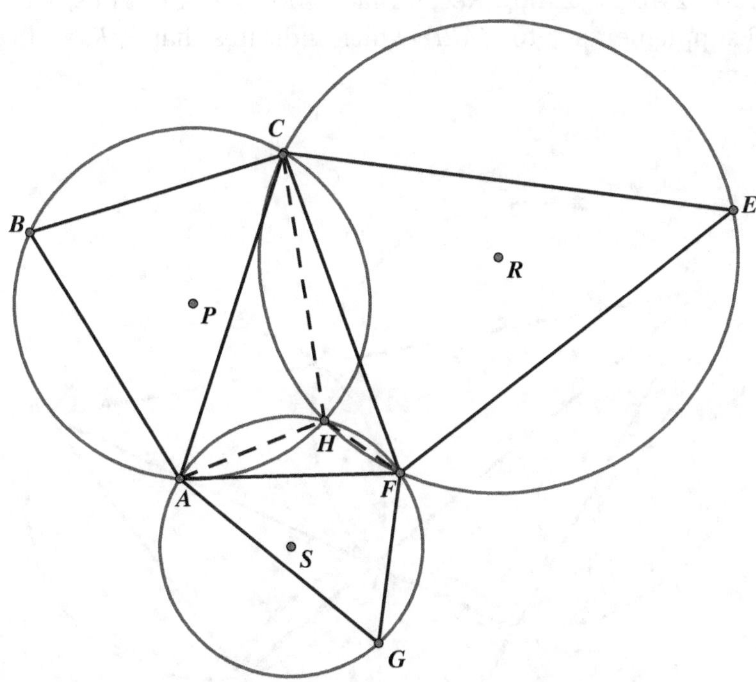

Figure 6-42

Geometrick 42: An Unexpected Collinearity

Begin by drawing line segment *AR*, as shown in Figure 6-43. The opposite angles of the cyclic quadrilateral are supplementary. Therefore, in cyclic quadrilateral *ABPR* we have $\angle ARP$ supplementary to $\angle ABP$, which is supplementary to $\angle ABC$. Therefore, $\angle ARP = \angle ABC$. In the original cyclic quadrilateral, we have $\angle ABC$ supplementary to $\angle ADC$, which is supplementary to $\angle ADQ$. Then, $\angle ADQ = \angle ABC$. Within cyclic quadrilateral *ADQR*, $\angle ADQ$ is supplementary to $\angle ARQ$, and therefore, $\angle ADC = \angle ARQ$. Recall that $\angle ARP = \angle ABC$. Thus, we have $\angle ARQ$ supplementary to $\angle ARP$, which indicates that *P*, *R*, and *Q* are collinear.

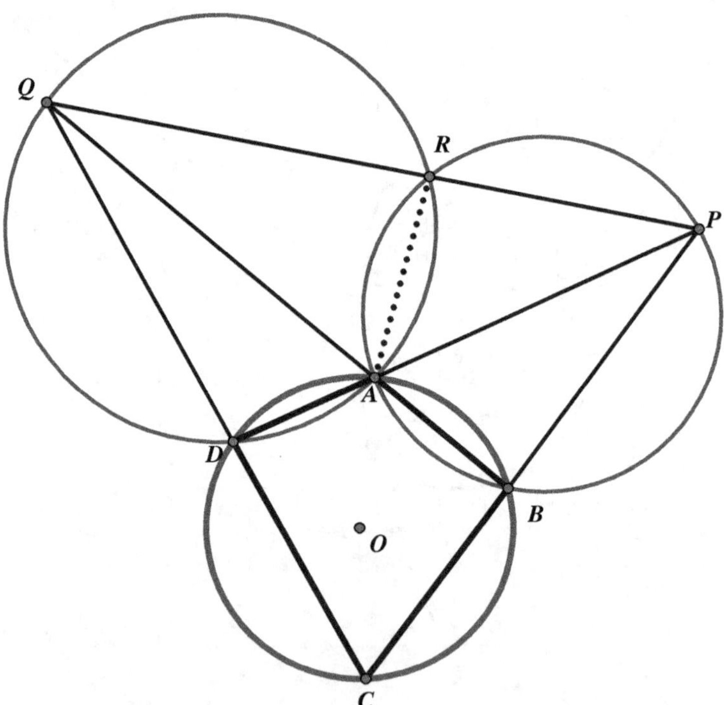

Figure 6-43

Geometrick 43: Four Congruent Concurrent Circles Define Another Circle

If we consider all the equal radii shown in Figure 6-44, we notice that each of the sides of quadrilateral *EFGH* are parallel to the sides of quadrilateral *ABCD*. As a result, the quadrilaterals are similar and their corresponding angles are equal. Furthermore, point *P* is equidistant from the vertices of quadrilateral *EFGH* because the circles all have equal radii. Therefore, quadrilateral *EFGH* is cyclic. Consequently, as quadrilateral *ABCD* is similar to quadrilateral *EFGH*, then *ABCD* is also cyclic.

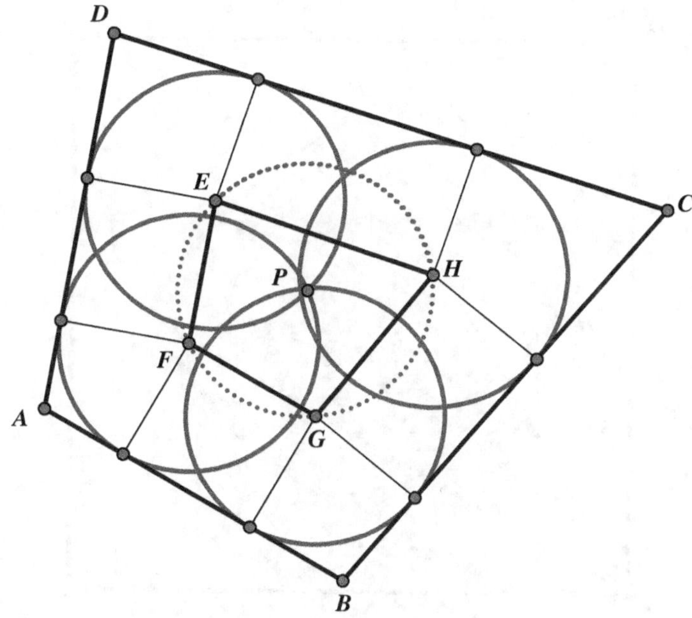

Figure 6-44

Geometrick 44: To Find the Common Area Determined by the Four Intersecting Circles

Analogously to the previous challenge, in order to find the area of the bold-line figure *EFGH* we need to get the area of four circular segments (for example, the area between line segment *GF* and arc *GF*) and add that to the area of the square *EFGH*. An essential piece of this process is to find the length of *GF*. Consider the triangle *AGF* in Figure 6-45, where $\angle GAF = \frac{1}{3}\angle DAB = 30°$. We draw the altitude *AK*, and we see that in right triangle *AKG*, we have $\text{Sin}\angle GAK = \frac{GK}{AG}$. Let the radius of the circles be *r*, so we then have $GK = r\text{Sin}15°$. To find the

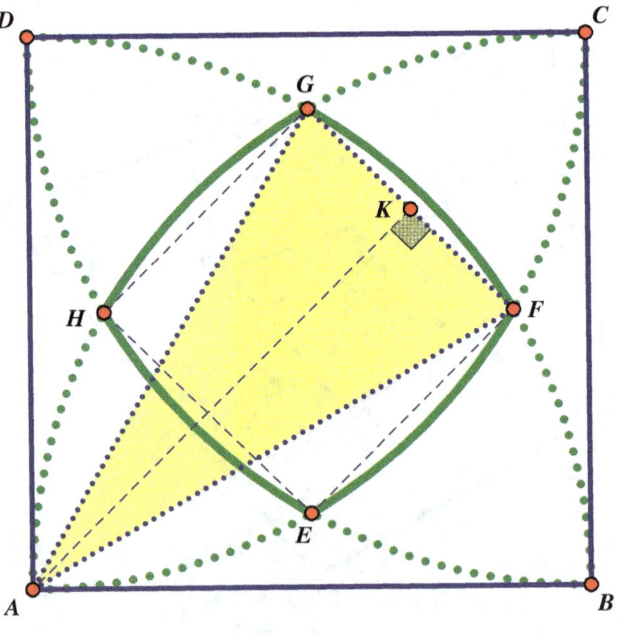

Figure 6-45

value of $\sin 15° = \sin(60° - 45°)$, we will use the trigonometric relationship $\sin(\alpha - \beta) = \sin\alpha\cos\beta - \cos\alpha\sin\beta$, which in this case provides us with the following:

$$\sin 15° = \sin 60°\cos 45° - \cos 60°\sin 45°$$

$$= \frac{\sqrt{3}}{2}\cdot\frac{\sqrt{2}}{2} - \frac{1}{2}\cdot\frac{\sqrt{2}}{2} = \frac{\sqrt{6}}{4} - \frac{\sqrt{2}}{4} = \frac{\sqrt{6}-\sqrt{2}}{4}.$$

Therefore, $GK = \frac{r(\sqrt{6}-\sqrt{2})}{4}$, and $GF = 2GK = \frac{r(\sqrt{6}-\sqrt{2})}{2}$. We then get the area of the square $EFGH$ as

$$\left(\frac{r\left(\sqrt{6}-\sqrt{2}\right)}{2}\right)^2 = \frac{r^2\left(4-2\sqrt{12}\right)}{4} = \frac{r^2\left(2-2\sqrt{3}\right)}{2} = r^2\left(1-\sqrt{3}\right).$$

The area of one sector, such as sector GAF, is $\frac{1}{12}\pi r^2$. The

$$Area\triangle GAF = \frac{1}{2}GF\cdot AK = \frac{1}{2}\left(\frac{r\left(\sqrt{6}-\sqrt{2}\right)}{2}\right)\cdot\left(\frac{r}{2}\sqrt{3+\sqrt{3}}\right).$$

The area of the segment formed by arc GF and line GKF is the area of the sector GAF minus the area of triangle GAF, which is $EFGH$. We then add four times this area to the area of the square to get the required area of the bold region. That is,

$$4\left[\frac{1}{12}\pi r^2 - \left(\frac{1}{2}\left(\frac{r\left(\sqrt{6}-\sqrt{2}\right)}{2}\right)\cdot\left(\frac{r}{2}\sqrt{3+\sqrt{3}}\right)\right)\right] + r^2\left(1-\sqrt{3}\right).$$

This expression can be easily simplified, and is left to the reader as an exercise.

Geometrick 45: Two Connected Squares Generate Another Square

We begin by constructing two parallelograms *BCEL* and *CDKG*, as shown in Figure 6-46. If we rotate parallelogram *BCEL* 90° in a clockwise direction about point *M*, we have placed it on parallelogram *CDKG*, since the two parallelograms are congruent. Consequently, congruent lines *MP* and *MQ* form an angle of 90°. When we rotate parallelogram *BCEL* 90° about point *N* in a counterclockwise direction, we find that congruent segments *NP* and *NQ* form an angle of 90°. Therefore, quadrilateral *NQMP* is a square.

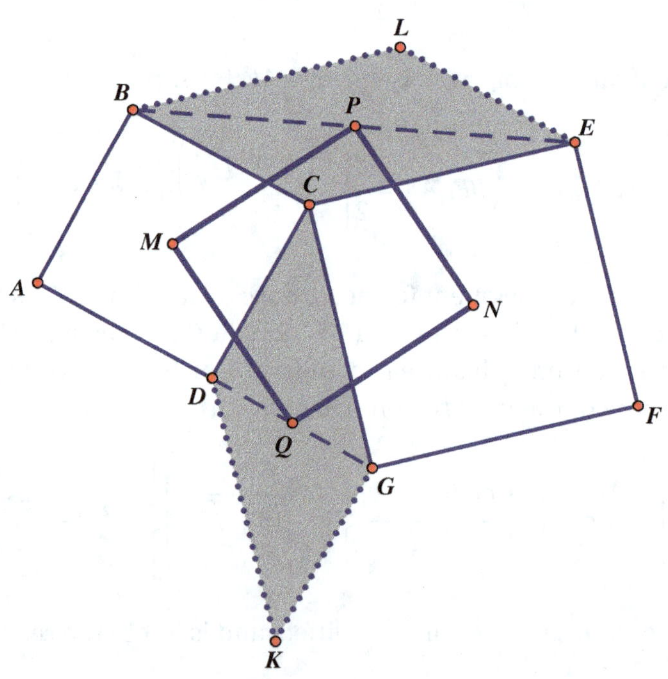

Figure 6-46

Geometrick 46: Generating Squares

We note in Figure 6-47 that *FAEP* is a cyclic quadrilateral on diameter *EF*. Since inscribed angles $\angle FAP = 45° = \angle EAP$, and the chords $EP = FP$, we know that quadrilateral *EFGH* is a square. A similar argument can be made to show that quadrilateral *JKLM* is also a square.

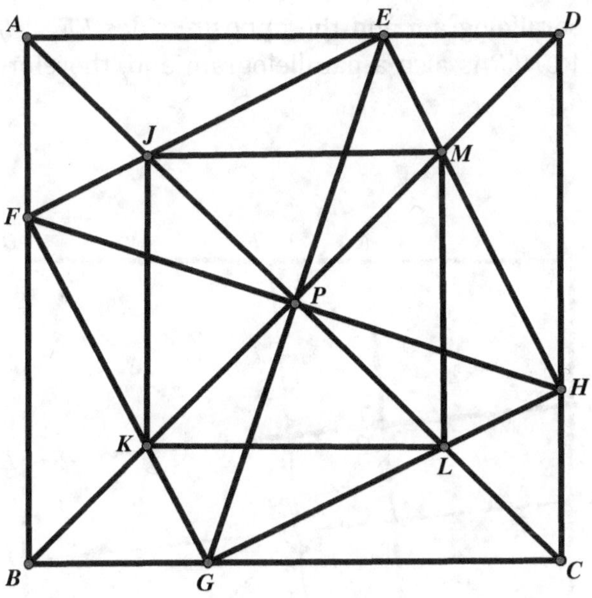

Figure 6-47

Geometrick 47: All Perpendicular Transversals in a Square Are Equal

In Figure 6-48, we have $NJ \perp GH$. We construct another set of perpendicular transversals, EF and RL, that intersect at the center of the square point P and are parallel to NJ and GH, respectively. We know from the previous Geometrick 46 that when two perpendicular transversals intersect at the center of the square, they are equal, so $EF = RL$. Consider quadrilateral $NFEJ$, whose opposite sides are parallel; therefore, it is a parallelogram and the opposite sides $EF = NJ$. Similarly, quadrilateral $GHLR$ is also a parallelogram and, therefore, $GH = RL$. Hence, $NJ = GH$.

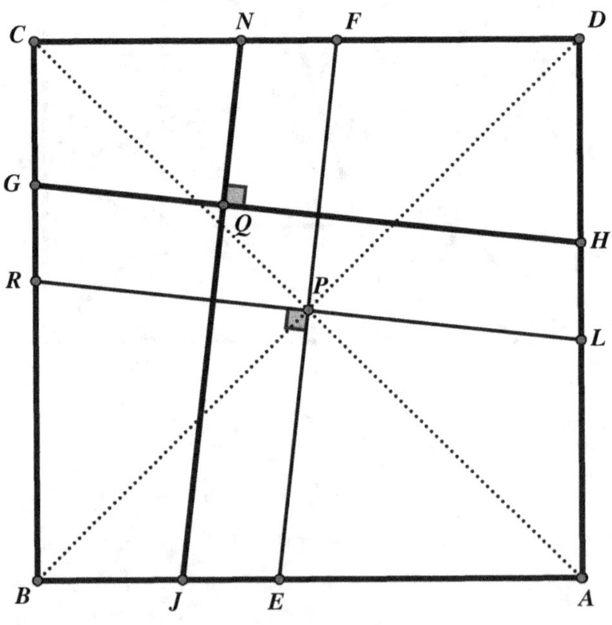

Figure 6-48

Geometrick 48: Relationship of a Square Within a Square

To show the comparison of the areas in Figure 6-49, we can conveniently let the area of square $EFHG = 4$, which means $EF = 2$. For isosceles triangle EDF, side $ED = \sqrt{2}$. For isosceles triangle AGE, with the leg $EG = 2$, we have $AE = 2\sqrt{2}$. With side $AD = \sqrt{2} + 2\sqrt{2} = 3\sqrt{2}$, the area of square $ABCD$ is $\left(3\sqrt{2}\right)^2 = 18$. Therefore, the ratio between the areas of the two squares is 2:9. In other words, square $ABCD$ has 4.5 times the area of square $EFHG$.

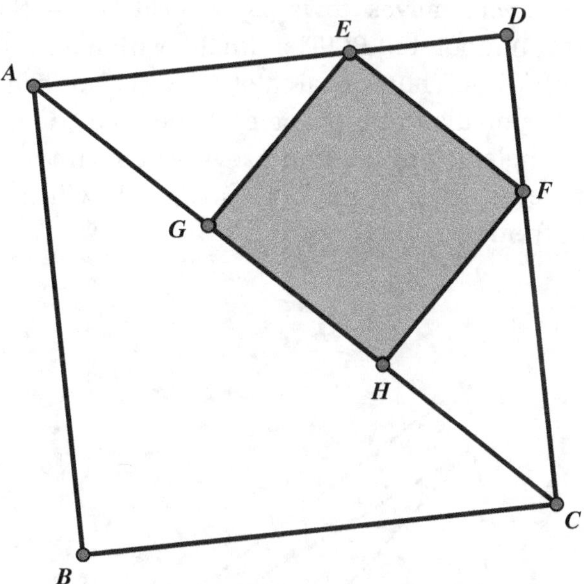

Figure 6-49

Geometrick 49: An Unexpected Area Determined

Using Figure 6-50, the typical approach to solving this problem is to recognize that the area of $\triangle BMC$ is one-half the area of $\triangle ABC$ (since the median partitions a triangle into two equal areas). Area $\triangle BMC =$ area $\triangle BMD +$ area $\triangle CMD =$ area $\triangle BMD +$ area $\triangle MPD$, which equals area $\triangle BPD = \frac{1}{2}$ area $\triangle ABC$. This rests on the property that when the vertices of two triangles lie on a line parallel to a common base, their areas are equal.

This problem can be considerably simplified by carefully considering an extreme case. We select point P at an extreme position, either at point M or point A. Suppose P is selected at point A. Notice that as P moves along BA towards A, we find that MD, which must stay parallel to PC, moves towards a position so that point D approaches the midpoint of BC. That final position for D then has AD as a median of $\triangle ABC$. Thus, the area of $\triangle PBD$ is one-half the area of $\triangle ABC$, since the median of a triangle divides the triangle into two equal area triangles. This solution provides us with an interesting example where we must watch all movements when we move a point to an extreme position.

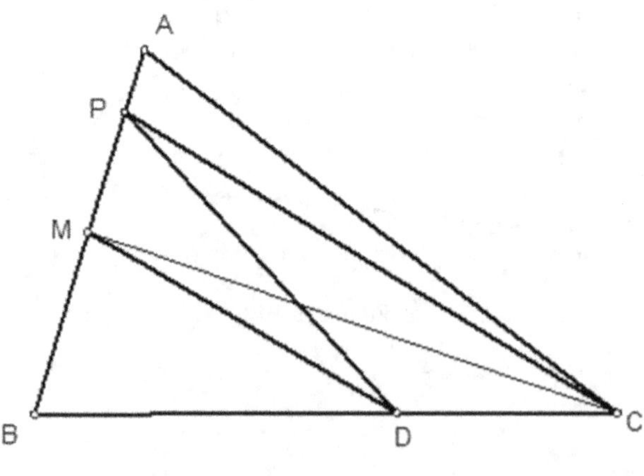

Figure 6-50

Geometrick 50: A Curious Point in the Interior of an Angle

In Figure 6-51, we see that points E, B, C, and D are concyclic with ED as the diameter. The secants generate $AB \cdot AE = AC \cdot AD$ and the cross chords generate $PB \cdot PD = PC \cdot PE$. Also, with secants AE and CE as well as AD and DB, we get $AD \cdot DC = DP \cdot DB$, and $AE \cdot BE = EP \cdot EC$.

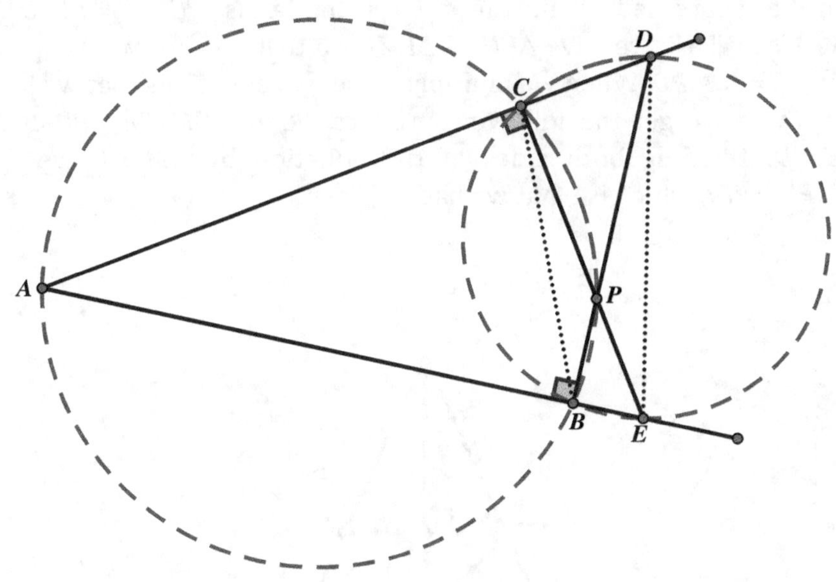

Figure 6-51

Geometrick 51: Finding Unusual Segment Products

We recognize that the area of *CNMB* is one-half the area of triangle *ABC*, and the same is true for triangle *CBD*, since the altitude divides the triangle into equal parts, as shown in Figure 6-52. These two equivalent surface areas have a common area, namely, quadrilateral *CPMD*. Therefore, the remaining parts of these two equivalent surface areas are also equal, that is, $Area\triangle DPM = Area\triangle CPN$. This can be expressed from the area formula as $PC \cdot NE = PD \cdot DM$. Since $NEF \parallel AB$, we have $\triangle PNE \sim \triangle PMD$, so that $\frac{DM}{NE} = \frac{PD}{PE}$, whereupon, $DM \cdot PE = NE \cdot PD$. When we multiply these two equations that we just generated we get the following product: $PC \cdot NE \cdot DM \cdot PE = PD \cdot DM \cdot NE \cdot PD$. Dividing both sides of the equation by $DM \cdot NE$, we get $PC \cdot PE = PD^2$, which is what we had to prove.

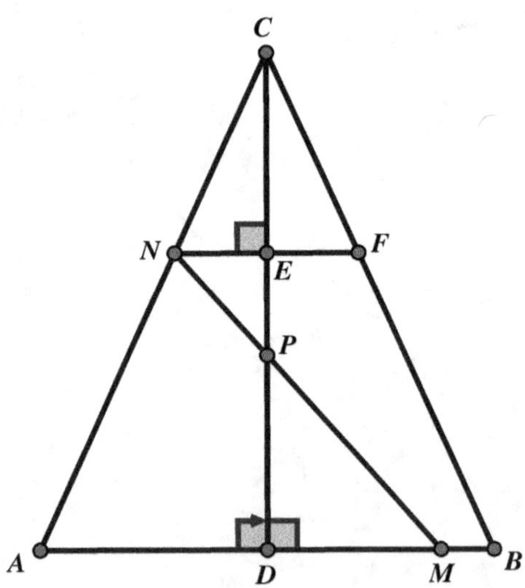

Figure 6-52

Geometrick 52: A Perpendicular Line that Is also a Bisector

In Figure 6-53, we have a series of angle equalities. One such are vertical angles $\angle BGF = \angle DGE$. Since the two angles are complementary to angle ABG, we have $\angle BGF = \angle ABC$. However, the two angles measured by $\frac{1}{2}\overset{\frown}{BC}$ are $\angle ABC = \angle BDC$. Therefore, $\angle BDC = \angle DGE$, which establishes isosceles triangle EDG, where $DE = GE$. A similar argument can be made to show that triangle CGE is also isosceles and $CE = GE$. Thus, $CE = DE$.

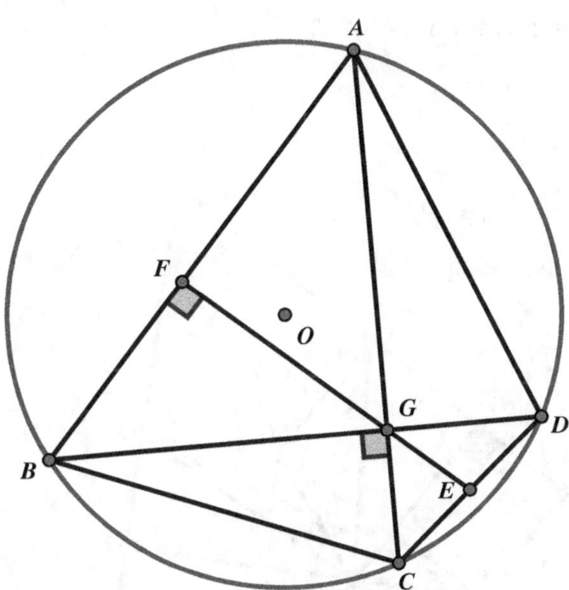

Figure 6-53

Geometrick 53: An Expected Concurrency in a Cyclic Quadrilateral

In order to prove the concurrency, we refer to Figure 6-54, where we locate point M, the intersection of the lines joining the midpoints of the opposite sides of the quadrilateral. From the circle's center O, we find that $OE \perp AB$ and $OG \perp CD$. When we extend OM its own length, we reach point N, where we find that quadrilateral $EOGN$ is a parallelogram because its diagonals bisect each other. We know that GI and EO are parallel as they are two sides of the parallelogram $EOGN$, and are, therefore, perpendiculars as expected. Similarly, we find $OH \perp AD$ and $OF \perp BC$ so that $OHNF$ is also a parallelogram, and line segments HJ and FL also contain point N. Thus, we have shown that EK, FL, GI, and HJ are concurrent at point N.

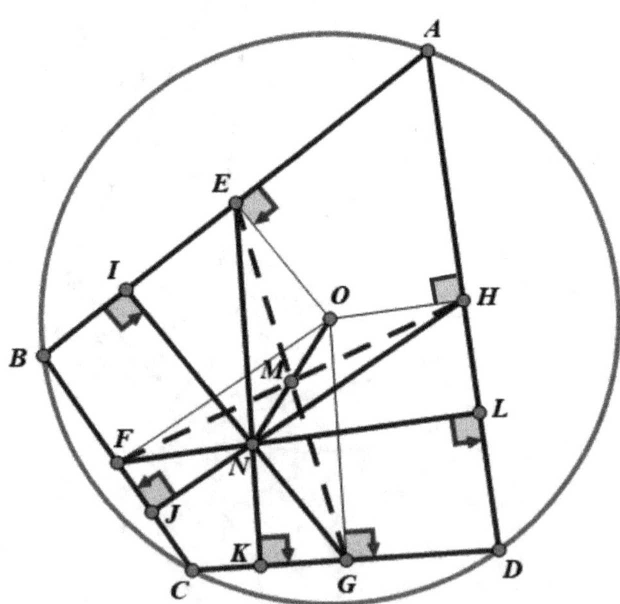

Figure 6-54

Geometrick 54: The Relationship of a Point on the Circumcircle of a Quadrilateral

Begin by drawing lines MD and MB to two opposite vertices, as seen in Figure 6-55. With the two line segments ME and MG perpendicular to a pair of opposite sides, we have two right triangles, MBE and MDH, which have a pair of equal angles $\angle MDH$ and $\angle MBA$, since both are measured by $\frac{1}{2}\widehat{AM}$. Therefore, $\triangle MBE \sim \triangle MDH$, which gives us $\frac{ME}{MH} = \frac{MB}{MD}$. Analogously, we find that right triangles MDG and MBF are also similar, so that $\frac{MG}{MF} = \frac{MD}{MB}$. Now multiplying these two equations, we get $\frac{ME}{MH} \cdot \frac{MG}{MF} = \frac{MB}{MD} \cdot \frac{MD}{MB} = 1$, which implies that $ME \cdot MG = MH \cdot MF$.

There is an analogous situation when two vertices of the cyclic quadrilateral converge to form a triangle, which we will see in the proof of Geometrick 55.

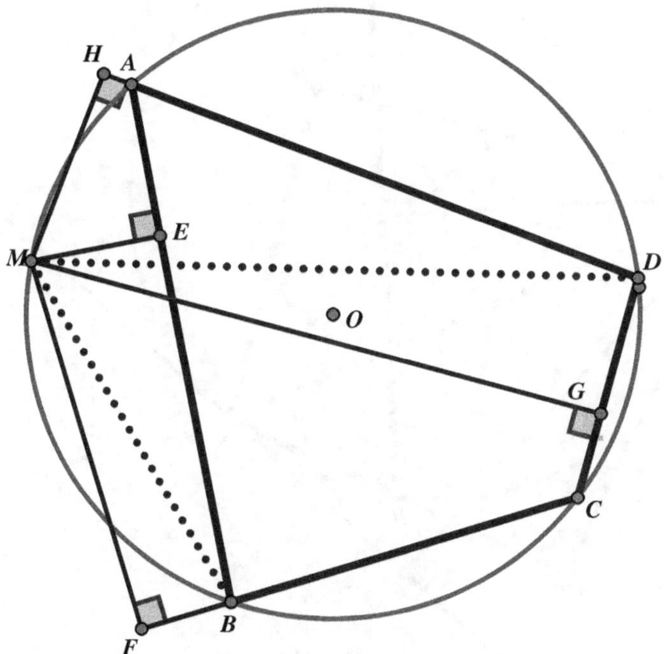

Figure 6-55

Geometrick 55: The Relationship of a Point on the Circumcircle of a Triangle

Begin by drawing lines MA and MC as shown in Figure 6-56. Triangles MGC and MHA are similar because they are both right triangles and $\angle MCG = \angle MAC$, as they are both measured by $\frac{1}{2}\overgroup{MBC}$. Therefore, $\frac{MA}{MC} = \frac{MH}{MG}$. Also, right triangles MAE and MFC are similar because $\angle MAE = \angle MCF$, as they are both measured by $\frac{1}{2}\overgroup{MB}$. Therefore, $\frac{MC}{MA} = \frac{MF}{ME}$. Once again, we multiply the two equations to get $\frac{MH}{MG} \cdot \frac{ME}{MF} = \frac{MA}{MC} \cdot \frac{MC}{MA} = 1$. Therefore, we get $ME \cdot MG = MH \cdot MF$.

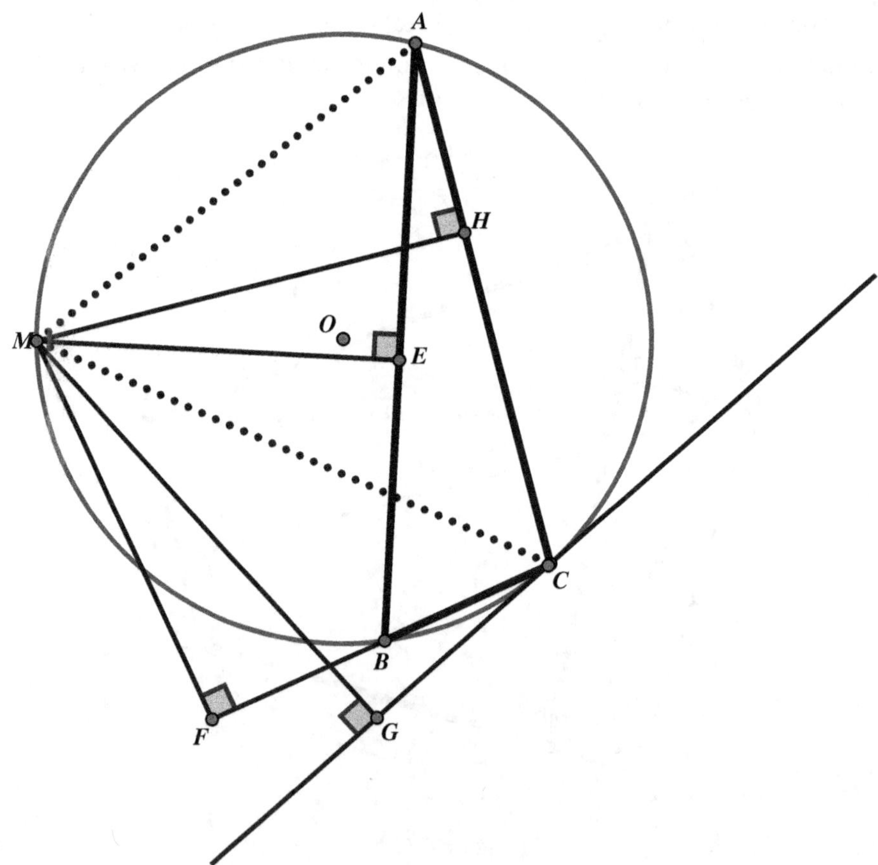

Figure 6-56

Geometrick 56: Determining the Radius of a Circle

To make some sense out of this rather confusing situation, we extend *BD* to meet the circle at point *E*, as shown in Figure 6-57. The diameter of the circle contains points *O* and *D* and intersects the circumfrence at points *G* and *F*. We shall let the radius of the circle be represented by *r*. Recall that when a tangent and a secant from an external point are drawn to a circle, the square of the tangent is equal to the secant times its external segment. Here, we have $BC \cdot BE = AB^2$, and therefore, $3(DE+6)=36$, and $DE=6$. Furthermore, since the product of segments of intersecting chords are equal, we have $DE \cdot DC = DF \cdot DG$. We then have $6 \cdot 3 = (r-2)(r+2)$, and $18 = r^2 - 4$, so that $r = \sqrt{22}$.

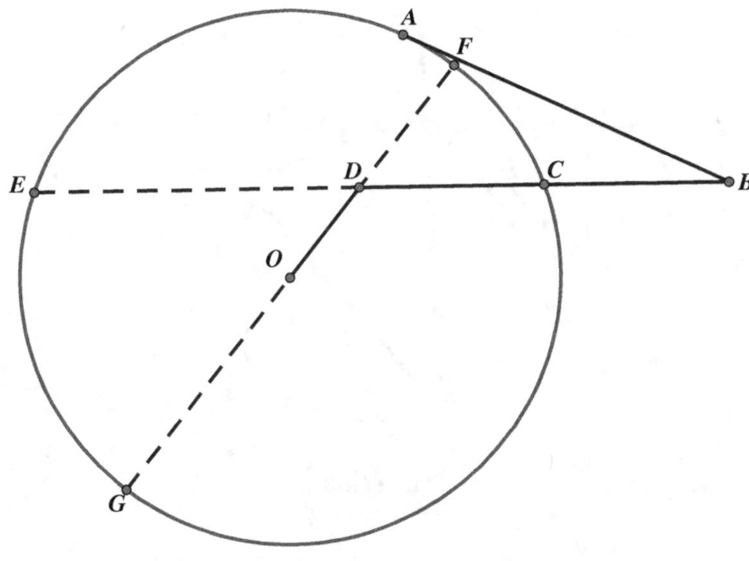

Figure 6-57

Geometrick 57: A Proportion Produced by Tangents and the Secant

In Figure 6-58, angles BAM and ADB are both measured by $\frac{1}{2}\overset{\frown}{AB}$. Therefore, $\angle BAM = \angle ADB$, which enables us to have $\triangle MAB \sim \triangle MAD$, so $\frac{AB}{AD} = \frac{MB}{MA}$. Similarly, $\triangle MBC \sim \triangle MCD$, which gives us $\frac{BC}{DC} = \frac{MB}{MC}$. Since $MA = MC$, we have $\frac{AB}{AD} = \frac{BC}{DC}$.

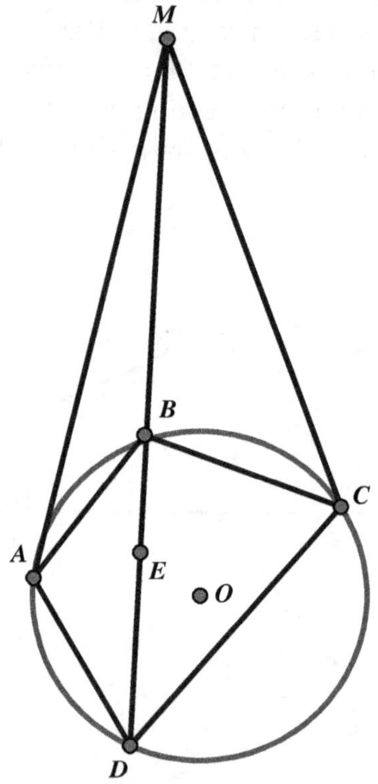

Figure 6-58

Geometrick 58: Parallelogram Comparisons

A common approach to solve the problem would be to look for congruent relationships that lead to equal areas. This method will lead nowhere. A clever method, although "off the beaten path," is to draw the line segment *PD*, as shown in Figure 6-59.

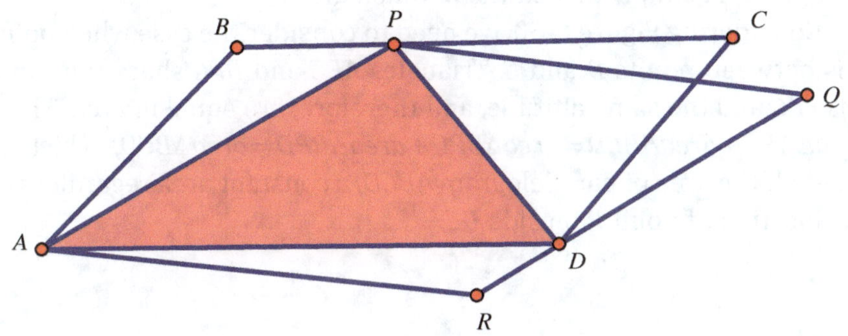

Figure 6-59

Notice that triangle *APD* is one-half the area of paralellograms *ABCD* and *APQR*, since it shares a base with each of the parallelograms as well as the related height. Although this is a clever approach to a rather challenging problem, there is an even more elegant way to approach this proof. When the problem was posed, we were told only that point *P* was on side *BC*, but not where along the side it was to be placed. We can consider an extreme case and place *P* on top of point *B*. Similarly, point *D*, which was to be placed on side *RQ*, could just as easily have been placed to overlap point *R*. Under these circumstances, which accord with the original problem's statement, the two parallelograms would overlap, and consequently would have the same area. Therefore, the area of parallelogram *APQR* is 18.

Geometrick 59: Constant Area of Two Triangles in a Parallelogram

The proof becomes trivial when we consider extreme cases: that is, when *P* takes on a position at either end of side *DC*. When that happens, one of the two triangles will disappear, and the sum of the triangles will simply be half the parallelogram.

Considering Figure 6-60, we need to consider the case when point *P* is between points *D* and *C*. Triangles *APC* and *BPC* share the same base *PC* and the same altitude, and therefore, are equal in area. Thus, $area\triangle APC + area\triangle BPD = area\triangle BPC + area\triangle BPD = area\triangle BCD$, which is one-half the area of parallelogram *ABCD*, a constant area regardless of the location of point *P* on side *DC*.

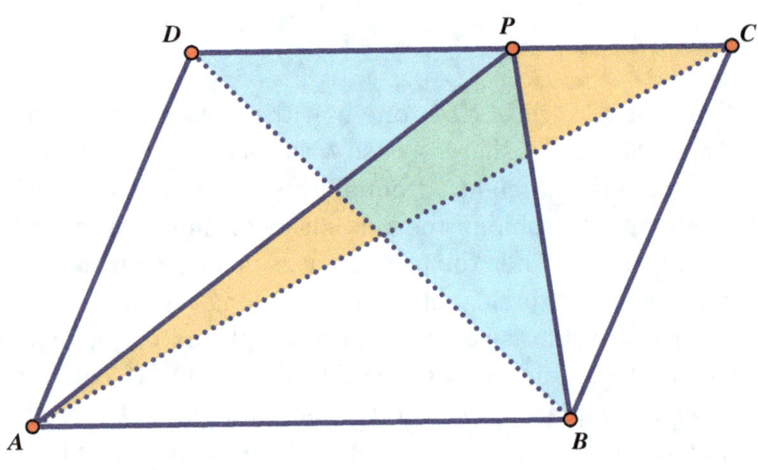

Figure 6-60

Geometrick 60: A Curious Area Difference Between Two Triangles

To find the difference of the areas required, we must create within triangle *PBE* an area equal to that of triangle *PDC*. When we draw $BG \parallel AC$, as shown in Figure 6-61, we have created $\triangle PDC \cong \triangle GBP$, since $CP = BP$ and the alternate interior angles created by the parallel lines are equal. Therefore, triangle *EBG* has an area of *K*, which is the difference of the areas of the two original triangles.

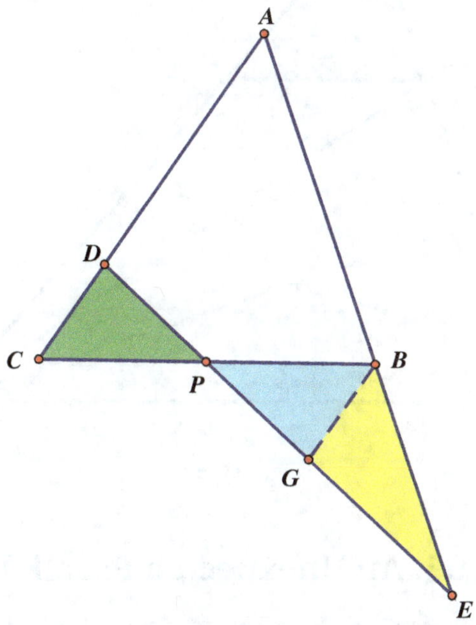

Figure 6-61

Geometrick 61: An Unusual Area Relationship in a Triangle

In Figure 6-62, we note that triangle DBC and triangle DCE share the same altitude DH. Since their bases are $EC = \frac{2}{3} BC$, the $area\triangle DEC = \frac{2}{3} area\triangle DBC$. However, since triangle DBC and triangle ABC share the common altitude CK and their bases $BD = \frac{1}{3} AB$, the $area\triangle DBC = \frac{1}{3} area\triangle ABC$. Thus, $area\triangle DEC = \frac{2}{3} \cdot \frac{1}{3} area\triangle ABC = \frac{2}{9} area\triangle ABC$.

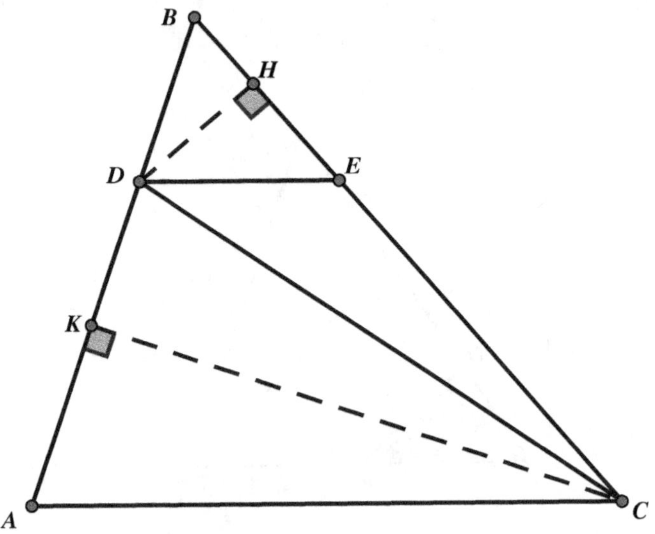

Figure 6-62

Geometrick 62: An Unexpected Parallelogram

In Figure 6-63, because $AC = AB$ in isosceles triangle ABC, we have $\angle ACB = \angle ABC$. Angle ABC is supplementary to angle ABD and angle AMD is supplementary to angle ABD (opposite angles of the cyclic quadrilateral). Therefore, $\angle AMD = \angle ABC = \angle ACB$. Also, $\angle MDB = \frac{1}{2} \overset{\frown}{BAM}$, and $\angle CAM = \frac{1}{2} \overset{\frown}{ABM}$. Since $\overset{\frown}{BAM} = \overset{\frown}{ABM}$, we have $\angle MDB = \angle CAM$.

Consider quadrilateral *ACDM*, where the sum of the angles is 360°. Therefore, $2\angle DMA + 2\angle CDM = 360°$ or $\angle DMA + \angle CDM = 180°$, which establishes that $AM \parallel CD$.

In a similar fashion, $\angle DMA + \angle CAM = 180°$, and $MD \parallel AC$. This proves that quadrilateral *ACDM* is a parallelogram.

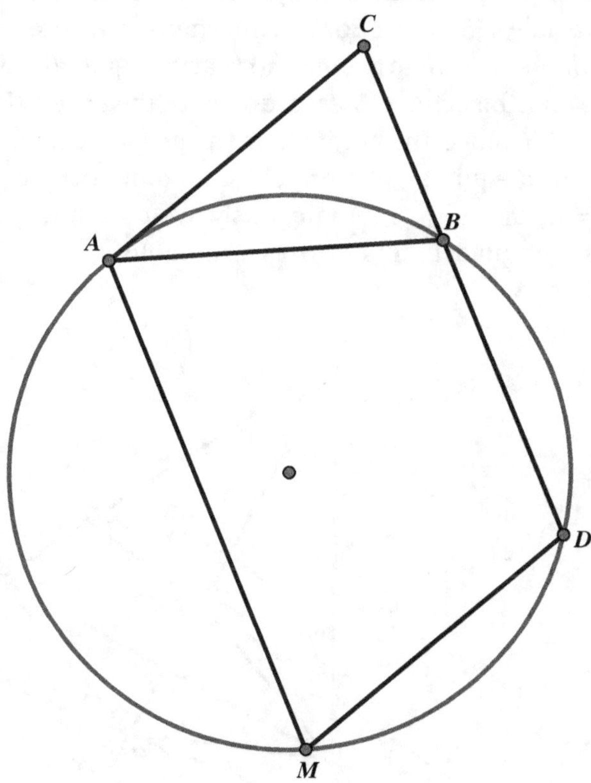

Figure 6-63

Geometrick 63: An Unexpected Rhombus

The basis of this proof will be to establish that triangle *FGH*, which is shown in Figure 6-64, is isosceles. Since $\angle CEG = \angle BEG$, the two triangles *DEH* and *BEG* have a pair of equal angles. Since quadrilateral *ABCD* is cyclic, we know that $\angle HDE$ is supplementary to $\angle GBA$. However, $\angle GBA$ is supplementary to $\angle GBE$. Therefore, $\angle HDE = \angle GBE$, and then the third angle of the two triangles *DEH* and *BEG* are also equal, namely, $\angle DHE = \angle BGE$. Furthermore, we note that the vertical angles *BGE* and *HGF* are equal, so that triangle *FGH* is isosceles. We have the angle bisector's intersection with the base *HG* at point *M*, making $FMJ \perp GH$. Since any point along the perpendicular bisector of a line segment is equidistant from the endpoints of a line segment, we have $GI = HI$ and $GJ = HJ$. Analogously, we can show that $GI = GJ$. The result is that quadrilateral *GIHJ* is a rhombus.

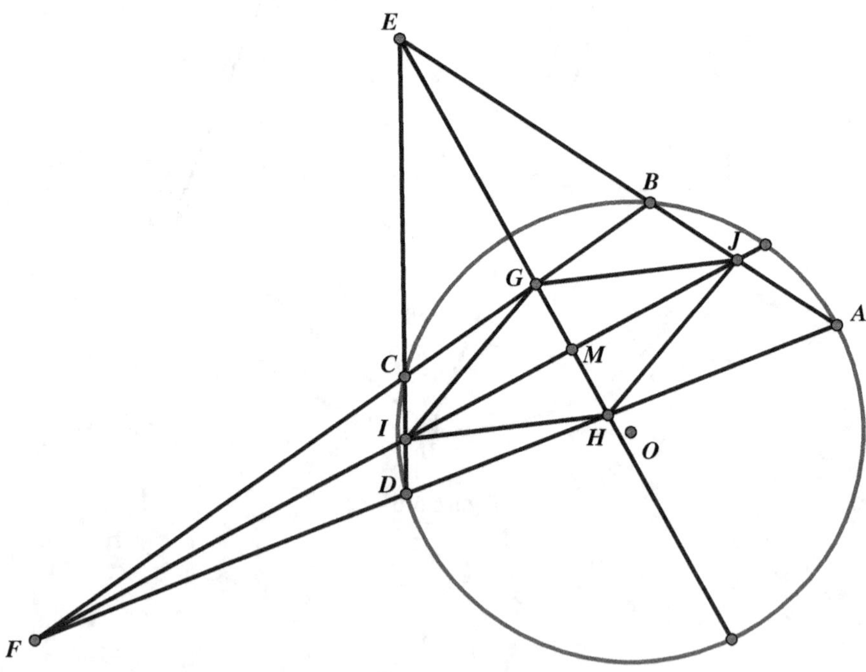

Figure 6-64

Geometrick 64: The Varignon Parallelogram

To prove this relationship, we rely on the basic theorem that the lines joining the midpoints of two sides of a triangle are parallel to the third side and one-half its length. Therefore, in Figure 6-65, in triangle ABD we have $KN \parallel BD$ and $KN = \frac{1}{2} BD$, and in triangle BCD we have $LM \parallel BD$ and $LM = \frac{1}{2} BD$. Thus, $KN \parallel LM$ and $KN = LM$, making $KLMN$ a parallelogram because one pair of opposite sides are equal and parallel.

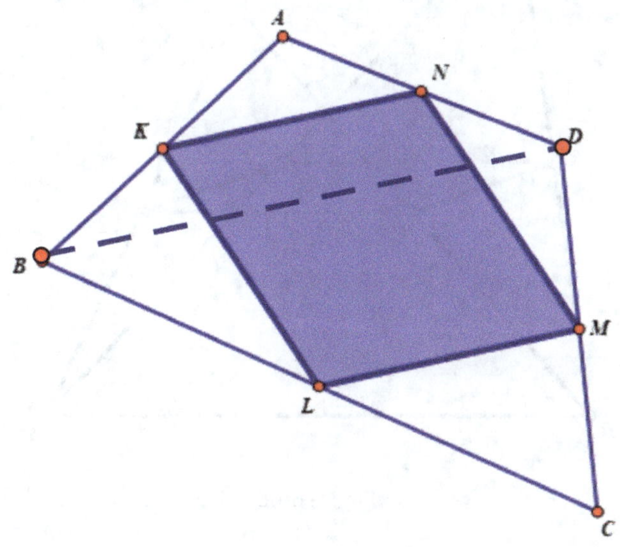

Figure 6-65

Moving along to the Varignon parallelogram, we can examine an alternative way of creating this shape. We know that the length of a line joining the midpoints of two sides of triangle is equal to one-half the length of the third side and parallel to the third side. Considering Figure 6-66, and applying this rule to triangles ABC and DBC, respectively, we get the following: $RQ \parallel BC$ and $PS \parallel BC$, and therefore, $RQ \parallel PS$. Also, $RQ = \frac{1}{2}BC$ and $PS = \frac{1}{2}BC$; therefore, $RQ = PS$. Thus, quadrilateral $PQRS$ is a parallelogram because one pair sides are equal and parallel.

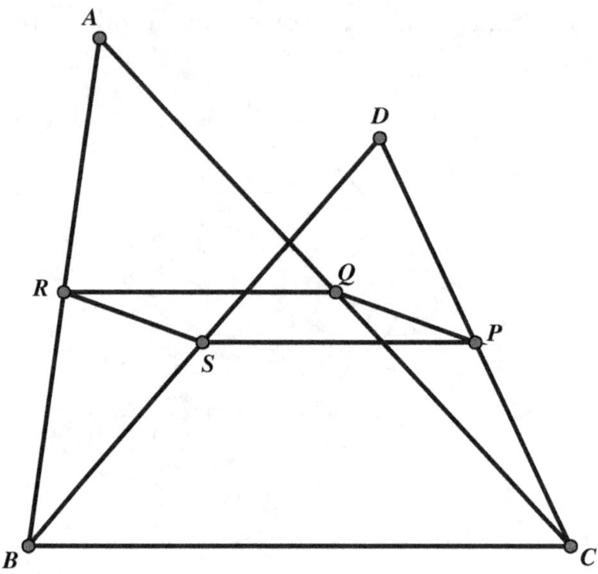

Figure 6-66

Geometrick 65: The Varignon Parallelogram's Special Properties

Using the relationships mentioned in Geometrick 64, we see in Figure 6-67 that $KN + LM = BD$. Similarly, we have $KL + MN = AC$, so that we can conclude that the perimeter of the parallelogram $KLMN$ is equal to the sum of the diagonals of the original quadrilateral. Symbolically, $KN + LM + KL + MN = BD + AC$.

We can also show that the Varignon parallelogram's area is one-half the area of the original quadrilateral. We once again refer to Figure 6-67, where $\triangle AKN \sim \triangle ABD$ and where the $area\triangle AKN = \frac{1}{4}area\triangle ABD$. Similarly, we have $\triangle LCM \sim \triangle BCD$ and $area\triangle LCM = \frac{1}{4}area \triangle BCD$. Therefore, $area\triangle AKN + area\triangle LCM = \frac{1}{4}area\triangle ABD + \frac{1}{4}area\triangle BCD = \frac{1}{4}areaABCD$. When we repeat this procedure using diagonal AC, we find that $area\triangle KBL + area\triangle NDM = \frac{1}{4}area\triangle ABC + \frac{1}{4}area\triangle ACD = \frac{1}{4}areaABCD$. Therefore, $area\triangle AKN + area\triangle LCM + area\triangle KBL + area\triangle NDM = \frac{1}{2}areaABCD$, which is to say that parallelogram $KLMN$ is one-half the area of the original quadrilateral.

This also demonstrates that the lines joining the midpoints of a quadrilateral's opposite sides bisect each other, as they are also the diagonals of the Varignon parallelogram, thus determining the common midpoint P of KM and LN, as shown in Figure 6-67.

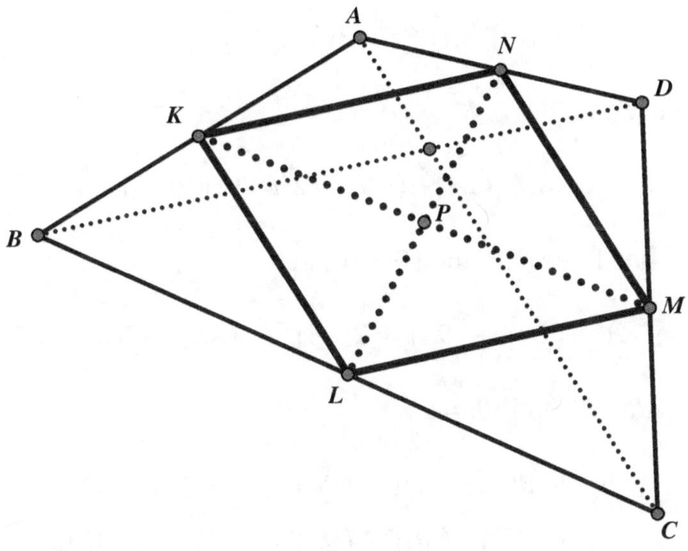

Figure 6-67

Geometrick 66: An Unexpected Relationship Between the Sides and the Diagonals of a Parallelogram

In the proof of Stewart's Theorem [see page 43, lines (II) and (IV)], we established the following relationships, which we apply to parallelogram $ABCD$ with $BF \perp AFEC$ in Figure 6-68. Consider $\triangle ABE$, where

$$(AB)^2 = (BE)^2 + (AE)^2 - 2(AE)(FE), \tag{I}$$

and for

$$\triangle EBC, (BC)^2 = (BE)^2 + (EC)^2 + 2(EC)(FE). \tag{II}$$

Since the diagonals of $ABCD$ bisect each other, $AE = EC$. Therefore, by adding equations (I) and (II), we get

$$(AB)^2 + (BC)^2 = 2(BE)^2 + 2(AE)^2. \tag{III}$$

Similarly, in

$$\triangle CAD, (CD)^2 + (DA)^2 = 2(DE)^2 + 2(CE)^2. \tag{IV}$$

By adding lines (III) and (IV), we get

$$(AB)^2 + (BC)^2 + (CD)^2 + (DA)^2 = 2(BE)^2 + 2(AE)^2 + 2(DE)^2 + 2(CE)^2.$$

Since $AE = EC$ and $BE = ED$,

$$(AB)^2 + (BC)^2 + (CD)^2 + (DA)^2 = 4(BE)^2 + 4(AE)^2,$$
$$(AB)^2 + (BC)^2 + (CD)^2 + (DA)^2 = (2BE)^2 + (2AE)^2,$$
$$(AB)^2 + (BC)^2 + (CD)^2 + (DA)^2 = (BD)^2 + (AC)^2.$$

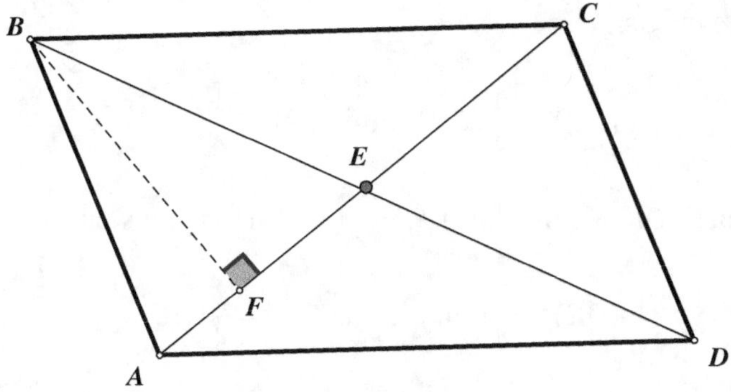

Figure 6-68

Geometrick 67: The Relationship Between the Sides and Diagonals of any Quadrilateral

In Figure 6-69, points P, Q, R, and S are the midpoints of the sides AD, AB, BC, and DC of quadrilateral $ABCD$. A line joining the midpoints of two sides of a triangle is one-half the length of the third side. Therefore, for triangle DAB we have $PQ = \frac{1}{2}(DB)$, and $SR = \frac{1}{2}(DB)$.

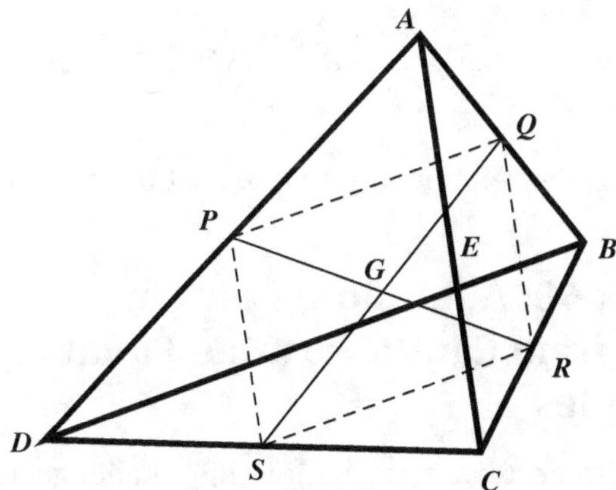

Figure 6-69

This gives us

$$(PQ)^2 = \frac{1}{4}(DB)^2 \text{ and } (SR)^2 = \frac{1}{4}(DB)^2. \tag{I}$$

Similarly, $QR = \frac{1}{2}(AC)$, and $PS = \frac{1}{2}(AC)$. This gives us

$$(QR)^2 = \frac{1}{4}(AC)^2 \text{ and } (PS)^2 = \frac{1}{4}(AC)^2. \tag{II}$$

Applying the relationship presented in the previous challenge to parallelogram *PQRS* gives us

$$PQ^2 + QR^2 + RS^2 + SP^2 = PR^2 + QS^2. \tag{III}$$

We now make the appropriate substitutions of (I) and (II) into (III):

$$\frac{1}{4}(DB)^2 + \frac{1}{4}(DB)^2 + \frac{1}{4}(AC)^2 + \frac{1}{4}(AC)^2 = (PR)^2 + (QS)^2$$
$$= \frac{1}{2}(DB)^2 + \frac{1}{2}(AC)^2 = (PR)^2 + (QS)^2$$
$$= (DB)^2 + (AC)^2 = 2\left[(PR)^2 + (QS)^2\right].$$

This justifies the relationship that we set out to prove.

Geometrick 68: A Random Point in a Parallelogram Can Determine Equal Area Triangles

Perhaps the simplest way to justify this challenge is to consider the parallel lines shown in Figure 6-70, with the area of each small triangular region marked with *m*, *n*, *p*, and *q*. Through point *P* two lines are drawn parallel to the sides of the parallelogram, partitioning *ABCD* into four parallelograms, each of which is divided in half by the four lines emanating from point P. $Area\triangle APB + Area\triangle DPC = (n+q)+(m+p)=m+n+p+q$, and similarly, $Area\triangle BPC + Area\triangle APD = (p+q)+(m+n)=m+n+p+q$. Therefore, $Area\triangle APB + Area\triangle DPC = Area\triangle BPC + Area\triangle APD$.

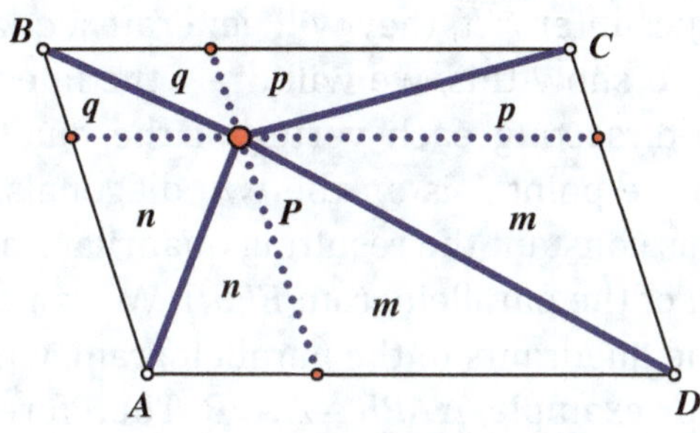

Figure 6-70

Geometrick 69: Equal Area Parallelograms in a Parallelogram

It is clear that the diagonal of a parallelogram divides a parallelogram into two equal area triangles, as shown in Figure 6-71. Consequently, we have $\triangle ABD \cong \triangle CBD$. However, this is true for the two pairs of non-shaded triangles as well, so we have $\triangle BEP \cong \triangle BDP$, and $\triangle PGD \cong \triangle PFD$.

Therefore, $Area\triangle BCD - Area\triangle BEP - Area\triangle PFD = Area\triangle ABD - Area\triangle BDP - Area\triangle PGD$, which leaves the two shaded parallelograms equal.

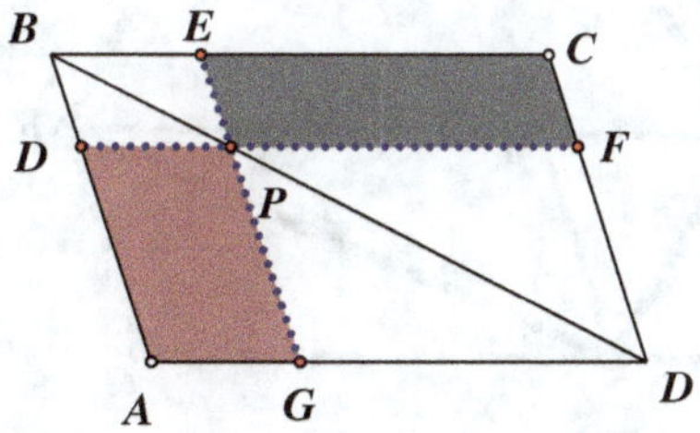

Figure 6-71

Geometrick 70: Detecting a Constant Quadrilateral

Suppose you are given two lines *AC* and *BD*, shown in Figure 6-72, which intersect at a given angle *APB*. We can show that no matter where the two lines intersect, they will generate a quadrilateral with a constant area. To show this, we will draw the lines parallel to the two diagonals intersecting each vertex of the quadrilateral *ABCD*. Regardless of where point *P* is on the two diagonals, as long as the angle *APB* remains constant the resulting quadrilateral will always be one-half the area of the parallelogram *EFGH*. We can easily show this by focusing on the quadrants in the parallelogram, where each one is divided in half; for example, $\triangle APB \cong \triangle AFB$. Therefore, the quadrilateral will always have half the area of the parallelogram, which is fixed regardless of where the diagonals of the quadrilateral intersect, provided that they remain at a constant length and intersect at the common angle *APB*.

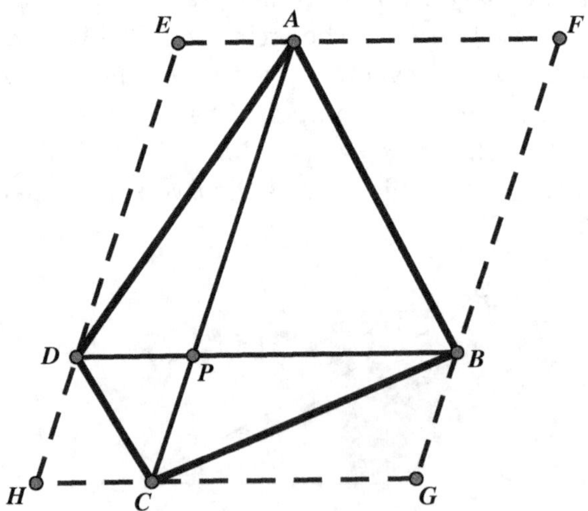

Figure 6-72

Geometrick 71: A Surprising Constant Product Created by a Parallelogram

In Figure 6-73, we notice that $\triangle ABM \sim \triangle NCM \sim \triangle ADN$, which generates the proportion $\frac{BM}{AB} = \frac{AD}{DN}$, and which gives us $BM \cdot DN = AB \cdot AD$. This is a constant product since they are the given sides of the parallelogram.

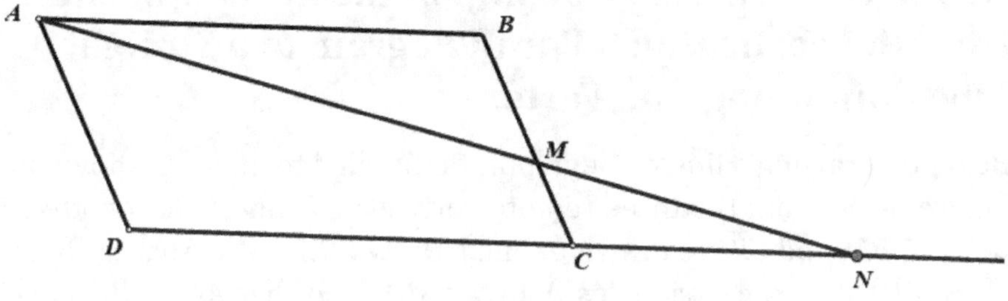

Figure 6-73

Geometrick 72: Finding the Largest Triangle in a Particular Parallelogram Setting

We know that $DM = CM = AK = BK$, as seen in Figure 6-74. Also, $\frac{JG}{GO} = \frac{1}{2}$. We draw a perpendicular line through G to produce similar triangles: $\triangle MJG \sim \triangle KGL$. Because the medians trisect each other, $\frac{MG}{GL} = \frac{GJ}{GD+DK} = \frac{1}{2+3} = \frac{1}{5}$, or $GL = \frac{5}{6} ML$. We know that the $areaABCD = ML \cdot AB$, while

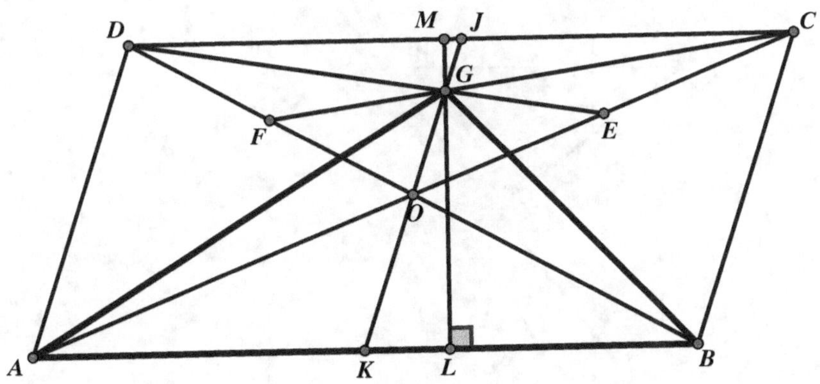

Figure 6-74

$area\triangle AGB = \left(\frac{1}{2}AB\right)\cdot\left(\frac{5}{6}ML\right) = \frac{5}{12}areaABCD$. Similarly, $area\triangle DGC = \frac{1}{6}areaABCD$, while $area\triangle AGD = area\triangle BGC = \frac{1}{4}areaABCD$. Hence, the largest triangle with vertex G and a side of the parallelogram is triangle $AGB = \frac{5}{12}\cdot 120 = 50$.

Geometrick 73: Relationship of the Perpendiculars from the Vertices of a Parallelogram to a Straight Line Containing One Vertex

Begin by drawing a line through point C parallel to line l, as shown in Figure 6-75. From the three remote vertices A, B, and C, perpendiculars AN, BM, and CR are drawn to line l. Draw $CE \perp BM$ so that we have $CR = ME$. Also, right triangles $\triangle AND \cong \triangle DEC$, so that $AN = BE$. Since $BM = BE + EM$, we get $AN + CR = BM$.

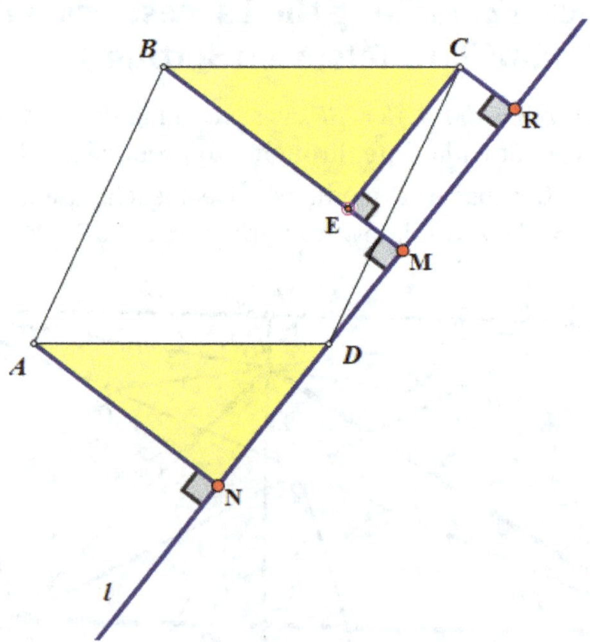

Figure 6-75

Geometrick 74: Isosceles Right Triangles on the Sides of a Parallelogram Create a Square

In Figure 6-76, we immediately have $\triangle AHD \cong \triangle BFC$, as they are both isosceles right triangles with the same hypotenuse, and so we have $AH = BF$. Clearly, $AE = BE$, and $\angle HAE = 360° - 45° - 45° - \angle BAD = 270° - \angle BAD$. Now consider $\angle EBF = 45° + 45° + \angle ABC = 90° + (180° - \angle BAD) = 270° - \angle BAD$. Therefore, $\angle HAE = \angle EBF$, and so $\triangle HAE \cong \triangle EBF$, and $HE = EF$. Now consider $\angle AEB = 90° = \angle AEF + \angle FEB = \angle AEF + \angle HEA = \angle HEF$. In an analogous fashion, we can show $HE = HG$, $HG = FG$, and $FG = EF$, which establishes a rhombus. However, we can also show in a similar way that $\angle EFG = \angle FGH = \angle GHE = 90°$. Therefore, we have established a square: $EFGH$.

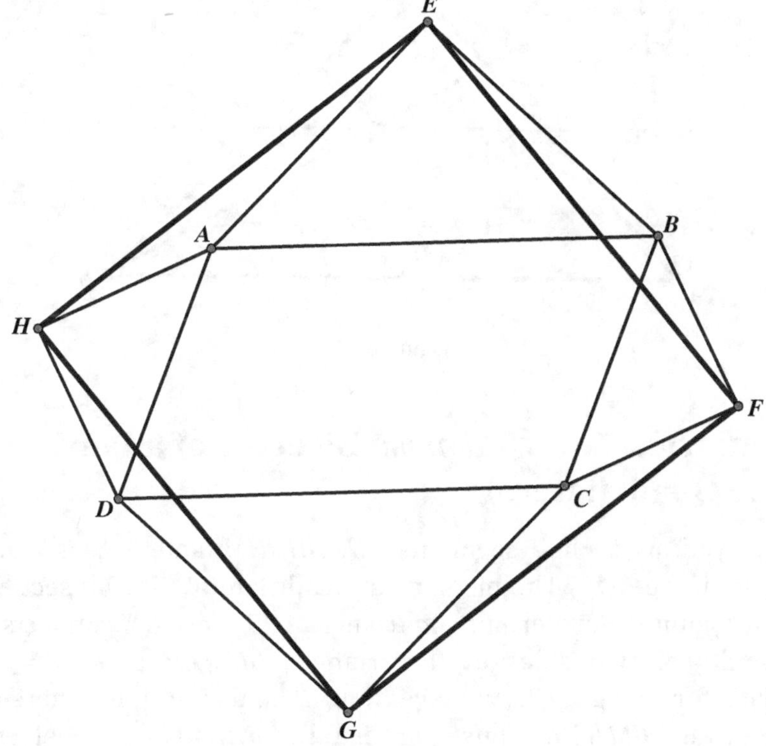

Figure 6-76

Geometrick 75: A Relationship Between the Midpoints of the Diagonals and the Bases of the Trapezoid

Begin by letting G be the midpoint of AB, as shown in Figure 6-77. For triangle ABC we have $GE \parallel BC \parallel AD$, since a line joining the midpoints of two sides of a triangle is parallel to the third side. If GE is extended it must pass through point F because $GF \parallel AD$ and G bisects AB. Therefore, in triangle ABD, we have $GF = \frac{1}{2}AD$, and for triangle ABC, we have $GE = \frac{1}{2}BC$. However, $EF = GF - GE = \frac{1}{2}(AD - BC)$.

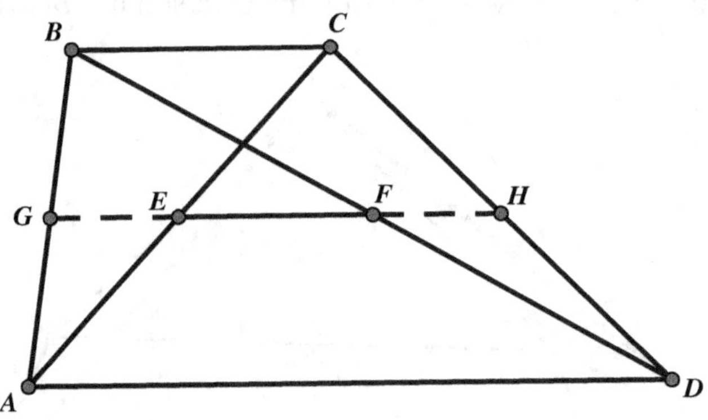

Figure 6-77

Geometrick 76: A Surprise Bisector of a Line in the Quadrilateral

Begin by drawing line segments GN, NH, HM, and GM, as can be seen in Figure 6-78. The medians of quadrilateral $ABCD$ bisect each other at point P. Furthermore, in triangle ADB, since GM connects the midpoints of two sides of the triangle, $GM \parallel AB$, and $GM = \frac{1}{2}AB$. Similarly for triangle ABC, we have $HN \parallel AB$, and $HN = \frac{1}{2}AB$. Therefore, $GM = HN$ and $GM \parallel HN$. Thus, quadrilateral $GNHM$ is a parallelogram where the diagonals bisect each other, indicating that P is the midpoint of MN.

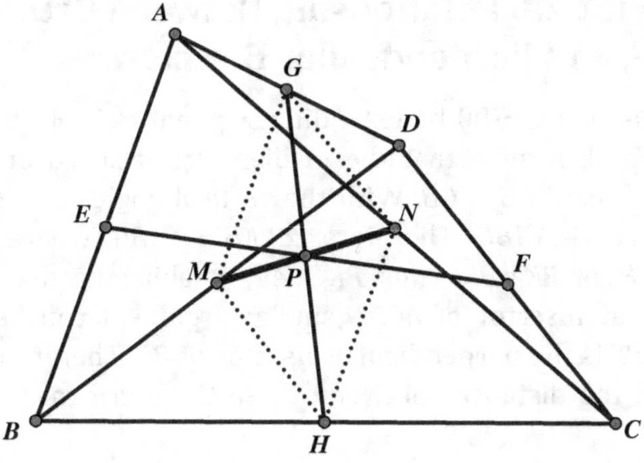

Figure 6-78

Geometrick 77: Constructing a Cyclic Quadrilateral

Using the angle measurements in Figure 6-79, consider triangle ABG, where $y = 180° - (s + t)$ and for triangle DCH, we have $x = 180° - (w + z)$. Therefore, $y + x = 180° - (s + t) + 180° - (w + z) = 360° - (s + t + w + z) = 360° - 180° = 180°$. Because we now have a pair of supplementary opposite angles for quadrilateral $EGFH$, the quadrilateral must be cyclic.

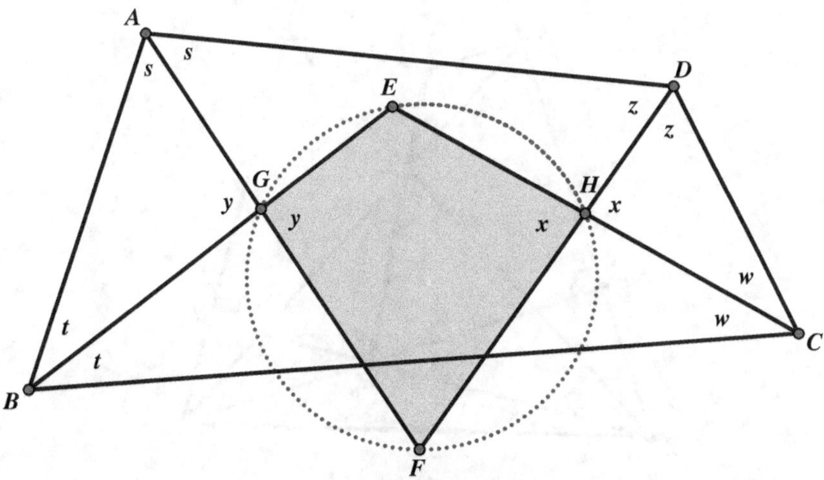

Figure 6-79

Geometrick 78: Relationship Between Orthocenter, Centroid, and Perpendicular Bisectors

We begin in Figure 6-80 by extending segment *HG* half its length to point *P*. We also know that the medians trisect each other at the centroid, so that $2KG = GB$. With the vertical angles $\angle PGK = \angle HGB$, we have $\triangle PKG \sim \triangle HBG$. Therefore, $\angle PKG = \angle HBG$, where it follows that $PK \parallel BH$, or $PK \parallel BE$. Thus, $PK \perp AC$, establishing that PK is the perpendicular bisector of *AC*. A similar argument can be made to show that *PF* is the perpendicular bisector of *BC*. Therefore, we have established the distances between these three critical points were $GH = 2\,PG$.

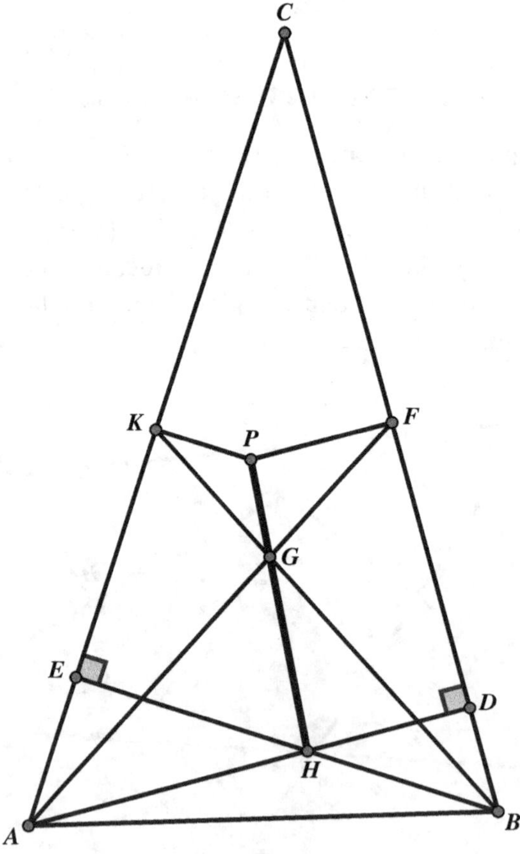

Figure 6-80

Geometrick 79: A Surprise Perpendicularity

Angles measured by equal arcs are equal, therefore, $\angle BCD = \angle DAE$. This equality generates a cyclic quadrilateral $ACGF$, shown in Figure 6-81. When we draw line AC, we find that $\angle ACB = 90°$, as it is inscribed in a semicircle. Within this quadrilateral, two other angles are measured by $\overset{\frown}{AF}$. Therefore, $\angle AGF = \angle ACB = 90°$, which is what we set out to prove.

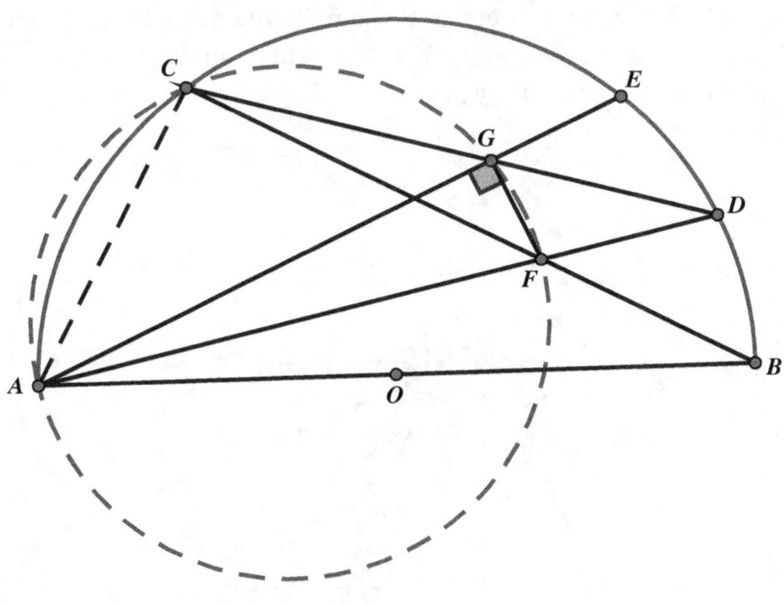

Figure 6-81

Geometrick 80: An Unexpected Cyclic Quadrilateral

For quadrilateral *BCGH* in Figure 6-82 we have $\angle BCG + \angle BHG = 180°$, and for quadrilateral *ABHE* we have $\angle BAE + \angle BHE = 180°$. Since $\angle BHG + \angle BHE + \angle GHE = 360°$, by substitution we get $\angle BCG + \angle BAE + \angle GHE = 360°$. The same argument can then be made to show that $\angle DCG + \angle DAE + \angle GFE = 360°$. However, in cyclic quadrilateral *ABCD*, the opposite angles are supplementary, namely, $\angle BCD + \angle BAD = 180°$, which is essentially $\angle BCG + \angle DCG + \angle BAE + \angle DAE = 180°$. When we subtract these pairs of angles from the above 360° sums, we get $\angle GHE + \angle GFE = 180°$, which allows us to conclude that quadrilateral *EFGH* is cyclic.

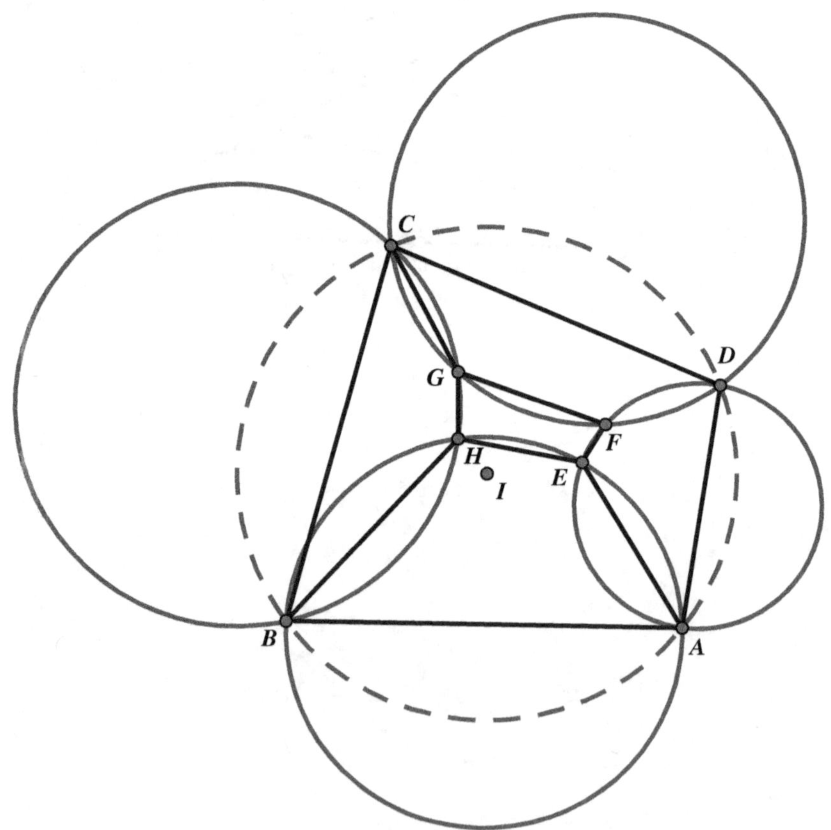

Figure 6-82

Geometrick 81: Another Surprise Cyclic Quadrilateral

We begin by drawing lines *PD*, *PE*, and *PF*, as shown in Figure 6-83. Since points *D*, *A*, *B*, and *M* are concyclic, $\angle AMB = \angle ADB$. Also, since points *C*, *E*, *B*, and *M* are concyclic, $\angle BMC$ is supplementary to $\angle BEC$, which in turn is supplementary to $\angle BEP$. This enables us to have $\angle BMC = \angle BEP$. We now recognize that $\angle AMB + \angle BMC = \angle AMC$. Therefore, $\angle AMC = \angle BEP + \angle ADB$. In triangle *APC*, we have $\angle P = 180° - (\angle BEP + \angle ADB) = 180° - \angle AMC$. Thus, we can conclude that $\angle P$ is supplementary to $\angle AMC$ and, therefore, the points *A*, *P*, *C*, and *M* are concyclic.

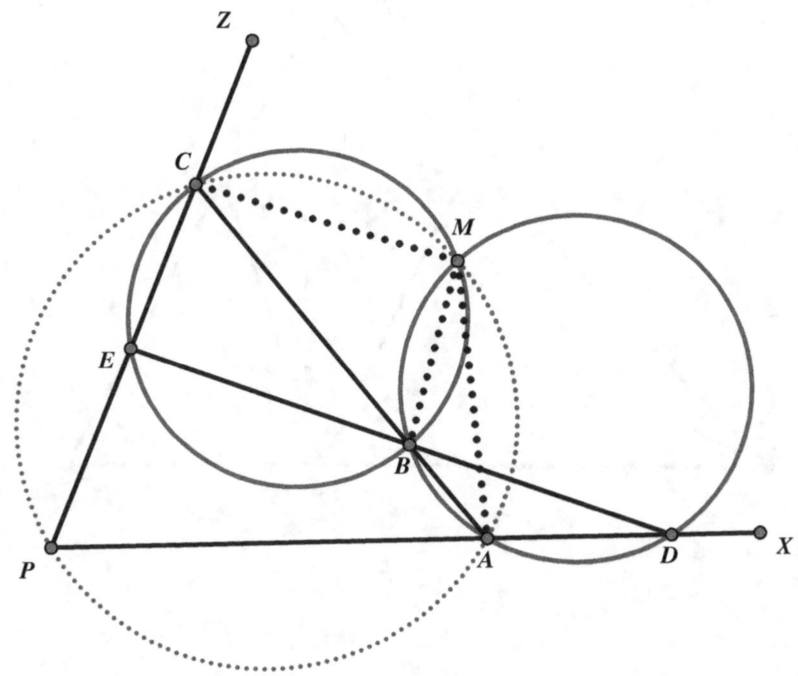

Figure 6-83

Geometrick 82: Three Intersecting Circles

In Figure 6-84, for cyclic quadrilateral *FODA*, angle *FOD* is supplementary to angle *A*. In cyclic quadrilateral *DOEB*, angle *DOE* is supplementary to angle *B*. The three angles $\angle FOD + \angle DOE + \angle FOE = 360°$, and along with the three angles of triangle *ABC*, the total sum is $360° + 180° = 540°$. Now $(\angle A + \angle AOB) + (\angle B + \angle DOE) = 180° + 180° = 360°$. Therefore, $\angle FOE + \angle C = 180°$, which determines that quadrilateral *CFOE* is cyclic and the circumference of the circle containing points *F*, *C*, and *E* must contain *O*. Consequently, the three circles share a common point *O*.

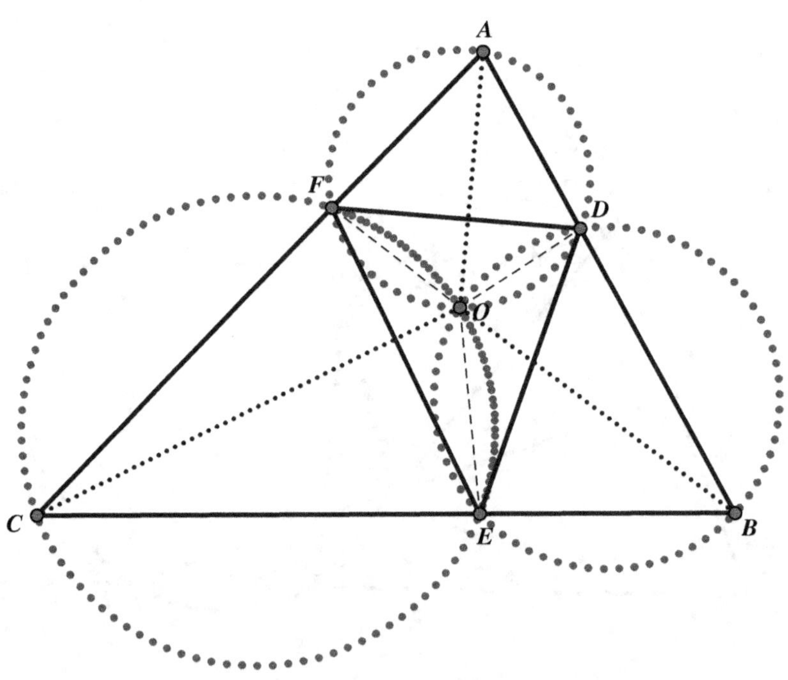

Figure 6-84

Geometrick 83: More with Three Intersecting Circles

In Figure 6-85, we have cyclic quadrilateral *FOEC*, where $\angle CFE = \angle COE$, as they are both inscribed in $\overset{\frown}{CE}$. Similarly, in cyclic quadrilateral *DOEB* we have $\angle BDE = \angle BOE$. Since $\angle BOC = \angle COE + \angle BOE$, we have $\angle BOC = \angle CFE + \angle BDE$. But $\angle CFE = 180° - (\angle FEC + \angle C)$, and $\angle BDE = 180° - (\angle BED + \angle B)$. Therefore, $\angle BOC = [180° - (\angle B + \angle C)] + [180° - (\angle FEC + \angle BED)]$. This essentially gives us $\angle BOC = \angle A + \angle DEF$. In a similar fashion, we can prove that $\angle AOC = \angle B + \angle DFE$, and $\angle AOB = \angle C + \angle FDE$.

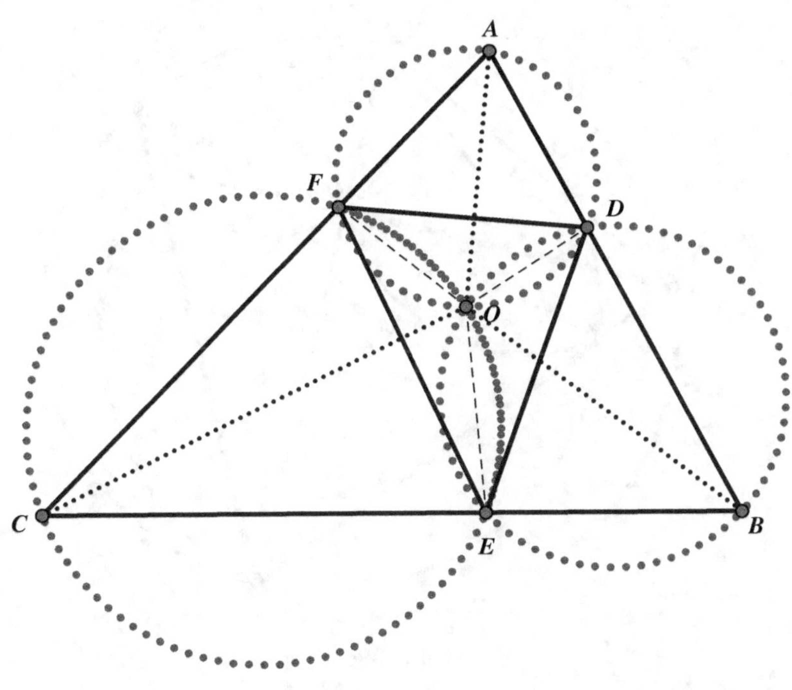

Figure 6-85

Geometrick 84: Three Intersecting Circles Creating a Straight Line

We begin by drawing lines *PD*, *PE*, and *PF*, as shown in Figure 6-86. Since angle *PDC* is inscribed in the semicircle with center *S*, we have *PD* ⊥ *DCB*; similarly, angle *PEA* and angle *PFC* are inscribed in semicircles with centers *Q* and *S*, respectively. Therefore, *PE* ⊥ *AB*, and *PF* ⊥ *AC*. We now can apply Simson's theorem to triangle *ABC* (See page 40) where the points *D*, *F*, and *E* are feet of the perpendiculars generated from point *P* and are therefore collinear.

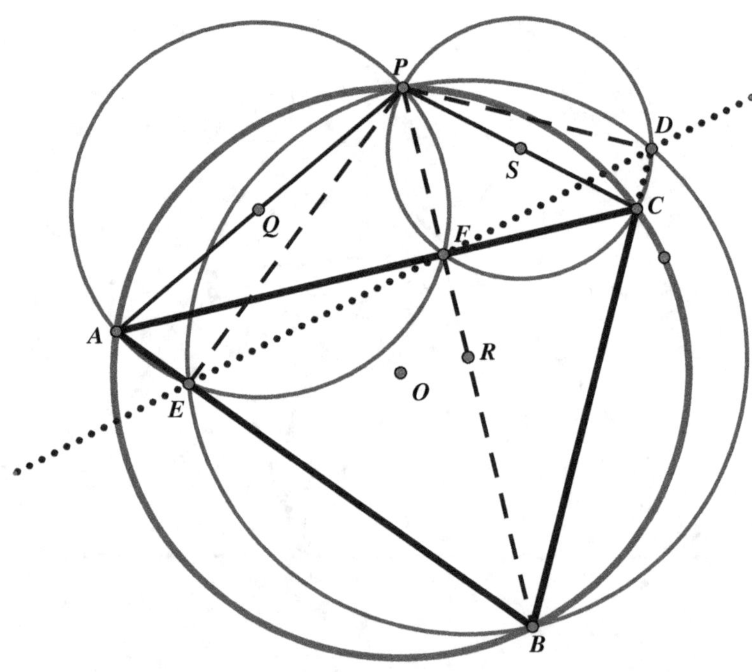

Figure 6-86

Geometrick 85: A Curiosity on a Decagon

In Figure 6-87, the measure of any one of the ten equal arcs is 36°. We note that inscribed $\angle DCM = \frac{1}{2} \cdot \overparen{DFH} = \frac{1}{2} \cdot 4 \cdot 36° = 72°$, and the angle formed by two cords $\angle CMD = \frac{1}{2} \cdot (\overparen{DC} + \overparen{AH}) = \frac{1}{2} \cdot 4 \cdot 36° = 72°$. Therefore, $\angle DCN = \angle CMD$ and triangle DMC is isosceles, with $CD = MD$. We can also show $\triangle DMC \sim \triangle AOM$, since vertical angles $\angle DMC = \angle AMO$, and inscribed angles measured by equal arcs yield $\angle CDM = \angle DAF$. Thus, triangle AOM is also isosceles, and $MA = AO$. Hence, $AD = CD + MA = CD + AO$.

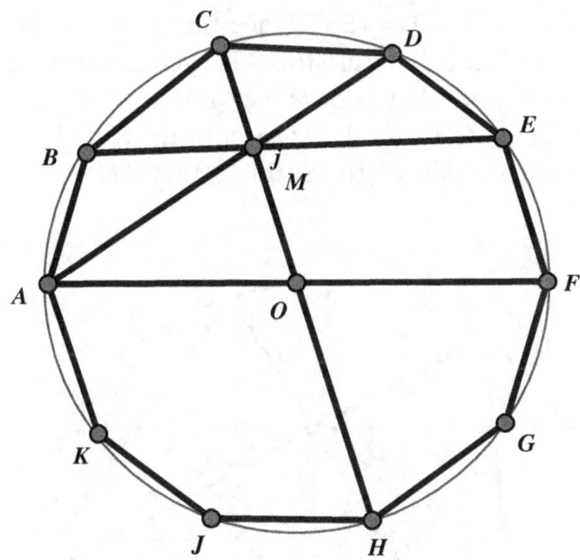

Figure 6-87

Geometrick 86: The Famous Butterfly Relationship

We begin by drawing the auxiliary lines $FG \parallel AB$ and $MN \perp FG$, as well as the lines GM, GQ, and GD, as shown in Figure 6-88. As $MN \perp AB$, we know that MN contains the center O of the circle. Since MN is also the perpendicular bisector of FG, and we have $MF = MG$, it follows that $\triangle MFN \cong \triangle MNG$ and $\angle FMN = \angle NMG$. Furthermore, we have $\angle AMF = \angle GMB$, since they are complements of the equal angles $\angle FMN = \angle NMG$. Since $FG \parallel AB$, we can conclude that $\overset{\frown}{AF} = \overset{\frown}{GB}$. The angle formed by two intersecting chords is one-half the sum of the intercepted arcs, and we obtain $\angle GMB = \angle AMF = \frac{1}{2} \cdot (\overset{\frown}{AF} + \overset{\frown}{BE}) = \frac{1}{2}(\overset{\frown}{GB} + \overset{\frown}{BE})$. We know that $\angle EDG = \frac{1}{2} \cdot \overset{\frown}{EAG}$, and by addition, $\angle GMB + \angle EDG = \frac{1}{2} \cdot (\overset{\frown}{GB} + \overset{\frown}{BE} + \overset{\frown}{EAG}) = \frac{1}{2} \cdot 360° = 180°$. Since this means $\angle GMQ + \angle QDG = 180°$, we see that quadrilateral $MGDQ$ is cyclic, and we get $\angle QDM = \angle QGM$. We then find that $\angle QGM = \angle QDM = \angle EDC = \angle EFC = \angle MFP$ (note that $\angle EDC = \angle EFC$, as they are both measured by one-half of arc FGD), which enables us to establish that $\triangle MPF \cong \triangle MGQ$ (ASA). Thus, $MP = MQ$.

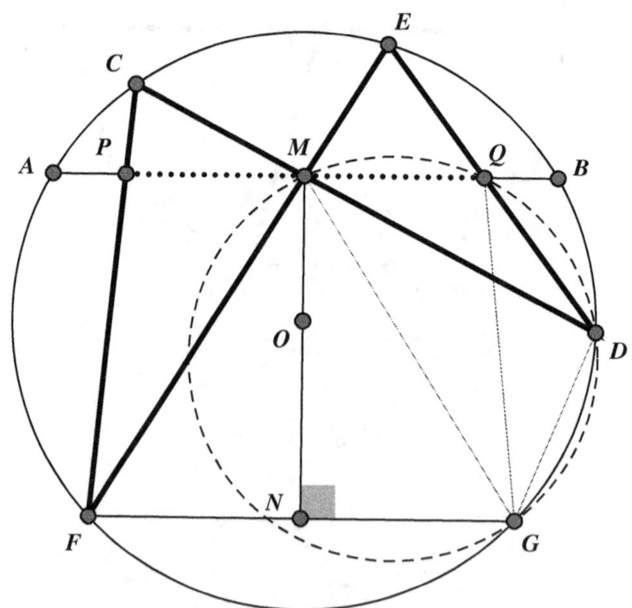

Figure 6-88

Geometrick 87: A Surprise Tangent

We begin by drawing a line through point *B* that is parallel to diameter *EF*, intersects *AC* at point *H*, and intersects *TC* at point *L*, as can be seen in Figure 6-89. Focusing on triangle *BCH*, we find that *HL* = *BL*. We also know that *AJ* = *BJ*, since a perpendicular from the center of the circle to a chord bisects the chord. Therefore, *JL*, which connects the midpoints of two sides of triangle *ABH*, is parallel to *AH*. We then have ∠*LJB* = ∠*GAC* = ∠*BTC* because angles *GAC* and *BTC* are both measured by $\overset{\frown}{BFC}$. Consequently, the points *L*, *J*, *T*, and *B* are concyclic; in other words, quadrilateral *LJTB* is cyclic. Furthermore, ∠*JGO* = ∠*LBJ* = ∠*OTJ*, because angles *LBJ* and *OTJ* are measured by $\overset{\frown}{LJ}$. Thus, ∠*OTG* = ∠*OJG* = 90°, since they are both inscribed in $\overset{\frown}{GO}$, and this proves that *GT* is tangent to circle *O* because it is at a point on the circle where the radius is perpendicular to the tangent.

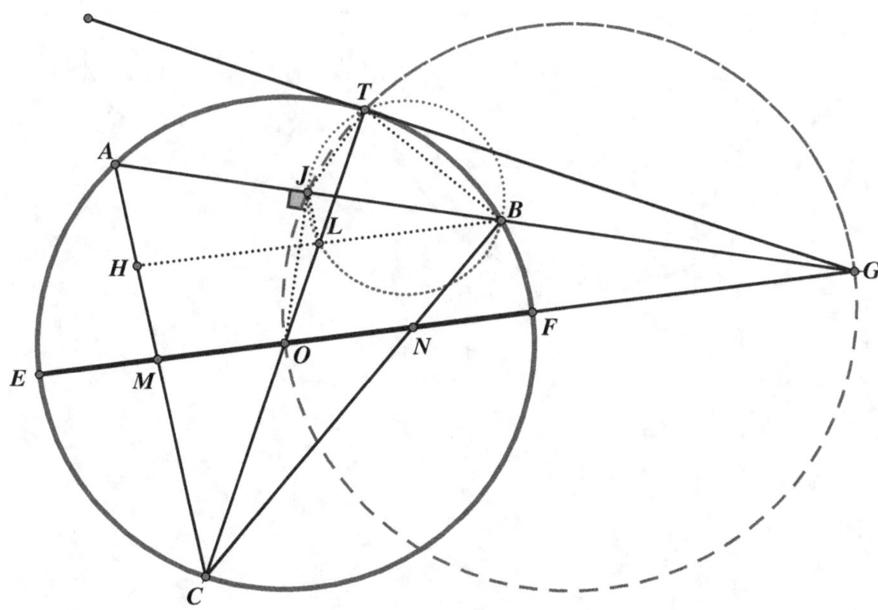

Figure 6-89

Geometrick 88: Equality of Common Tangents

Using Figure 6-90, we simplify matters by letting $AB = CD = x$, $AE = EP = y$, and $FQ = FD = z$. The basic relationship that we will use in this proof is that tangents drawn from an external point to the same circle are of equal length. Therefore, we get $EB = EQ = x - y$, and $FC = FP = x - z$. We also have $EQ + QF = FP + EP$. This can be written as $x - y + z = x - z + y$, which reduces to $z = y$. We also have $EF = EQ + QF = x - y + z = x$. Therefore, we have shown that $AB = EF$. A similar argument shows that $CD = EF$ as well.

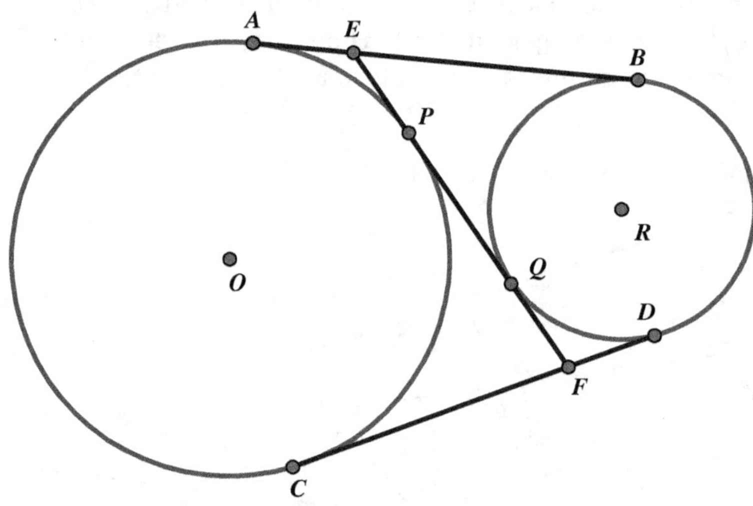

Figure 6-90

Geometrick 89: Appearance of an Equilateral Triangle in an Equilateral Triangle

Besides the given equal segments along the sides of the equilateral triangle in Figure 6-91, we also have $AD = BF = CE$ and the vertex angles of the equilateral triangle are equal. Therefore, $\triangle ABD \cong \triangle ECA \cong \triangle BCF$ (by SAS) and $AE = BD = CF$. Also, $AD = BF = CE$, and by subtraction $DC = BE = AF$. We will also need to know that $\angle ABD = \angle BCF = \angle CAE$. We then have $\triangle AFC \cong \triangle BCD \cong \triangle ABE$ (by SAS). The exterior angles of these three congruent triangles are also equal, namely, $\angle ADK = \angle BFH = \angle CEG$. We now have enough to establish that $\triangle ADK \cong \triangle BFH \cong \triangle CEG$ (by ASA). Since $\angle AKD = \angle BHF = \angle CGE$, their respective vertical angles are also equal, namely, $\angle HKG = \angle KHG = \angle KGH$, which makes triangle GHK equiangular, which is what we needed to prove it's equilateral.

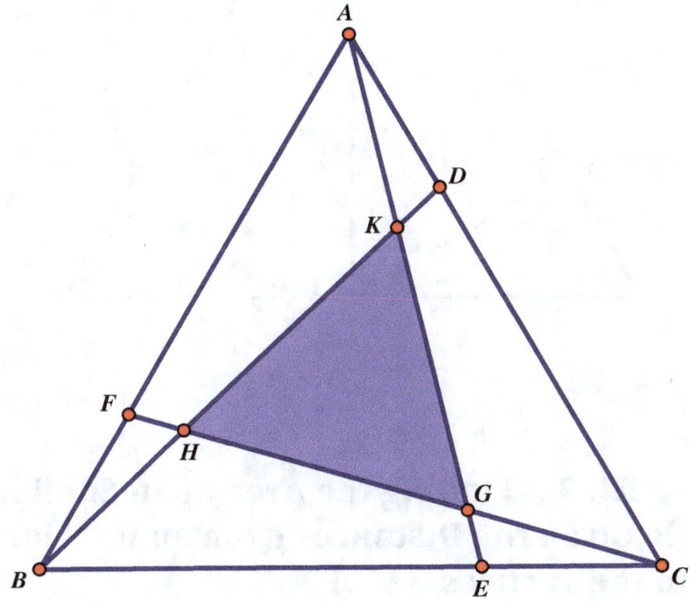

Figure 6-91

Geometrick 90: How a Circle Can Partition an Equilateral Triangle

Symmetry indicates that AM is a diameter of the circle that makes $\angle ADM = 90°$, as shown in Figure 6-92. We also know that $\angle B = 60°$, which makes triangle BDM a $30° - 60° - 90°$ triangle. Therefore, $\frac{BD}{BM} = \frac{1}{2}$. Since $BM = \frac{BC}{2}$, we then have $\frac{BD}{BC} = \frac{BD}{AB} = \frac{1}{4}$. Thus, $\frac{BD}{AD} = \frac{1}{3}$.

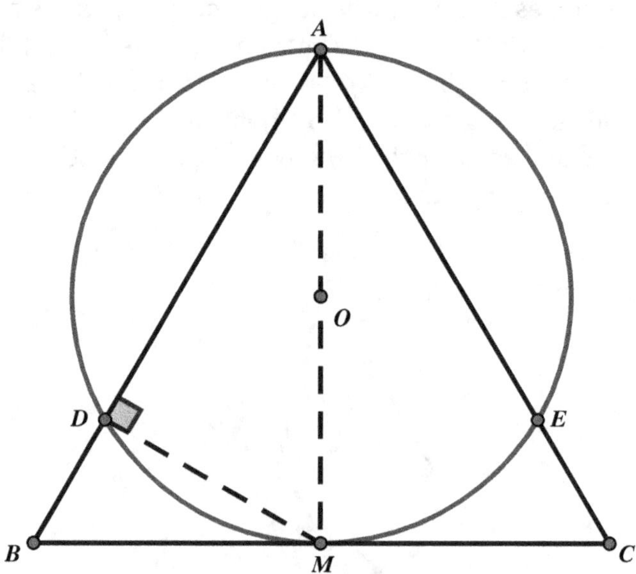

Figure 6-92

Geometrick 91: Finding the Area of an Equilateral Triangle Given the Distances from an Interior Point to the Vertices

Begin by constructing equilateral triangles *PAD*, *PBE*, and *PCF* upon the segments *PA*, *PB*, and *PC*, as we show in Figure 6-93. Now draw *AF*, *BD*, and *CE* to complete the hexagon *ADBECF*. Although this hexagon is not a regular hexagon, the opposite sides are congruent and parallel, as we see in the figure. First, consider $\triangle BPC \cong \triangle AFC$, since $PC = FC$

and $BC = AC$, and we also have $\angle ACF = \angle PCF - \angle PCA = \angle ACB - \angle PCA = \angle PCB$, which establishes the aforementioned triangle congruence (SAS). Similarly, we can establish that $\triangle CPA \cong \triangle ADB$ and $\triangle APB \cong BEC$. Keep in mind that $area\triangle ABC = area\triangle BPC + area\triangle CPA + area\triangle APB$. Therefore, the area of hexagon $ADBECF$ is twice the area of triangle ABC. However, the area of hexagon $ADBECF$ equals the sum of the areas of $\triangle ADP + \triangle DBP + \triangle BEP + \triangle ECP + \triangle CFP + \triangle FAP$. We note that triangles ADP, BEP, and CFP are equilateral with side lengths 5, 6, and 7, respectively. The areas of these three triangles can be easily calculated with the formula $\frac{s^2\sqrt{3}}{4}$ so that we have $\frac{5^2\sqrt{3}}{4} + \frac{6^2\sqrt{3}}{4} + \frac{7^2\sqrt{3}}{4} = \frac{110\sqrt{3}}{4} = \frac{55\sqrt{3}}{2}$. We can also quickly establish that $\triangle ECP \cong \triangle DBP \cong \triangle FAP$ using the above criteria we have established. Heron's formula for calculating the area of a triangle whose sides have lengths a, b, and c, and semiperimeter s is $area = \sqrt{s(s-a)(s-b)(s-c)}$. We can use Heron's formula for any one of these triangles, say, $area\triangle ECP = \sqrt{9(9-5)(9-6)(9-7)} = 6\sqrt{6}$. The area of the three triangles is then $18\sqrt{6}$. Therefore, the area of the hexagon $ADBECF$ is equal to the sum of the areas of the six triangles, which is $\frac{55\sqrt{3}}{2} + 18\sqrt{6}$. Thus, the area of triangle ABC is equal to $\frac{55\sqrt{3}}{4} + 9\sqrt{6}$.

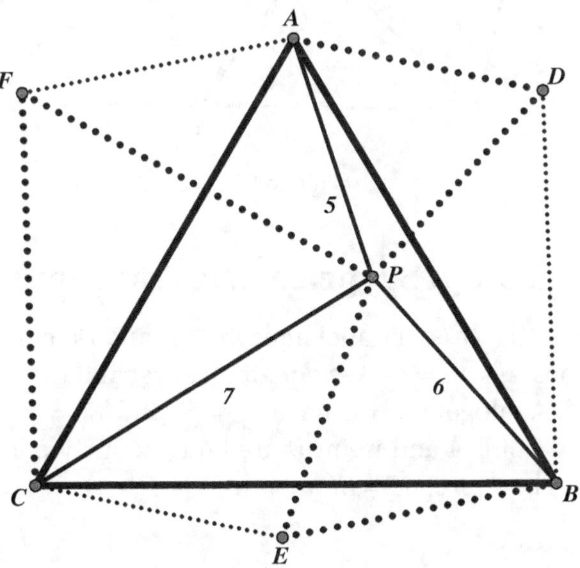

Figure 6-93

Geometrick 92: Unexpected Isosceles Triangles Generate Concyclic Points

The median in a right triangle drawn to the hypotenuse is half the length of the hypotenuse. For triangle ANC in Figure 6-94, this relationship gives us $AC = NE$, and for triangle AMB, sides $AF = MF$. Therefore, $\angle NAE = \angle ANE$, and $\angle AMF = \angle MAF$. We can clearly see that $\angle MAF + \angle NAE + \angle EAF = 180°$. We also note that in triangle MNP we have $\angle AMF + \angle ANE + \angle P = 180°$; therefore, $\angle P = \angle EAF$. For parallelogram $AEDF$, we have $\angle EDF = \angle EAF$. Thus, $\angle P = \angle EDF$. This allows point P to be situated on the circumscribed circle of triangle DEF, as they will be two inscribed angles measured by the same arc EF. Thus, points D, P, E, and F are concyclic.

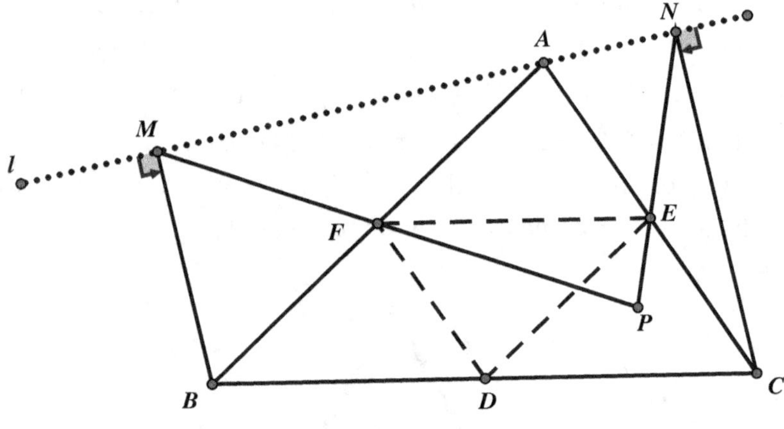

Figure 6-94

Geometrick 93: A Surprise in a Trapezoid

In Figure 6-95, the three parallel lines AB, EF, and DC emit the following proportions: $\frac{GF}{DC} = \frac{BF}{BC} = \frac{AE}{AD}$. We equate the first and third fractions to get $GF = \frac{DC \cdot AE}{AD}$. Analogously, we have $\frac{EG}{AB} = \frac{ED}{AD}$, and $EG = \frac{AB \cdot ED}{AD}$. Since the tangents from point A and point D are equal pairs, we have $AB = AE$ and $ED = DC$. Therefore, $GF = EG$.

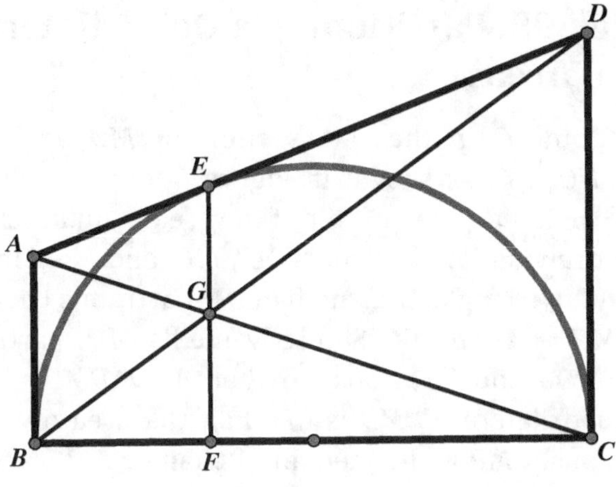

Figure 6-95

Geometrick 94: Another Surprise in a Trapezoid

In Figure 6-96, when we draw AO and DO we essentially create a right triangle AOD, because AO bisects angle EOB, DO bisects angle EOC, the angles are supplementary, and one-half their sum is 90°. Thus, $\angle AOD = 90°$. Furthermore, a radius is perpendicular to a tangent at the point of tangency, so $EO \perp AD$. The similar triangles EOA and EOD then yield the proportion $\frac{AE}{EO} = \frac{EO}{DE}$, which yields $AE \cdot DE = EO^2$.

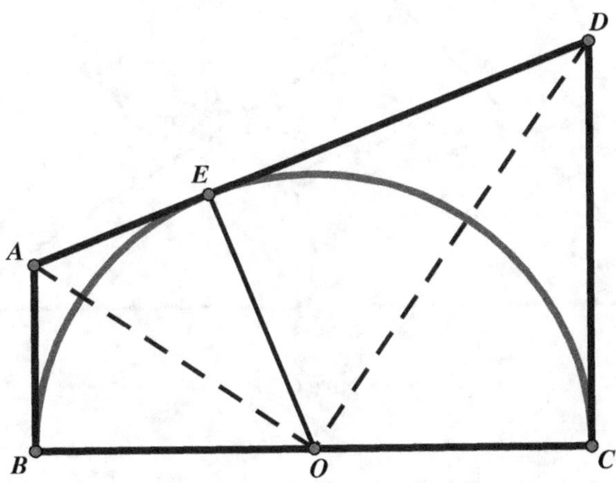

Figure 6-96

Geometrick 95: Partitioning a Quadrilateral into Four Equal Areas

Recall that point O is the intersection of MO and NO, where $MO \parallel BD$, and $NO \parallel AC$, as shown in Figure 6-97. We will prove that quadrilateral $OFCG$ has one quarter of the area of quadrilateral $ABCD$. We begin by drawing line segments MB, MD, and MG. Since M is the midpoint of AC in triangle ABC, median BM partitions triangle ABC so that $Area\triangle BMC = \frac{1}{2} Area\triangle ABC$. Similarly, median DM partitions triangle ABC into two equal regions, so that $Area\triangle DMC = \frac{1}{2} Area\triangle ABC$. Therefore, quadrilateral $CBMD$ is one-half the area of quadrilateral $ABCD$. Analogously, MF is the median of triangle AMC, and MG is the median of triangle DMC, so that we have $Area\triangle MFC = \frac{1}{2} Area\triangle BMC$ and $Area\triangle CMG = \frac{1}{2} Area\triangle DMC$. Therefore, quadrilateral $MFCG$ is half the area of quadrilateral $CBMD$, which then makes quadrilateral $MFCG$

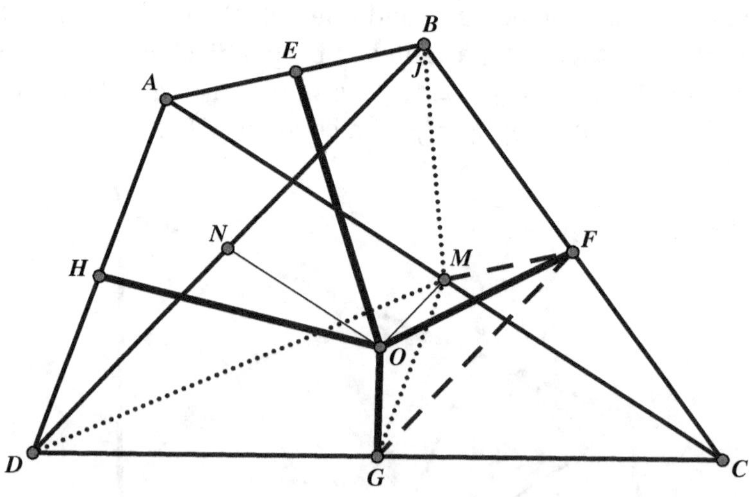

Figure 6-97

one-quarter the area of quadrilateral *ABCD*. Since $FG \parallel BD \parallel MO$, we find that $Area\triangle MFO = Area\triangle MOG$, as they have the same base, *MO*, and corresponding vertices on the line *FG* parallel to their base. Therefore, the area of quadrilateral *FMGC* is equal to the area of quadrilateral *OFCG*, that is, one quarter of the area of quadrilateral *ABCD*. This procedure can then be applied to the other quarter areas at vertices *A*, *B*, and *D*.

Geometrick 96: The Measure of an Angle in the Interior of a Triangle

Begin by noting in Figure 6-98 that triangle *PBC* has $\angle P = 180° - (\frac{\angle PBC}{2} + \frac{\angle PCB}{2})$. Considering triangle *ABC*, we get $\angle A = 180° - (\angle PBC + \angle PCB)$. Therefore, $90° - \frac{\angle A}{2} = (\frac{\angle PBC}{2} + \frac{\angle PCB}{2})$, and so when we substitute this into the previous equation, we get $\angle P = 180° - (90° - \frac{\angle A}{2}) = 90° + \frac{\angle A}{2}$.

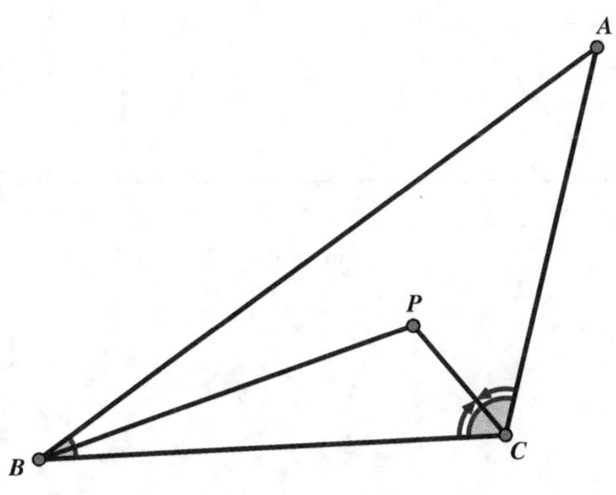

Figure 6-98

Geometrick 97: The Strange Angle Between an Altitude and an Angle Bisector

In triangle ABC, shown in Figure 6-99, where we know that the sum of the angles is $180°$, we can get half of angle A by $\frac{\angle A}{2} = \frac{180° - (\angle B + \angle C)}{2}$. Also, $\angle DAE = \frac{\angle A}{2} - \angle EAC$, and $\angle EAC = 90° - \angle C$.

Then, $\angle DAE = \frac{180° - (\angle B + \angle C)}{2} - 90° - \angle C = \frac{180° - (\angle B + \angle C) - 180° + 2\angle C}{2} = \frac{\angle C - \angle B}{2}$.

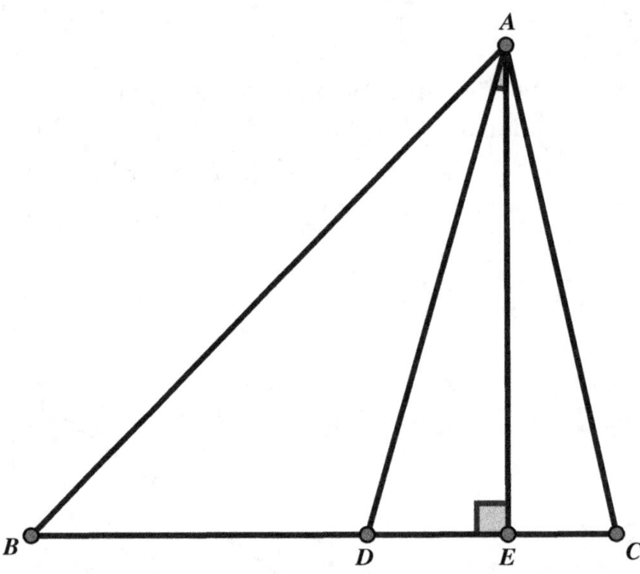

Figure 6-99

Geometrick 98: Another Curiosity Formed by the Altitude and Angle Bisector

In Figure 6-100, we have $\angle B + \angle ADB = 180° - \angle BAD$, and $\angle C + \angle ADC = 180° - \angle DAC$. But $\angle BAD = \angle DAC$, therefore, $\angle B + \angle ADB = \angle C + \angle ADC$. This can be expressed as $\angle C - \angle B = m - n$. We apply the relationship that we proved in Geometrick 97 to find $\angle DAE = \frac{m-n}{2}$. In triangle DAE, we see that $\angle DAE = 90° - n$, therefore, $\frac{m-n}{2} = 90° - n$, so $m - n = 180° - 2n$, whereupon it follows that $m + n = 180°$, which is consistent with what we see in the diagram as a straight line.

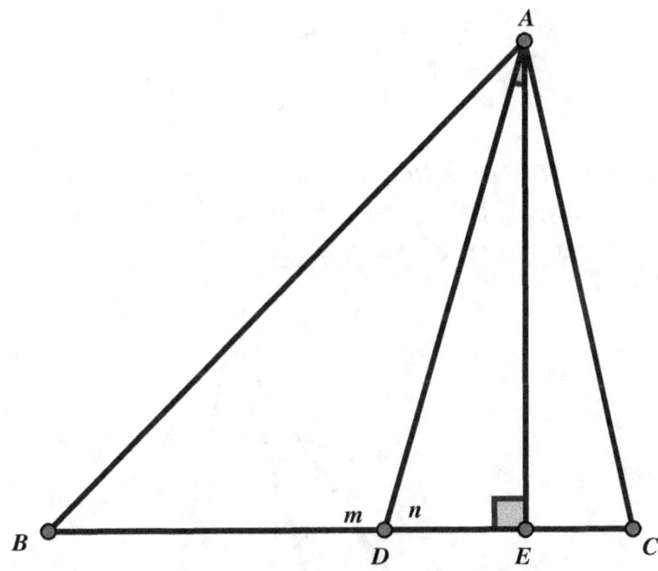

Figure 6-100

Geometrick 99: Comparison Between the Distance from the Vertices to a Point in a Triangle to the Sum of the Two Longest Sides of the Triangle

Consider triangle *ABC* in Figure 6-101, where the two longest sides are *AC* and *BC* and $AC > BC$. We need to prove that $AC + BC > AP + BP + CP$. We begin by drawing $DPE \parallel AB$, as can be seen in the Figure. Keep in mind that $DC > EC > DE$ and $DC > CP$. The sum of two sides of the triangle is always greater than the third side:

$$AD + DP > AP$$

$$BE + EP > BP$$

$$DC > CP$$

$$EC > DE.$$

By addition: $AD + DP + BE + EP + DC + EC > AP + BP + CP + DE$.
When we subtract $DE = DP + EP$ from both sides, we are left with $AD + DC + BE + EC > AP + BP + CP$, which is $AC + BC > AP + BP + CP$.

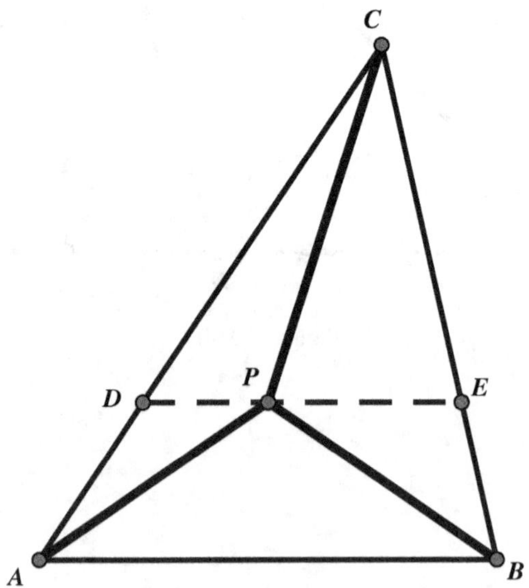

Figure 6-101

Geometrick 100: Results of Perpendiculars from a Random Point in the Triangle

We begin by drawing lines AO, BO, and CO in Figure 6-102. Applying the Pythagorean theorem to triangle AFO, we get $FO^2 = AO^2 - AF^2$, and then applying the Pythagorean theorem to triangle BFO, we get $FO^2 = BO^2 - BF^2$. Therefore, $AO^2 - AF^2 = BO^2 - BF^2$ or $AO^2 - BO^2 = AF^2 - BF^2$. We repeat this process with the remaining two sides of the triangle: for side BC we get $BO^2 - CO^2 = BD^2 - CD^2$, and for side AC we get $CO^2 - AO^2 = CE^2 - AE^2$. When we add these three equations, the left side has the sum 0, and the right side has the sum $AF^2 - BF^2 + BD^2 - CD^2 + CE^2 - AE^2 = 0$. Therefore, $AF^2 + BD^2 + CE^2 = BF^2 + CD^2 + AE^2$, which are the sums of the squares of alternate segments along the sides of the triangle.

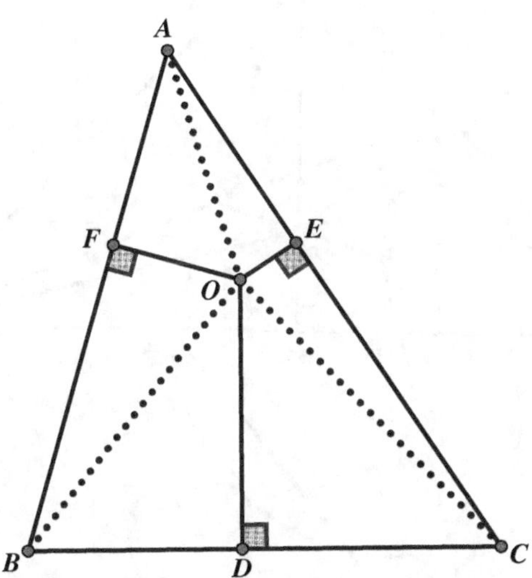

Figure 6-102

Geometrick 101: The Wonders Created by the Altitudes and a Median of a Triangle

Since $BK \parallel AC$ and $BH \perp AC$, then $BK \perp BH$ and similarly $CK \perp CH$. Because right angles are inscribable in a semicircle, we have, as seen in Figure 6-103, concyclic points $BPHC$ with diameter HK. Similarly, we have concyclic points $BCEF$ with diameter BC, and we also have concyclic points $EFHP$ with AH as the diameter. Now consider the common chords of the three circles: $BPHC$ and $BCEF$ share BC, $EFHP$ and $BPHC$ share PH, and $BCEF$ and $EFHP$ share EF. These chords are concurrent at Q, and therefore, EF extended contains point Q. We know that $AD \perp GQ$ and $PQ \perp AG$. Thus, H is the orthocenter of triangle AGQ.

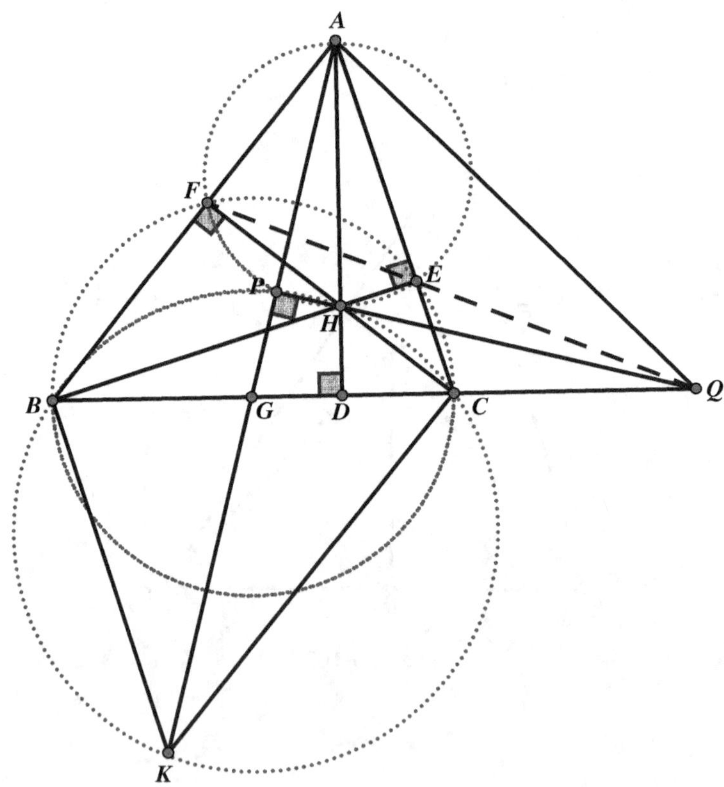

Figure 6-103

Geometrick 102: More Wonders Created by the Altitudes and a Median of a Triangle

Notice in Figure 6-104 that $\angle CBP = \frac{1}{2}\overset{\frown}{CHP} = \angle CKP$. Because $AB \parallel CK$, the alternate interior angles $\angle BAP = \angle CKA = \angle CKP$, therefore, $\angle CBP = \angle BAP$. Similarly, $\angle BCP = \frac{1}{2}\overset{\frown}{CHP} = \angle BKP$ and $\angle CAP = \angle BKA = \angle BKP$, therefore, $\angle BCP = \angle CAP$.

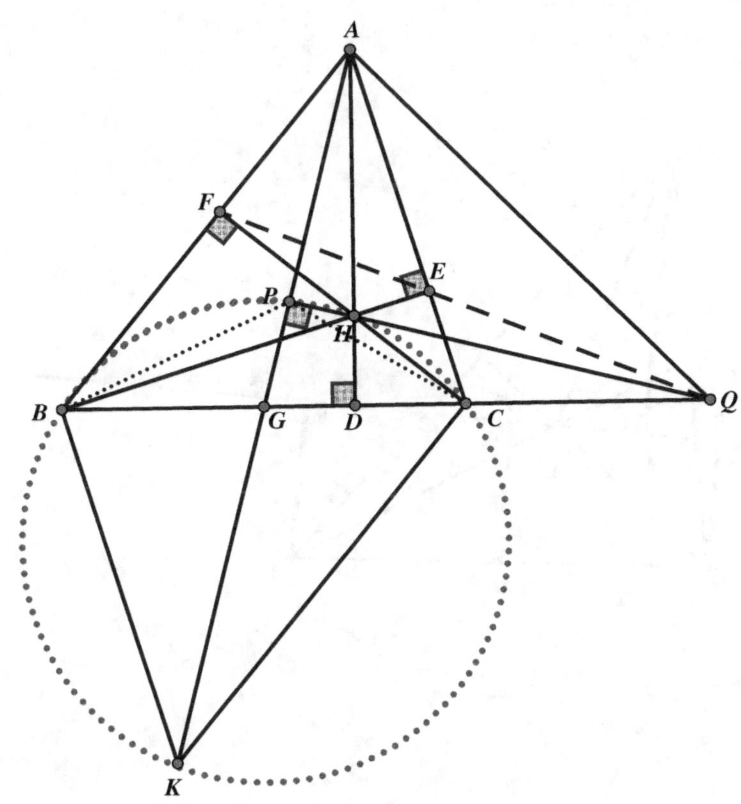

Figure 6-104

Geometrick 103: The Amazing Line Through a Triangle's Centroid

We begin by recognizing that quadrilateral *BCDH* in Figure 6-105 is a trapezoid since $DC \parallel BH$. Because *M* is the midpoint of *BC* and $FM \perp DH$, we find that *FM* is the median of the trapezoid, and therefore, $2FM = CD + BH$. The right triangles that share vertex angles are similar: $\triangle AEG \sim \triangle MFG$. Because the centroid trisects the median, we have $\frac{AE}{MF} = \frac{2}{1}$, so that $2MF = AE$. Therefore, $AE = CD + BH$, which is what he set out to prove.

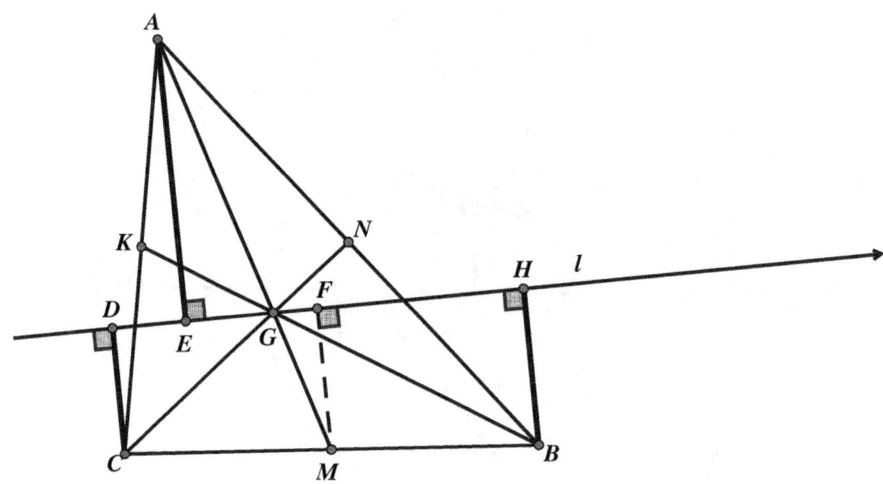

Figure 6-105

Geometrick 104: A Surprising Relationship Between the Distances to a Random Line from the Vertices of the Triangle

Consider the line drawn through point G that is parallel to l, as shown in Figure 6-106. In Geometrick 103, we established the relationship that $CS = AN + BR$. We could also write this equation as $CS - AN - BR = 0$. We then add $3g$ to both sides of the equation to get $g + CS + g - AN + g - BR = 3g$. This can be written as $CI + AG + BL = 3g$, or as $a + b + c = 3g$, which we set out to demonstrate.

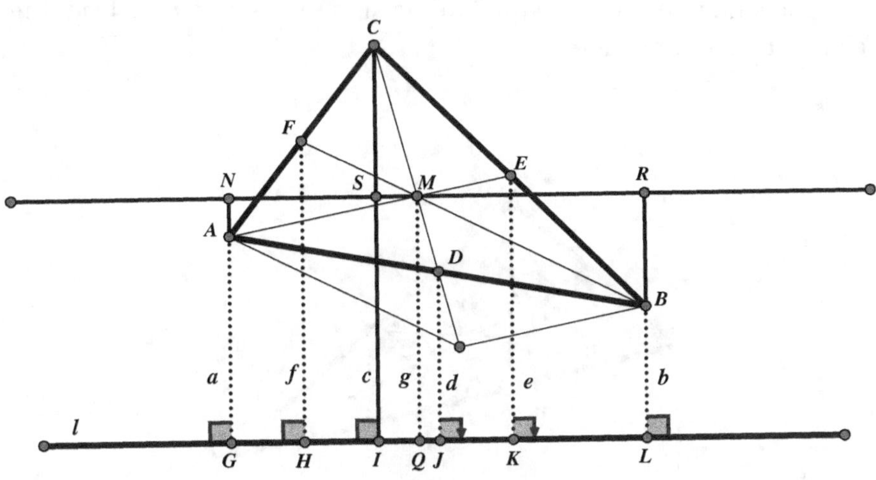

Figure 6-106

Geometrick 105: An Unusual Relationship Between the Midpoints and the Vertices of the Triangle

In Figure 6-107, the midpoints of the sides of triangle *ABC* are points *D*, *E*, and *F*. We represent the distance to the line *l* from vertices *A*, *B*, and *C* as *a*, *b*, and *c*, respectively, and the distances to the line *l* from the midpoints *D*, *E*, and *F* as *d*, *e*, and *f*, respectively. Since the median of a trapezoid is one-half the sum of the bases, we get the following: for trapezoid *ABLG*, we have $d = \frac{a+b}{2}$; for trapezoid *CBIL*, we have $e = \frac{b+c}{2}$; and for trapezoid *ACIG*, $f = \frac{a+c}{2}$. We add these three equations to get $d + e + f = a + b + c$, which proves that the sum of the distances from the side midpoints is equal to the sum of the distances from the vertices to the same line.

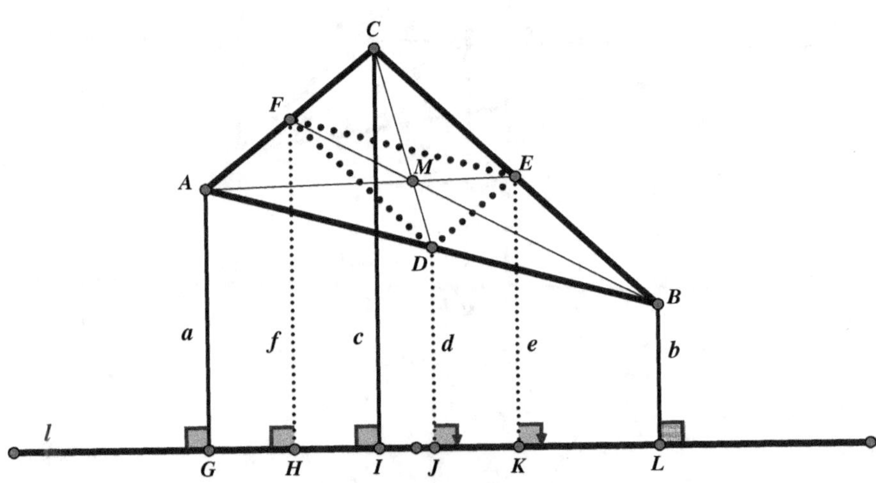

Figure 6-107

Geometrick 106: The Surprising Relationship Between the Circumcenter and Orthocenter of a Triangle

In Figure 6-108, begin by drawing diameter *COK*, *BK*, and *AK*. We then have *KB* ⊥ *BC* since angle *KBC* is inscribed in a semicircle. Since *AH* ⊥ *BC*, then *KB* ∥ *AH*. Similarly, *AK* and *BH* are both perpendicular to *AC*, and therefore, *AK* ∥ *BH*. Quadrilateral *AKBH* is thus a parallelogram, where *AH* = *KB*. In the triangle *KBC*, points *O* and *F* are midpoints of *KC* and *BC*, respectively. Therefore, *KB* = 2*OF*, and so *AH* = 2*OF*. Naturally, this relationship holds for the other altitudes as well.

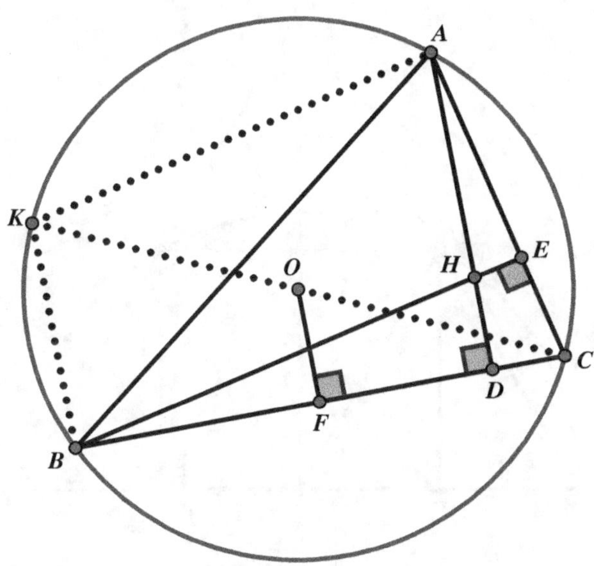

Figure 6-108

Geometrick 107: An Equilateral Triangle and Semicircle Create a Square

Draw altitude *AO* to intersect *DE* at point *K*, and then draw *DO*, as can be seen in Figure 6-109. We can then prove that $\triangle BFD \cong \triangle DOF \cong \triangle DKO$. Since triangle *BOD* is isosceles (with equal radii *PO* and *DO*), and $\angle BOD = 60°$, we have triangle *BOD* as equilateral; therefore, $\angle BDF = 30°$. Furthermore, we can also show the triangle *ADE* is equilateral and $\angle ADE = 60°$. Thus, $\angle EDF = 90°$, which establishes quadrilateral *FDEG* as a rectangle.

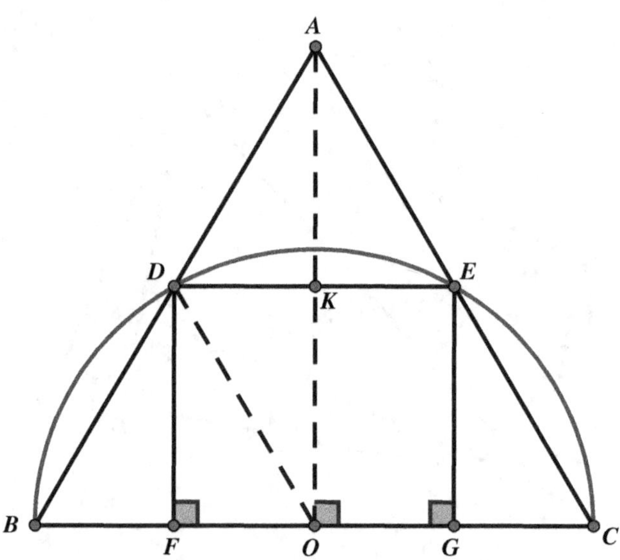

Figure 6-109

Geometrick 108: Introducing the Orthic Triangle

In Figure 6-110, quadrilateral *AEDB* is cyclic since ∠*AEB* and ∠*ADB* are right angles inscribed in a semicircle. Therefore, ∠*EAB* is supplementary to ∠*EDB* (opposite angles of a cyclic quadrilateral). However, ∠*EDC* is also supplementary to ∠*EDB*. Therefore, ∠*EAB* = ∠*EDC*. Thus, △*ABC* ~ △*DEC*, since both triangles also share ∠*ECD*. Simply repeat this procedure with cyclic quadrilateral *ECBF* to prove △*ABC* ~ △*AEF*, and with cyclic quadrilateral *AFDC* to get △*ABC* ~ △*DBF*.

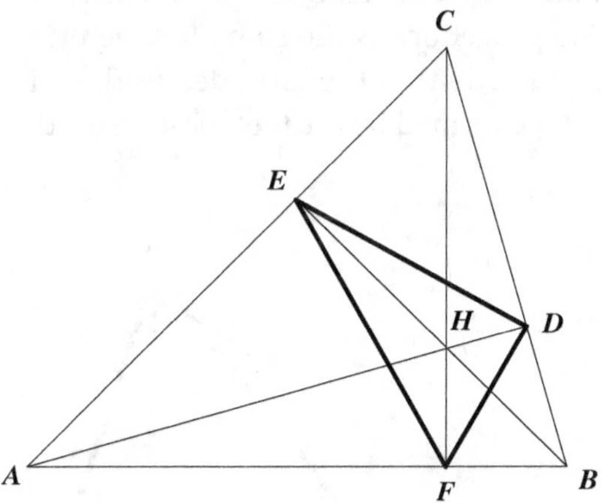

Figure 6-110

The property of an orthic triangle established in Geometrick 108 leads us to even more intriguing properties about an orthic triangle. Consider a triangle whose vertices lies on the sides of a second triangle. This is called an *inscribed triangle* of the second triangle. Now consider the possible inscribed triangles that a given acute triangle can have. Of these, the one with the shortest perimeter is the orthic triangle.

Geometrick 109: Triangle Altitudes as Angle Bisectors

In Figure 6-111, consider the quadrilaterals *BDGF* and *CEGD*. Each has a pair of opposite angles which are right angles, and are, therefore, cyclic quadrilaterals. Thus, for quadrilateral *BDGF* we have $\angle GDF = \angle GBF$, and for quadrilateral *CEGD* we have $\angle ECG = \angle EDG$. Consider the two right triangles $\triangle ABE$ and $\triangle ACF$, where we see that $\angle ABE = \angle ACF$ because both angles are complementary to $\angle BAC$. Therefore, $\angle ABF = \angle ACF$, and as a result, $\angle EDG = \angle GDF$. Altitude *AD* is a bisector of angle *EDF*. The procedure is the same for the other two altitudes, whereupon we conclude that the altitudes of the triangle bisect the angles of the triangle formed by the feet of the altitudes of the original triangle.

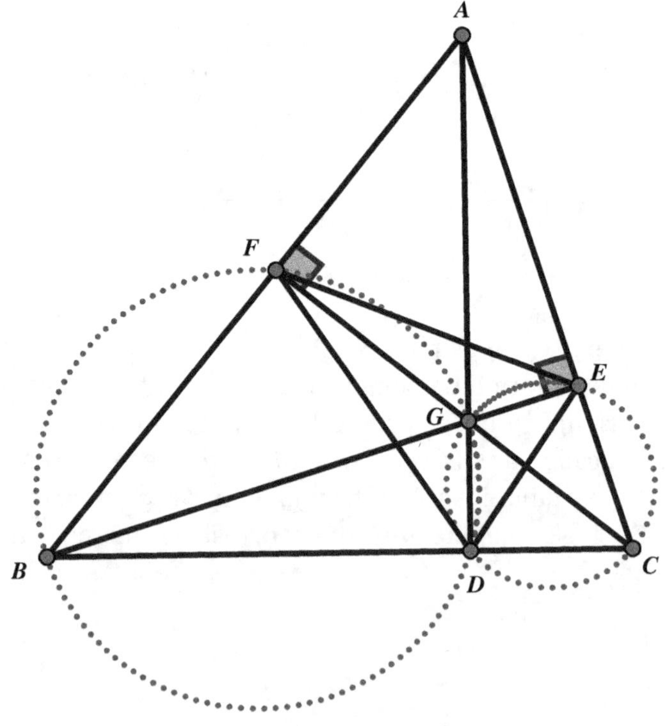

Figure 6-111

Alternate Proof: In Figure 6-112, consider $\triangle ABC$, where CD is the altitude to AB and P is any point on DC. Also, AP intersects CB at Q, and BP intersects CA at R. The line containing C and parallel to AB is intersected by the extensions of DR and DQ at points G and H, respectively.

$$\triangle CGR \sim \triangle ADR, \text{ so that } \frac{CR}{RA} = \frac{GC}{AD}. \tag{I}$$

$$\triangle BDQ \sim \triangle CHQ, \text{ so that } \frac{BQ}{QC} = \frac{DB}{CH}. \tag{II}$$

We now apply Ceva's Theorem (see Chapter 2) to $\triangle ABC$ to get

$$\frac{CR}{RA} \cdot \frac{AD}{DB} \cdot \frac{BQ}{QC} = 1. \tag{III}$$

Substituting (I) and (II) into (III) gives us $\frac{GC}{AD} \cdot \frac{AD}{DB} \cdot \frac{DB}{CH} = 1,$ or $\frac{GC}{CH} = 1.$ This implies that $GC = CH$. Thus, CD is the perpendicular bisector of GH. Hence, $\triangle GCD \cong \triangle HCD$, and therefore, $\angle RDC = \angle QDC$.

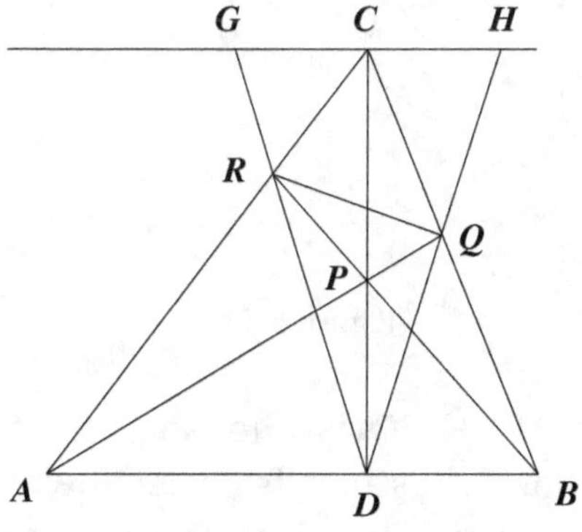

Figure 6-112

Geometrick 110: The Inscribed Triangle with the Shortest Perimeter

Before we start the proof, we should consider the following: The shortest distance that connects a given point to a given line to another given point on the same side of the line is the path that forms congruent angles with the given line. For example, consider the points A and B on the same side of line ℓ, shown in Figure 6-113. Let A' be the reflection of A in line ℓ, so that $AA' \perp \ell$ and $AR = A'R$. The intersection of $A'B$ and ℓ is point P. We can show that $AP + PB$ is the shortest distance from A to ℓ to B. Also, $\triangle ARP \cong A'RP$, and $\angle APR = \angle A'PR$. Thus, $\angle APR = \angle BPS$, which is an important property of this minimum perimeter.

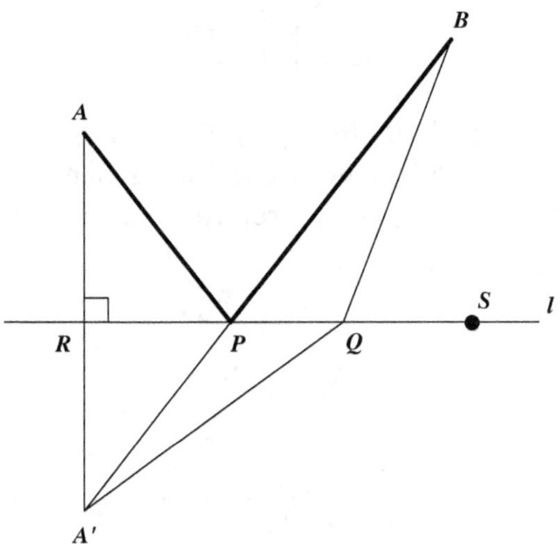

Figure 6-113

To prove that $AP + PB$ is the shortest required distance, select point Q, on line ℓ, distinct from P. Now $A'B < A'Q + QB$, or $A'P + PB < A'Q + QB$, which verifies the selection of point P as the point determining the minimum distance, as required. We are now ready to prove the statement of Geometrick 110.

In Geometrick 108, we established that the orthic triangle determined three triangles similar to the original triangle. From this, we show in Figure 6-114 that $\angle AEF = \angle CED$, $\angle CDE = \angle BDF$, and $\angle AFE = \angle BFD$. Therefore, the shortest path from E to AB to D is $EF + FD$. Similarly, the shortest path from E to CB to F is $ED + DF$, and the shortest path from D to AC to F is $DE + EF$. This implies that $\triangle EDF$ is the minimum perimeter inscribed triangle of acute $\triangle ABC$. Were we to compare the perimeter of orthic $\triangle DEF$ to that of any other inscribed triangle of $\triangle ABC$ we could, in this way, easily show that $\triangle DEF$ has a smaller perimeter.

From the congruent angles we established earlier, another interesting property of an orthic triangle evolves. Notice that since $\angle AFE = \angle BFD$, and $\angle EFC$ is complementary to $\angle AFE$, and $\angle DFC$ is complementary to $\angle DFB$, we have $\angle EFC = \angle DFC$. The general case is stated is the next Geometric.

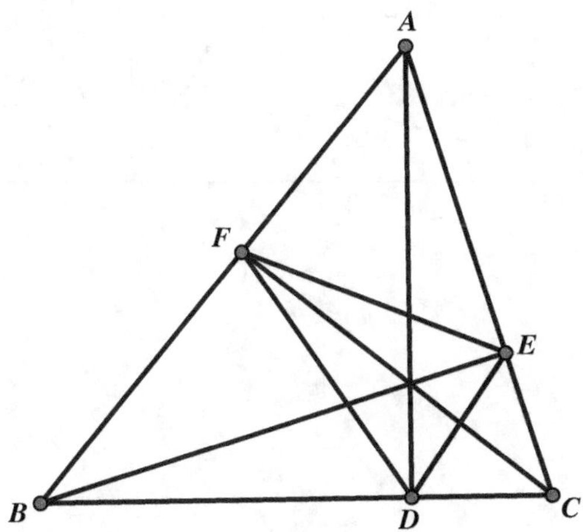

Figure 6-114

Geometrick 111: The Orthocenter in Relation to Each Altitude

From an earlier proof we know that since $\triangle CDH \sim \triangle AFH$, we have $\frac{CH}{AH} = \frac{HD}{HF}$, as shown in Figure 6-115. This can be rewritten as $(CH)(HF) = (AH)(HD)$. The proof is completed by using other pairs of similar right triangles in the same manner as above.

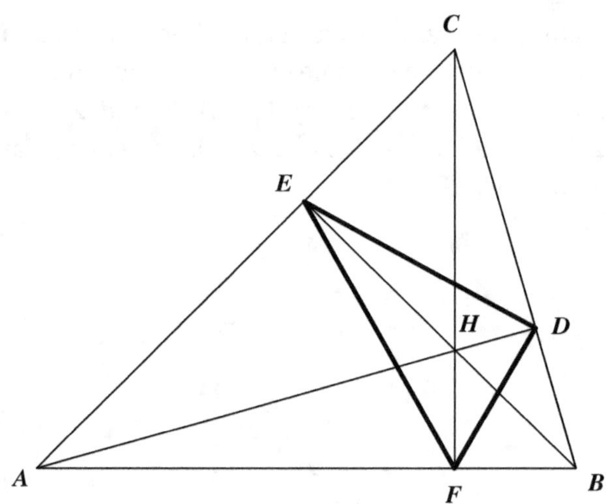

Figure 6-115

Geometrick 112: An Unexpected Bisection

We begin with $\angle CSB = \angle CAB$, since both angles are inscribed in the same circle and intercept the same arc, $\overset{\frown}{BC}$, as shown in Figure 6-116. In $\triangle ACF$, $\angle ACF$ is complementary to $\angle CAF$. In $\triangle CEH$, $\angle ECH$ is complementary to $\angle CHE$. But vertical angles $\angle BHF = \angle CHE$. Therefore, $\angle CAF = \angle BHF$. Since both $\angle CSB$ and $\angle BHF$ are equal to $\angle CAB$, they are equal to each other. Thus, $\triangle HBS$ is isosceles, and $\triangle HFB \cong \triangle SFB$. It then follows that $HF = SF$, which proves the theorem for one altitude. This can be repeated for the other altitudes.

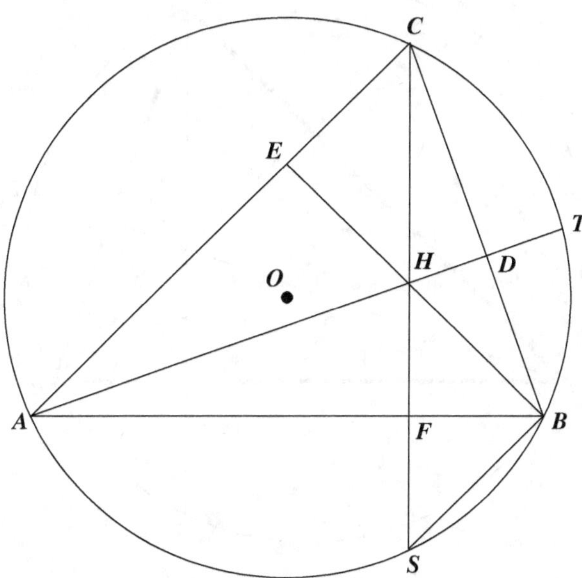

Figure 6-116

Geometrick 113: Triangle Vertices Bisect Arcs on the Triangle's Circumcircle

In Figure 6-117, quadrilateral *AFDC* is cyclic since $\angle AFC$ and $\angle ADC$ are right angles. Therefore, $\angle FAD = \angle DCF$ since both are measured by $\overset{\frown}{DF}$. It then follows that $\overset{\frown}{SB} = \overset{\frown}{TB}$ because equal inscribed angles in the same circle have equal intercepted arcs. What holds true for one pair of altitudes can also be shown to hold true for other pairs of altitudes.

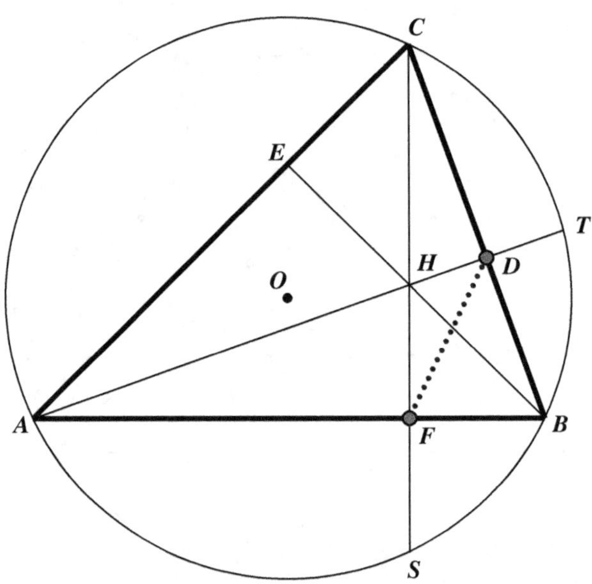

Figure 6-117

Geometrick 114: Proportional Areas

Since we notice in Figure 6-118 that triangles *ABC*, *ABG*, and *ABH* have the same base *AB*, we can compare their areas by comparing the altitudes. That is, we need to show that the altitude of triangle *ABG* is the mean proportional between the altitudes of triangles *ABC* and *ABH*, which is $\frac{DA}{DG} = \frac{DG}{DB}$, or $DG^2 = DA \cdot DB$, We note that right $\triangle ADH \sim \triangle CDB$, since $\angle DAH = \angle DCB$, as they are both complementary to the vertical angles at point *H*. Therefore, $\frac{DA}{DC} = \frac{DH}{DB}$, and $DA \cdot DB = DC \cdot DH$, which, from the earlier relationship, gives us $DG^2 = DC \cdot DH$, and thus, $\frac{DC}{DG} = \frac{DG}{DH}$. With the altitudes in proportion and the bases the same, the areas of triangles are defined by the relation of their altitudes. That is, the area of triangle *ABG* is a mean proportional between the areas of triangle *ABC* and triangle *ABH*.

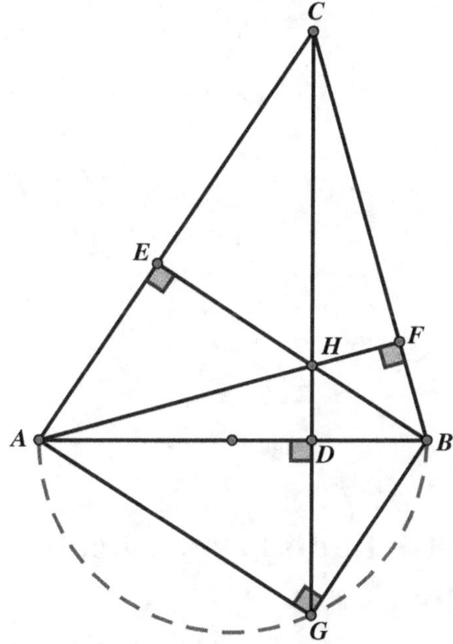

Figure 6-118

Geometrick 115: An Unexpected Similarity to the Orthic Triangle

We established in Geometrick 112 that $HF = SF$ and $HD = TD$, which can be seen in Figure 6-119. Therefore, in $\triangle HST$, DF is a midline parallel to ST. The same argument can be used to prove $EF \parallel US$ and $DE \parallel TU$. It then follows that $\triangle DEF \sim \triangle TUS$.

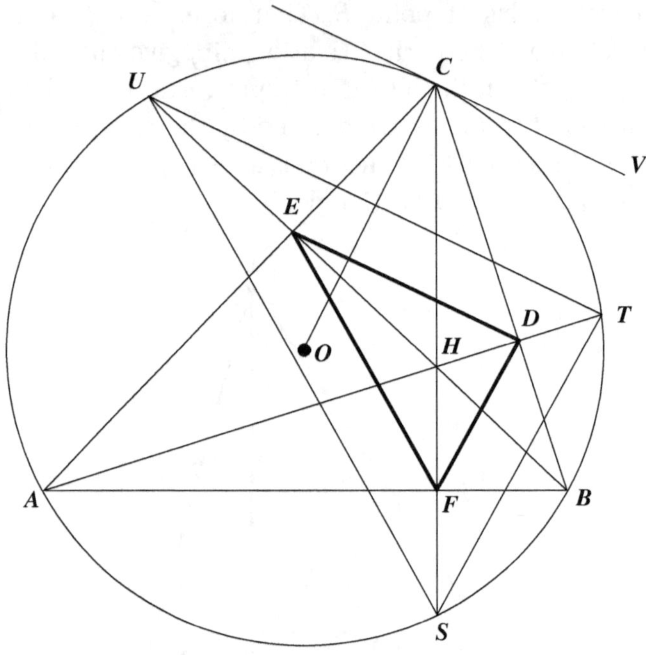

Figure 6-119

Geometrick 116: Radii Perpendicular to the Orthic Triangle

We shall prove this relationship for one of the radii in question and leave the proofs for the other two radii to the reader. In Figure 6-119, we see that $\overarc{UC} = \overarc{TC}$, and OC is a radius of the circumcircles of $\triangle ABC$ and $\triangle STU$. Earlier, we proved that $\overarc{UC} = \overarc{TC}$. Therefore, OC is the perpendicular bisector of TU. Since $OC \perp TU$, OC must also be perpendicular to DE because we already established $DE \parallel TU$.

Geometrick 117: Lines Parallel to the Orthic Triangle

Again, we shall prove this relationship for only one side of the orthic triangle. Radius OC is perpendicular to tangent VC, as can be seen in Figure 6-119. However, we earlier established that $OC \perp DE$. Therefore, $VC \parallel DE$. The same argument holds for the other sides of the orthic triangle.

Geometrick 118: Relationships of the Incenter, Circumcenter, and Orthocenter of a Triangle

The bisectors of the angles of a triangle are concurrent at the center of the inscribed circle. Therefore, points P, Q, and R are the midpoints arcs BC, CA, and AB, as seen in Figure 6-120. The exterior angle of

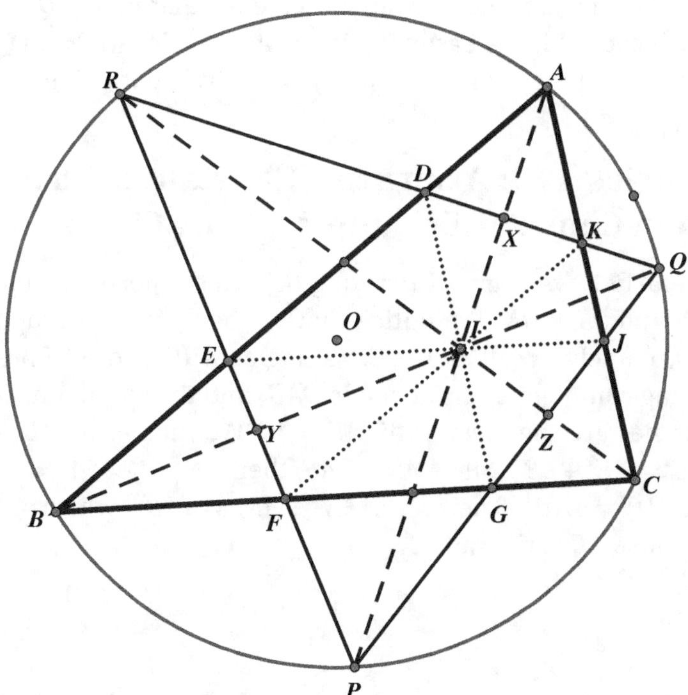

Figure 6-120

a triangle is equal to the sum of the two remote interior angles; therefore, in triangle *RXP*, we have $\angle PXQ = \angle PRX + \angle XPR$. This can be expressed as $\angle PXQ = \angle PRQ + \angle XPR = \angle PRC + \angle CRQ + \angle APR$, which is the sum of one-half the three angles of triangle *ABC*, which is 90°. This means we have $PX \perp QR$, and similarly, $QY \perp RP$, and $RZ \perp PQ$. Therefore, point *I* is the orthocenter of triangle *PQR*. We also get $\triangle AXD \cong \triangle AXK$ and $AD = AK$, and similarly, $BE = BF$ and $CG = CJ$. Since *I* is the orthocenter of triangle *PQR*, we have $AX = XI$, which we proved in Geometrick 112, or recall the fact that $\triangle RXA \cong \triangle RXI$. Hence, quadrilateral *ADIK* is a rhombus, as are quadrilaterals *BFIE* and *CJIG*. Also, $IJ \parallel GC$ and $IE \parallel BF$, so that *IE* and *IJ* are both parallel to *BC*, which makes *EIJ* a straight line. Similarly, *DIG* and *KIF* are also straight lines parallel to sides *AC* and *AB* of triangle *ABC*. With regard to rhombus *ADIK*, line *QR* is a perpendicular bisector of *AI*. Similarly, *RP* and *QP* are perpendicular bisectors of *BI* and *CI*, respectively. Therefore, points *P*, *Q*, and *R* are the circumcenters of triangles *BIC*, *CIA*, and *AIB*, respectively. The last relationship indicated is that points *Q*, *R*, *Y*, and *Z* are concyclic, which is easily justified since angles *RYQ* and *QCR* are right angles, and therefore, inscribable in a circle with diameter *RQ*.

Geometrick 119: A Surprise Constant from a Common Chord of Two Intersecting Circles

In Figure 6-121, we begin by drawing *BC* and the perpendicular lines $BD \perp MN$ and $CE \perp MN$. We notice that we have two congruent right triangles, namely, $\triangle ADB \cong \triangle ACE$, as angles *ABD* and *EAC* are equal (both being complementary to angle *DAB*) and the hypotenuses of the two triangles are the equal radii $AB = AC$. We then have $AD^2 + BD^2 = AB^2 = AC^2 = AE^2 + CE^2$. Since $AN = 2AE$, then $AN^2 = (2AE)^2 = 4AE^2$ and $AM^2 = (2AD)^2 = 4AD^2$. Therefore, $AN^2 + AM^2 = 4AB^2$, which is a constant because *AB* is the radius.

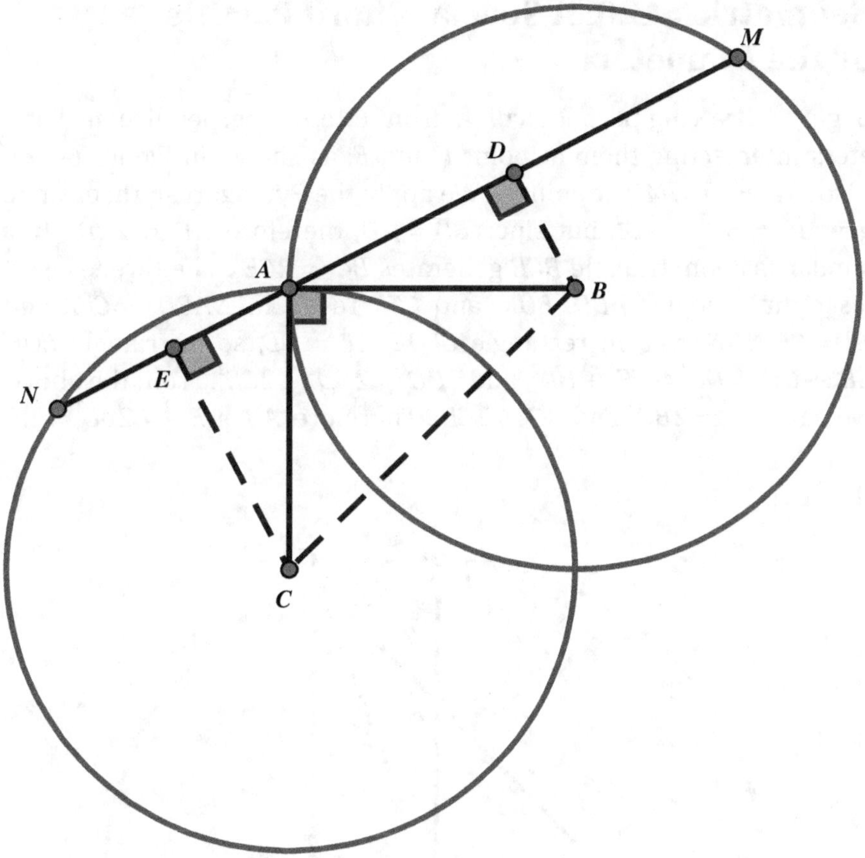

Figure 6-121

Geometrick 120: A Special Chord Partitioned by the Diameter

Begin by drawing perpendiculars from C to the perpendicular diameters intersecting them at point F and E, as shown in Figure 6-122. Also draw $AD \perp MN$ at point D. We apply the Pythagorean theorem to get $AB^2 = BD^2 + AD^2$, but since $BD = AD$, therefore, $AB^2 = 2BD^2$. In a similar fashion, triangle BCE generates $BC^2 = 2CE^2$. The three isosceles right triangles ABD, BOK, and CFK result in $\triangle ABD \cong \triangle CFK$ and $BD = CF$. However, in rectangle $CFOE$, $CF = OE$, so in triangle COE, $CE^2 + OE^2 = OC^2 = CE^2 + BD^2$, and $2OC^2 = 2BD^2 + 2CE^2$. From the above, we have $AB^2 = 2BD^2$ and $BC^2 = 2CE^2$. Therefore, $AB^2 + BC^2 = 2OC^2 = 2r^2$.

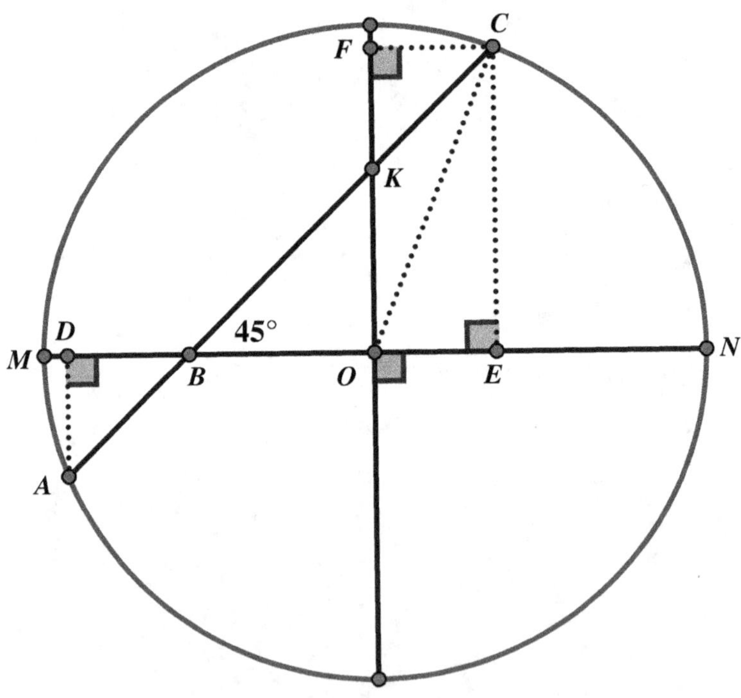

Figure 6-122

Geometrick 121: Perpendicular Chords Intersect with the Constant Sum

Draw diameter COE, and then draw segments AE, AC, and BD, as shown in Figure 6-123. We apply the Pythagorean theorem to triangle AMC to get $a^2 + c^2 = AC^2$, then apply the Pythagorean theorem to triangle BMD to get $b^2 + d^2 = BD^2$. We add these two equations to get $a^2 + b^2 + c^2 + d^2 = AC^2 + BD^2$. Because the chords are perpendicular, $90° = \angle AMC = \frac{1}{2}(\overarc{AC} + \overarc{BD})$, then $\overarc{AC} + \overarc{BD} = 180°$. Also, $\overarc{AC} + \overarc{AE} = 180°$. Therefore, $\overarc{BD} = \overarc{AE}$ and $BD = AE$. However, $AE^2 + AC^2 = CE^2$, which is equal to $BD^2 + AC^2 = CE^2$. We then have $a^2 + b^2 + c^2 + d^2 = AC^2 + BD^2 = CE^2 = (2r)^2 = 4r^2$.

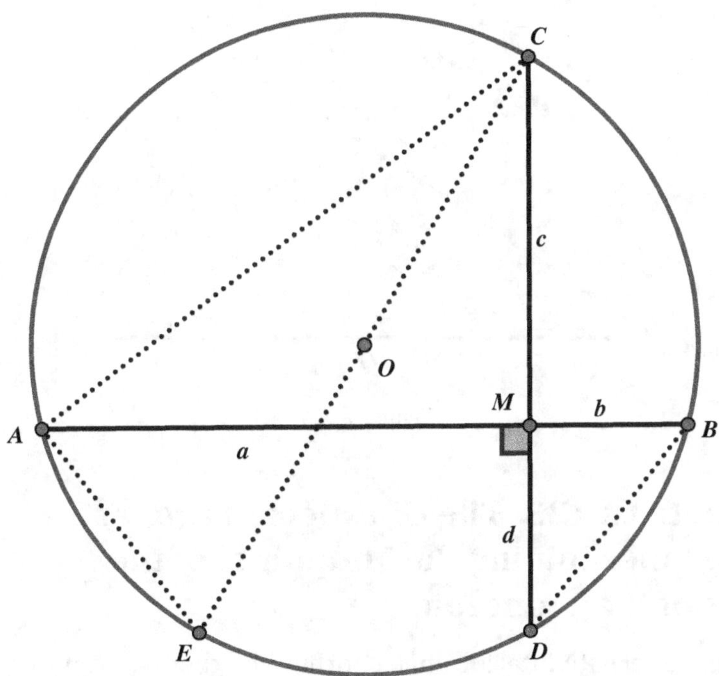

Figure 6-123

Geometrick 122: A Surprising Semicircular Path

In Figure 6-124, the right triangles *CDF* and *EDB* are similar because $\angle B = \angle F$, as they are both complementary to $\angle C$. Therefore, $\frac{CD}{DE} = \frac{DF}{DB}$, which can be written as $DE \cdot DF = DB \cdot CD$. When the point *P* is on the circumference of the semicircle, $\triangle CPB$ is a right triangle, and since the altitude to the hypotenuse of a right triangle is the mean proportional between the two segments along the hypotenuse, we have $DP^2 = DB \cdot CD$. This relationship holds for any point *P* that is on the semicircle.

The same argument can be made for the locus of points that the point *A* would make —that is, a path along the semicircle —under the condition that $DE \cdot DF = DB \cdot CD$.

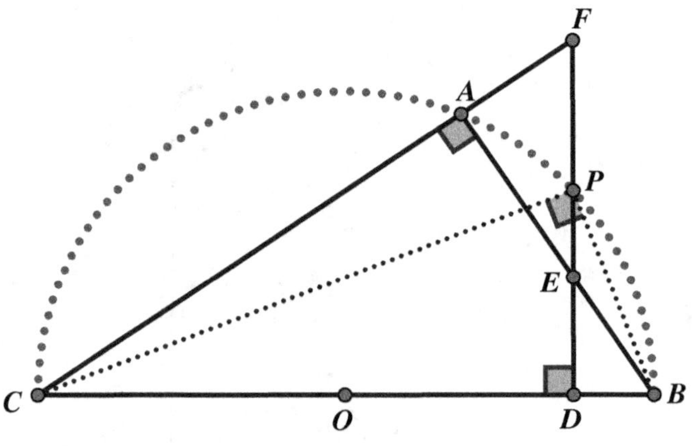

Figure 6-124

Geometrick 123: The Unexpected Property of the Lines Joining the Midpoints of the Bases of the Trapezoid

Since the alternate interior angles formed by the two transversals of parallel lines *AD* and *BC* are equal, we know that $\triangle AGD \sim \triangle BGC$, as can be seen in Figure 6-125. Line segment *GE* is a median of triangle *AGD*, and *GF* is the median of triangle *BGC*. Therefore, the medians of these similar triangles form the same angles with the sides of their respective triangles. Hence, $\angle AGE = \angle CGF$, which in turn indicates that *EGF* is a straight line concurrent with the diagonals of the trapezoid.

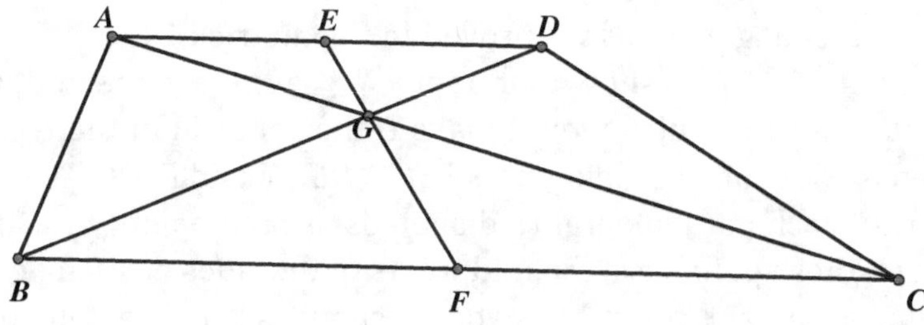

Figure 6-125

Geometrick 124: A Multitude of Properties in the Triangle

To prove that point *G* is the centroid of triangle *ABC*, we first notice that $AG = \frac{2}{3}AM$ because the parallel lines partition *AM* into thirds, as they do with the sides of triangle *ABC*. Since $BH = FL = CK$, and *AM* is trisected at point *G*, the line *AM* is the median of triangle *ABC*, as shown in Figure 6-126. Point *G* is the midpoint of *ED*, making *AG* a

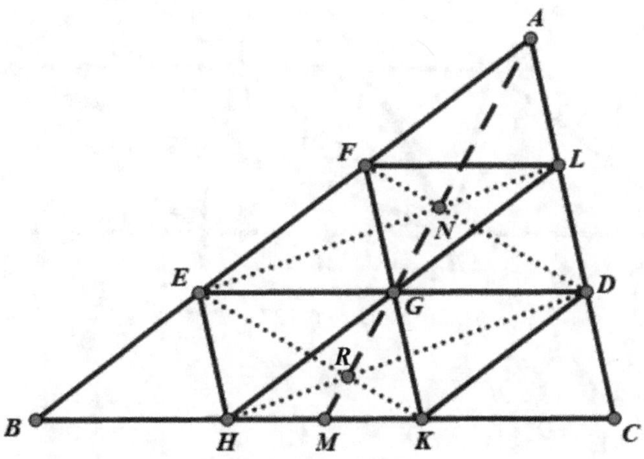

Figure 6-126

median of triangle *AED*; also, *DF* and *EL* are medians of triangle *AED*, and are thus concurrent at point *N*.

Now consider trapezoid *EDKH* and notice that the line joining the midpoints of the two bases, *GM*, must intersect the point of intersection, *R*, of the diagonals, as we proved in Geometrick 123.

In Figure 6-127, to show that points *X*, *Y*, and *Z* are the midpoints of *FH*, *DH*, and *DF*, which would make *G* the centroid of the triangle, we need to focus on parallelograms *FEHG*, *HGDK*, and *FLDG*. The intersections of each parallelogram's diagonals are the points *X*, *Y*, and *Z*; therefore, these points are the midpoints of the sides of triangle *FHD*. A similar argument could be made to show that *G* is a centroid of triangle *ELK*.

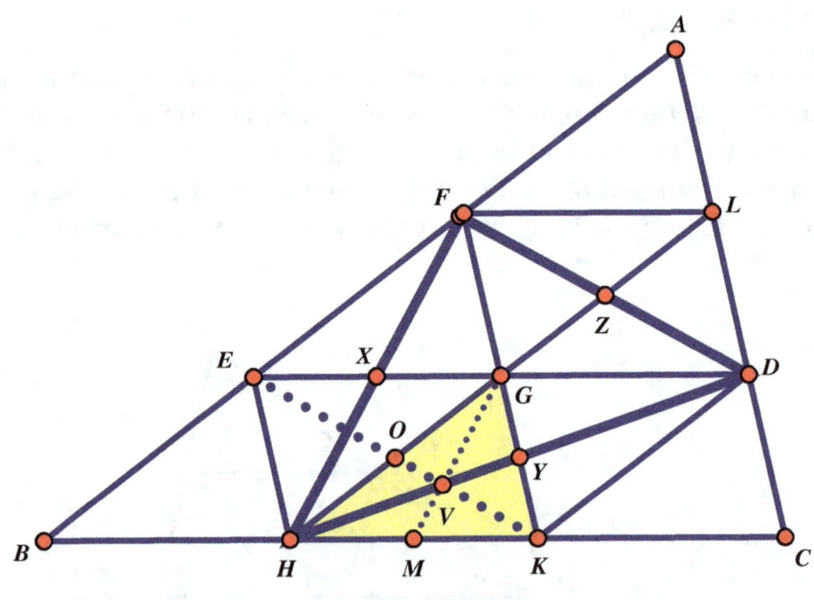

Figure 6-127

In Figure 6-127, line segment *EK* is a diagonal of parallelogram *EGKH* and is bisected by the other diagonal *GH* at point *O*. Similarly in parallelogram *HGDK*, *Y* is the intersection of the diagonals. Point *V* is then the centroid of triangle *HGK*.

In Figure 6-128, the centroids *T*, *S*, and *U* of triangles *EGF*, *LGD*, and *HGK* are on the medians of triangle *ABC*, and therefore, concurrent at point *G*. The symmetry in the figure enables us to see that when *TG*, *SG*, and *UG* are extended to meet the sides of triangle *TSU*, they bisect the sides, and are therefore the medians of the triangle.

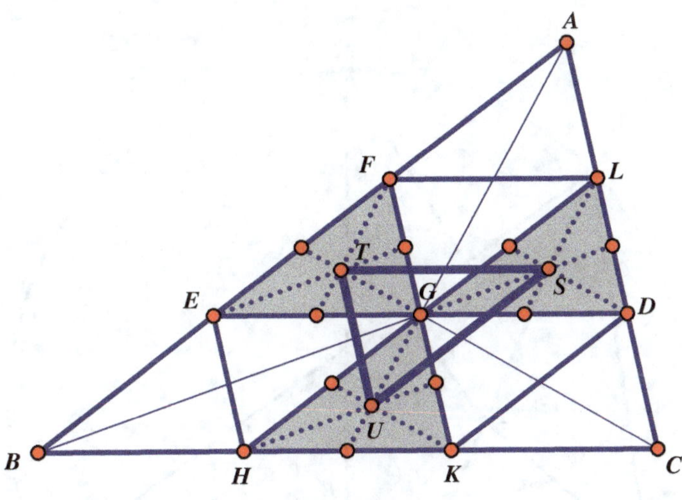

Figure 6-128

Geometrick 125: A Relationship Between the Circumradius and the Inradius

In Figure 6-129, the bisector of angle A intersects the circumscribed circle at point N. The perpendicular bisector of BC intersects the circumscribed circle at points M and N. With a common intercepted arc, we get $\angle NBC = \frac{1}{2}\overset{\frown}{NC} = \angle CAN$ and $\angle BMN = \frac{1}{2}\overset{\frown}{NB} = \angle BAN$, but $\overset{\frown}{NC} = \overset{\frown}{NB}$. Therefore, $\angle NBC = \angle CAN = \angle BMN = \angle BAN$. Angle BIN is the exterior angle of triangle ABI, then $\angle BIN = \angle BAI + \angle ABI = \angle NBI$, which establishes that triangle NBI is isosceles, so that $NI = NB$. Thus,

$$R^2 - d^2 = NI \cdot AI = NB \cdot AI = NM \cdot \frac{\overset{NB}{\overline{NM}}}{\underset{\overline{AI}}{\overline{IX}}} \cdot IX = NM \cdot \frac{\sin \angle BMN}{\sin \angle CAN} \cdot IX = NM \cdot IX = 2Rr.$$

Therefore, $d^2 = R^2 - 2rR$.

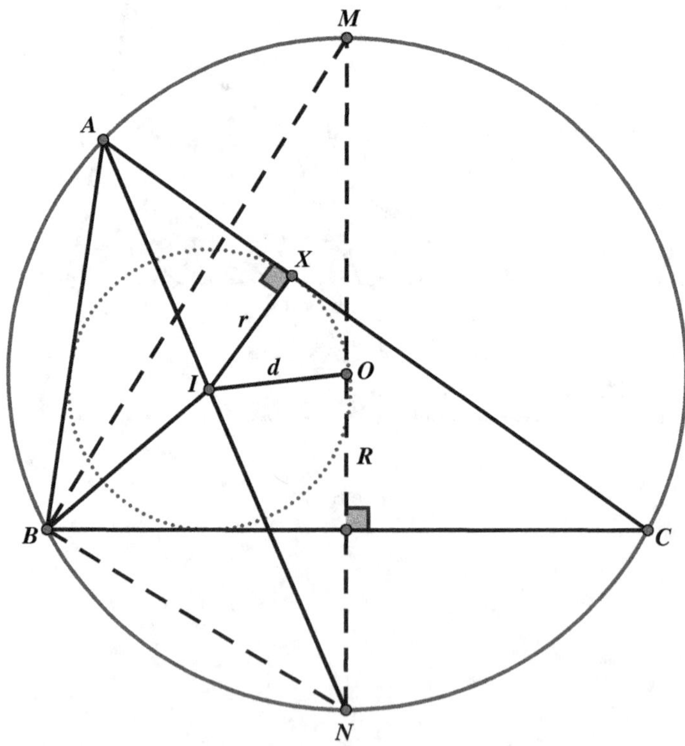

Figure 6-129

Index

www.ingramcontent.com/pod-product-compliance
Lightning Source LLC
Chambersburg PA
CBHW072207150726
48002CB00005B/1705